Advances in
MARINE BIOLOGY

VOLUME 52

Advances in
MARINE BIOLOGY

Edited by

DAVID W. SIMS

Marine Biological Association, The Laboratory, Citadel Hill,
Plymouth, United Kingdom

ELSEVIER

AMSTERDAM • BOSTON • HEIDELBERG • LONDON
NEW YORK • OXFORD • PARIS • SAN DIEGO
SAN FRANCISCO • SINGAPORE • SYDNEY • TOKYO
Academic Press is an imprint of Elsevier

Academic Press is an imprint of Elsevier
525 B Street, Suite 1900, San Diego, California 92101-4495, USA
84 Theobald's Road, London WC1X 8RR, UK

This book is printed on acid-free paper. ∞

For information on all Academic Press publications
visit our Web site at www.books.elsevier.com

ISBN-13: 978-0-12-373718-2
ISBN-10: 0-12-373718-4

PRINTED IN THE UNITED STATES OF AMERICA
07 08 09 10 9 8 7 6 5 4 3 2 1

CONTRIBUTORS TO VOLUME 52

KEIRON P. P. FRASER, *British Antarctic Survey, Natural Environment Research Council, Cambridge CB3 OET, United Kingdom*

ELENA GUIJARRO GARCIA, *Marine Research Institute, 101 Reykjavík, Iceland*

SALLY P. LEYS, *Department of Biological Sciences, University of Alberta, Edmonton, Alberta T6G 2E9, Canada*

GEORGE O. MACKIE, *Department of Biology, University of Victoria, Victoria V8W 3N5, Canada*

HENRY M. REISWIG, *Department of Biology, University of Victoria, Victoria V8W 3N5, Canada; Natural History Section, Royal British Columbia Museum, Victoria V8W 9W2, Canada*

ALEX D. ROGERS, *British Antarctic Survey, Natural Environment Research Council, Cambridge CB3 OET, United Kingdom*

CONTENTS

The Biology of Glass Sponges

Sally P. Leys, George O. Mackie and Henry M. Reiswig

The Northern Shrimp (*Pandalus borealis*) Offshore Fishery in the Northeast Atlantic

Elena Guijarro Garcia

Protein Metabolism in Marine Animals: The Underlying Mechanism of Growth

Keiron P. P. Fraser and Alex D. Rogers

SERIES CONTENTS FOR LAST TEN YEARS*

*The full list of contents for volumes 1–37 can be found in volume 38.

The Biology of Glass Sponges

Sally P. Leys,* George O. Mackie[†] and Henry M. Reiswig[†,‡]

*Department of Biological Sciences, University of Alberta,
Edmonton, Alberta T6G 2E9, Canada
[†]Department of Biology, University of Victoria,
Victoria V8W 3N5, Canada
[‡]Natural History Section, Royal British Columbia Museum,
Victoria V8W 9W2, Canada

0065-2881/07 $35.00
DOI: 10.1016/S0065-2881(06)52001-2

As the most ancient extant metazoans, glass sponges (Hexactinellida) have attracted recent attention in the areas of molecular evolution and the evolution of conduction systems but they are also interesting because of their unique histology: the greater part of their soft tissue consists of a single, multinucleate syncytium that ramifies throughout the sponge. This trabecular syncytium serves both for transport and as a pathway for propagation of action potentials that trigger flagellar arrests in the flagellated chambers. The present chapter is the first comprehensive modern account of this group and covers work going back to the earliest work dealing with taxonomy, gross morphology and histology as well as dealing with more recent studies. The structure of cellular and syncytial tissues and the formation of specialised intercellular junctions are described. Experimental work on reaggregation of dissociated tissues is also covered, a process during which histocompatibility, fusion and syncytialisation have been investigated, and where the role of the cytoskeleton in tissue architecture and transport processes has been studied in depth. The siliceous skeleton is given special attention, with an account of discrete spicules and fused silica networks, their diversity and distribution, their importance as taxonomic features and the process of silication. Studies on particle capture, transport of internalised food objects and disposal of indigestible wastes are reviewed, along with production and control of the feeding current. The electrophysiology of the conduction system coordinating flagellar arrests is described. The review covers salient features of hexactinellid ecology, including an account of habitats, distribution, abundance, growth, seasonal regression, predation, mortality, regeneration, recruitment and symbiotic associations with other organisms. Work on the recently discovered hexactinellid reefs of Canada's western continental shelf, analogues of long-extinct Jurassic sponge reefs, is given special attention. Reproductive biology is another area that has benefited from recent investigations. Seasonality, gametogenesis, embryogenesis, differentiation and larval biology are now understood in broad outline, at least for some species. The process whereby the cellular early larva becomes syncytial is described. A final section deals with the classification of recent and fossil glass sponges, phylogenetic relationships within the Hexactinellida and the phylogenetic position of the group within the Porifera. Palaeontological aspects are covered in so far as they are relevant to these topics.

1. INTRODUCTION

Glass sponges, Hexactinellida, are emerging as an important group of animals which, because of their ancient heritage, can shed light on fundamental questions such as the origin of multicellular animals, molecular evolution, and the evolution of conduction systems. Glass sponges are unusual animals with a skeleton of silicon dioxide whose triaxonal (cubic), six-rayed symmetry and square axial proteinaceous filament distinguishes them from other siliceous sponges. The fossil record suggests glass sponges were established by the Late Proterozoic, thrived during the middle Cambrian, diversified during the Jurassic when they formed vast reefs in the Tethys Sea and reached their maximum radiation and diversity during the Late Cretaceous. Estimated rates of molecular evolution place their origin even earlier at 800 million years ago (Ma).

However, it is their soft tissues that are really remarkable. They are interestingly different from all other animals (including other sponges) in having syncytial tissues that arise by fusion of early embryonic cells. The larva and adult have an elegant combination of multinucleate and cellular cytoplasmic regions unknown in any other animal. The continuity of this tissue not only allows food to be transported around the animal symplastically as in plants, it allows electrical signals to travel throughout the animal—in essence it functions analogously to the nervous system of other animals. These unusual features show how Nature has a few tricks up her sleeve none of us could have imagined 30 years ago.

What we know of these unusual animals has been constrained by limited access to their deep-water habitat. There are approximately 500 species world wide; the greatest diversity inhabit 300- to 600-m depths and only a few populations inhabit shallow (>15 m) water. Glass sponges were first sampled in the late nineteenth century during the Challenger Expedition, and though we owe much of our understanding of the taxonomy to that early descriptive work, use of modern techniques such as SCUBA, submersibles and remote operated vehicles (ROVs) have vastly extended our knowledge of the cytology and physiology of the group. Recently, use of submersibles and ROVs coupled with modern multibeam surveys have revealed modern reefs in the Northeast (NE) Pacific, and have now made it possible to study the ecology and even physiology of feeding *in situ*.

These are exciting times, but much remains to be done in several key areas. Reproductive individuals are hard to encounter, and our only detailed information comes from a tiny cave-dwelling species that is both accessible by SCUBA and is reproductive year round. How other species reproduce, whether their larvae are similar, and how they colonise new habitats and maintain the massive reef structures over time are areas we know nothing about. Syncytial tissues are remarkable structures that we little understand. How do nuclei arise in the multinucleated regions; how are the microtubule transport pathways nucleated—are there 'roaming' microtubule-organizing centres as there are in some protists? How does the tissue differentiate zones for fusing and how does fusion of the membrane occur? What is the protein and molecular make up of the unusual plugged junctions that separate functional regions of cytoplasm into 'cells' and to what extent do they function to regulate transjunctional traffic? The syncytial tissues can propagate action potentials, but intracellular recordings have yet to be achieved, and the basis of the impulse as a calcium spike requires verification by analysis at the membrane level. Glass sponges inhabit many sediment-rich regions, yet particulates in the incurrent water are one of the principal irritants that trigger feeding current arrests. What limits glass sponges to their present habitats and how robust are modern-day shallow populations in the face of growing human impact on this habitat?

This is the first review of glass sponge biology, and thus will likely be an important resource for those who continue to advance our knowledge of the group. Here we have tried to cover as much of the biology of the group as possible, but only lightly touched on palaeontology as it pertains to systematics and skeletal structure. The reader may find more information in other excellent papers (Steiner *et al.*, 1993; Reitner and Mehl, 1995; Mehl, 1996, Brasier *et al.*, 1997; Mehl-Janussen, 2000). Our ultimate goal is to stimulate interest and continued research into a fascinating and ancient group of animals.

2. GENERAL ORGANISATION

2.1. Gross morphology

Glass sponges are vase, plate or tube-shaped animals (never encrusting) that range in size from 0.5-cm diameter in *Oopsacas minuta* up to 2-m diameter in *Aphrocallistes vastus* or *Scolymastra joubeni* (Figure 1). The sponge body may arise directly from the substratum and be organised around a single atrial feeding cavity, or the sponge can be suspended on a tall stalk of glass spicules, in which case the atrial cavity is all but absent and numerous excurrent canals simply merge on an oscular 'field' at one side of the animal (e.g. *Hyalonema*). Budding is common in many non-stalked species.

Most hexactinellids require a hard substratum for settlement (either rocks or shell debris), but two groups, all Amphidiscophora (including *Hyalonema* and *Monorhaphis*) and Euplectellinae (including *Euplectella*) anchor in soft sediments with special basal spicules. Deviations from the vase-form in a few genera are thought to represent adaptations to the deep sea environment. In four species (in the genera *Polioplogon*, *Sericolophus*, *Platylistrum* and *Tretopleura*) the sponge body is bilaterally symmetrical, an adaptation which is thought to expose the dermal surface to currents and place the atrial chamber in the lee of currents. In some specimens of *Farrea oc̄ea* and another species (possibly *Chonelasma*), tubes and oscula alternate along a central axis in an apparent form of metamery (Tabachnick, 1991).

Glass sponges fall into two gross morphological categories based on the type of skeleton the animal produces (Figure 2): those with a loose spicule skeleton—lyssacine—that is held together by strands of the trabecular tissue, and those with a fused spicule skeleton—dictyonine—that forms a rigid three-dimensional scaffold. Sponges with a dictyonal framework form part of the Subclass Hexasterophora (Figure 55).

2.2. Structure of the body wall

The whole sponge is formed by a single continuous syncytial tissue, the trabecular reticulum (Section 3.4), which stretches from the outside or dermal membrane to the inside, or atrial membrane, and encloses cellular components of the sponge. The dermal and atrial tissues are called membranes by convention because they are two-dimensional syncytial extensions of the trabecular reticulum.

The body wall is divided into three regions (Figures 3 and 4). The major component occupying the centre of the body wall is the choanosome, which contains the flagellated chambers. The choanosome is bordered on either

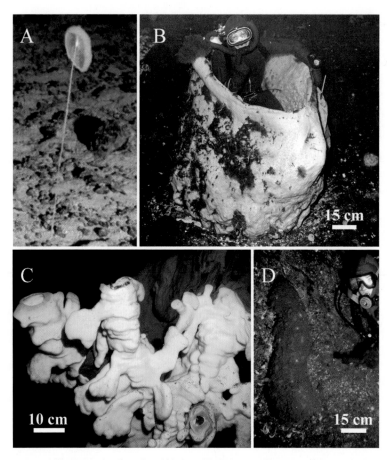

Figure 1 Glass sponge diversity. (A) A stalked glass sponge, possibly *Caulophacus* sp. photographed by submersible at an oligotrophic site in the North Pacific (courtesy of S. Beaulieu). (B) *Scolymastra joubini*, a vase-shaped sponge that grows large enough for a diver to perch inside the osculum during a 1967 dive at 54 m (180 ft) at McMurdo Sound, Antarctica (courtesy of P. K. Dayton). (C) *Aphrocallistes vastus*, a billowy series of cream-yellow tubes, photographed at 35 m depth in Saanich Inlet, British Columbia, Canada. (D) *Rhabdocalyptus dawsoni*, a tube-shaped sponge that hangs with osculum down from the near vertical fjord walls at 30 m in Barkley Sound, Canada (S. Leys, unpublished data).

side by a network of fine strands of the trabecular reticulum called the peripheral trabecular network (Reiswig and Mehl, 1991; Leys, 1999). The inner and outer trabecular networks occupy a similar amount of the body wall (5–12%) in both *F. occa* and *Rhabdocalyptus dawsoni*.

Water enters the sponge through pores in the dermal membrane (Figure 4A) that have a diameter of 5–22 μm (*F. occa*) and 4–30 μm (*Rhabdocalyptus*

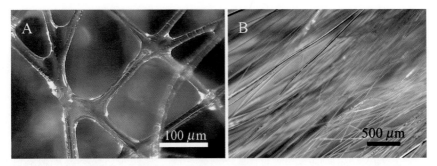

Figure 2 Cleaned spicule skeletons from glass sponges. (A) Dictyonine (fused) from *Aphrocallistes vastus*. (B) Lyssaccine (loose) from *Rhabdocalyptus dawsoni*.

dawsoni) (Mackie and Singla, 1983; Reiswig and Mehl, 1991). The size and shape of incurrent and excurrent canals have been determined both from sections of several species and from plastic casts of the canal systems in *Rhabdocalyptus dawsoni*, *Aphrocallistes vastus* and *Scolymastra joubini* (Leys, 1999; Bavestrello *et al.*, 2003). In *Rhabdocalyptus dawsoni*, the outer peripheral trabecular network lines a loose system of broad, interconnected channels. From this region, distinct incurrent canals (lined by the trabecular reticulum) lead into the body of the sponge and taper from a broad starting diameter of 1.25 mm just below the dermal membrane to approximately 0.5 mm in the centre of the body wall, whereupon they give rise to numerous narrow branches that terminate at individual flagellated chambers. Casts of the incurrent canals show them to be discrete channels throughout their length (Leys, 1999). Canals in *Aphrocallistes vastus* are also discrete channels, but are more uniform in diameter throughout their length. Incurrent and excurrent canals are interlacing pillars 550 and 350 μm in diameter, respectively. Branches off the principal incurrent canals 200 μm in diameter bifurcate once before terminating at flagellated chambers.

In *Rhabdocalyptus*, the fine branches of each incurrent canal terminate in several cup-shaped spaces that surround flagellated chambers. The water enters the chambers through over a hundred pores (prosopyles) formed in the trabecular reticulum, and because there is a second layer of the trabecular reticulum that fits snugly around each collar in the flagellated chamber, the water must be drawn through the collar mesh into the flagellated chamber (Figure 4B). Chambers are large, 55 to 70 μm internal diameter cavities in *Aphrocallistes vastus* and *Rhabdocalyptus dawsoni* and up to 100 μm internal diameter in *Oopsacas minuta*. The water exits the flagellated chambers through large openings (apopyles), 20–35 μm in diameter, directly into small 'collecting' excurrent canals 100–150 μm in diameter. These in turn empty into tubular excurrent canals that increase in diameter from 0.3 to 1.5 mm from the centre of the sponge body wall to the peripheral trabecular

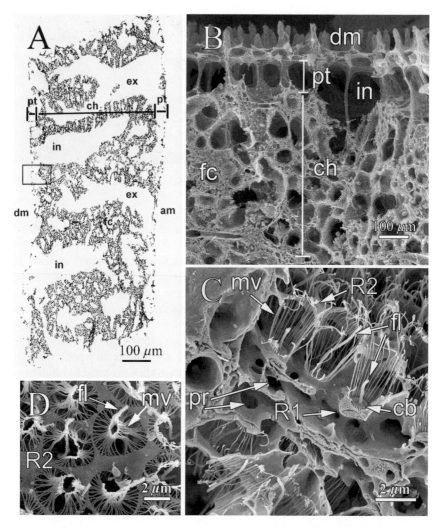

Figure 3 The canal system and tissue structure of the body wall in lysaccine (A, D) and dictyonine (B, C) glass sponges (A, Leys, 1999). (A) Composite of paraffin sections through the body wall of *Rhabdocalyptus dawsoni*. A region equivalent to that in the box is shown in B. (B) Scanning electron micrograph of a fracture through the outer wall of *Aphrocallistes vastus*. Discrete incurrent (in) canals lead via fine branches to flagellated chambers (fc) in the choanosome (ch). The outer zone of both incurrent and excurrent canals (ex) in both sponges consists of a loose network of the trabecular reticulum, the peripheral trabecular network (pt) that supports the dermal (dm) or atrial (am) membranes. (C) A fracture through the wall of a two adjacent flagellated chambers in *Aphrocallistes vastus*. Primary (R1) and secondary (R2) reticula, extensions of the trabecular reticulum, incurrent pores, prosopyles (pr), collar bodies (cb) branching from choanocytes, with collar microvilli (mv) and flagellum (f). (D) A view from the inside of the chamber of *Rhabdocalyptus dawsoni* showing the secondary reticulum (R2) surrounding the collar microvilli (mv) and flagellum that arise from each collar body.

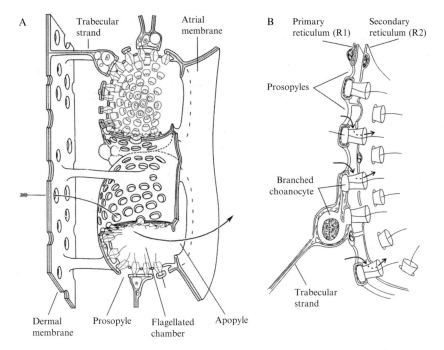

Figure 4   Hexactinellid soft tissue organisation. (A) *Farrea occa*: two flagellated chambers connected to the dermal membrane by trabecular strands. Arrow shows water flow (after Reiswig and Mehl, 1991, with kind permission of Springer Science and Business Media). (B) Section through the wall of a flagellated chamber, based on *Rhabdocalyptus dawsoni*. Water (arrows) is drawn through prosopyles in the primary reticulum and through the choanocyte collars by the beating of the flagella. The secondary reticulum surrounds the collars, forcing water through the collar microvilli (after Leys, 1999).

network on the atrial side. At the atrial side of the sponge the excurrent canals merge once again to form a loose network of broad interconnected canals. The water passing through these canals vents into the atrium of the sponge through openings 100 to 200 μm diameter.

3. CELLS AND SYNCYTIA

3.1. Definitions

Hexactinellids differ from other sponges in consisting largely of a single syncytial tissue, the trabecular syncytium, but cells are also present and some explanation is necessary as to how these words will be used in this review, as terminology has always been a problem in this group.

Eukaryotic cells may be defined as membrane-enclosed cytoplasmic bodies containing a single nucleus. If a number of such units fuse, or become multinucleate by internal nuclear division, they are referred to as syncytia. The trabecular reticulum of hexactinellids is a syncytium, containing thousands of nuclei within a single, ramifying, cytoplasmic domain that permeates all parts of the sponge. As noted in Section 2.2, it forms the dermal and atrial covering layers and the trabecular strands, and provides the structural support for the flagellated chambers, as well as the secondary reticulum that directs water flow through the choanocyte microvilli (Figure 4B). We also recognise a second syncytial tissue, here termed sclerosyncytium, that secretes the spicules.

Cells in animals, plants and fungi are frequently joined by specialised junctional structures such as gap junctions, plasmodesmata and perforate septa that permit some degree of direct exchange of materials between cells via aqueous pathways. This does not disqualify the nucleated domains on either side from consideration as cells in the sense of 'separate working units of protoplasm' (Dahlgren and Kepner, 1930). Accordingly, we also recognise cells in hexactinellids, even though they are typically connected to one another or to the trabecular syncytium by cytoplasmic bridges. Despite sharing a common plasmalemma, the cells have their own distinctive cytoplasmic components. The intercellular bridges may be open initially, but in the mature state they become 'plugged' with dense material. These plugs evidently provide partial barriers against the free exchange of components between the two sides, judging from the fact that the cytoplasm on either side of plugged junctions is often markedly different (Figure 5A). We recognise several categories of cells, chief of which are branched choanocytes (formerly termed 'choanosyncytium'), archaeocytes, cystencytes and gametocytes. Leys (2003b) further discusses the concept of syncytiality in hexactinellids and its phylogenetic significance.

The chief sources drawn in this section are Reiswig (1979a) and Leys (1999) for *Aphrocallistes vastus*; Mackie and Singla (1983) for *Rhabdocalyptus dawsoni*; Reiswig and Mehl (1991) for *Farrea occa*; Mehl *et al.* (1994) for *Schaudinnia rosea* (reported as *S. rosea*); Köster (1997) for species of *Rossella* and Boury-Esnault and Vacelet (1994), Perez (1996), Boury-Esnault *et al.* (1999) and Leys (2003a) for *Oopsacas minuta*.

3.2. Plugged junctions

These junctions were first described in *Rhabdocalyptus dawsoni* as 'perforate septal partitions' or 'plugs' (Mackie, 1981; Mackie and Singla, 1983). They have since been observed in many other hexactinellids studied by transmission electron microscopy and may be regarded as a defining feature of the group.

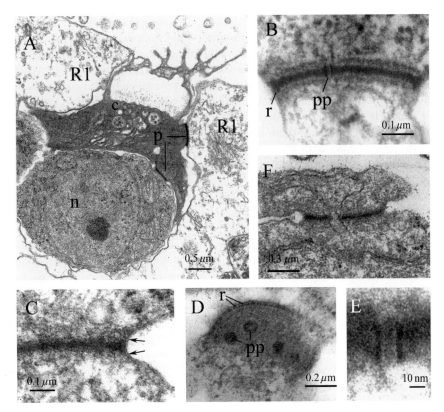

Figure 5 Plugged junctions and plug structure in *Rhabdocalyptus dawsoni* (A–D, F, from Mackie and Singla, 1983) and *Rossella racovitzae* (E, courtesy of J. Köster, 1997). (A) Developing choanocyte with a single collar body (c). The cytoplasm of the collar body is markedly darker than that of the nucleated body (n) and the trabecular syncytium (R1) to both of which it is connected by plugged junctions (p). (B) Section through a plug showing trilaminar structure revealed by chrome-osmium fixation. Rodlets (r) and pore particles (pp) are visible in the plug. (C) Section through a plugged junction showing continuity of the cell membrane (arrows) around the edge of the plug. (D) Horizontal section through a plug showing rodlets (r) and pore particles (pp). (E) Pore particle at high magnification. (F) Membrane-bounded vesicle apparently in transit through a plug.

In *Oopsacas minuta*, the equivalent structures were termed 'dense osmiophilic junctions' (Boury-Esnault *et al.*, 1999). In *Rhabdocalyptus dawsoni*, they are seen as flat, trilaminate plaques or discs inserted into bridges between cells (Figure 5B). The plaque is a sandwich whose outer layers consist of loose fibrous or granular material and the inner of finer, electron-dense material. The whole plaque is about 50 nm thick. No lipid bilayer has ever been observed within or enveloping the plug. In fact, the cell membrane can be seen going around the edges of the plug (arrows in Figure 5C) joining the cell membranes of the domains on either side. Peripheral rodlets form an orderly array around the perimeter of the plug, lying closely adjacent to the cell membrane (Figure 5B and D). Structures interpreted as pore particles are also visible within the plug. In *Rhabdocalyptus dawsoni*, these seem to be hollow cylinders about 50 nm in diameter, with a 7 nm central channel (Figure 5B and D). In *Rossella racovitzae*, the central channel is given as 11 nm (Figure 5E), and there are other minor differences in the dimensions of the plug components, but the overall picture is very similar. The term pore particle implies that the central channel is a true pore but in some images it contains electron-dense material similar to that of the surrounding plug material. Finally, membrane-bounded saccules or cisternae are frequently observed apparently crossing plugs (Figure 5F). Reiswig and Mehl (1991) call them 'transit vesicles'. They are probably not permanent fixtures as many plugs lack them, and it is not clear if the vesicles pass through expanded pore particles or insinuate themselves directly through the plug material.

As noted, the cytoplasm on the two sides of a plug often differs strikingly in terms of electron density and organellar content. Plugs evidently represent a barrier to the free passage of materials, allowing certain components to cross while blocking others, but no experimental work has been done to determine precisely what sort of materials can cross, or how they are selected. It is clear that plugs are not a barrier to the passage of ions, as electrical impulses can spread from the trabecular syncytium into choanocytes, bringing about flagellar arrests (Section 5.3.3).

Stages in the formation of plugs have been described in *Rhabdocalyptus dawsoni* (Pavans de Ceccatty and Mackie, 1982; Mackie and Singla, 1983; reviewed by Leys, 2003b). Plugs first appear as plaques of electron-dense material lying close to the nuclear membrane (Figure 6A and B). In these areas, pores in the nuclear membrane sometimes show material (presumably ribosome precursors) in transit to the cytoplasm. The plaques become associated with small vesicles in the vicinity of the Golgi complex (Figure 6B and C), but there does not appear to be any regular progression of the plaque material through the classic *cis–trans* Golgi sequence. The fate of the small vesicles is uncertain, but larger vesicles, probably of *trans*-Golgi origin, are later seen around the edges of the plaque (Figure 6D). These appear to fuse, forming a more or less complete, toroidal cisterna around the plug (Figure 6E). Pore particles can already be seen in the plaque at this stage. The plaque with its satellite vesicles or cisterna moves into a region where a narrow bridge will form connecting adjacent cytoplasmic domains. The cisternal membrane fuses with the cell membrane, lodging the plaque firmly into the equatorial region of the bridge. At this point the structure becomes a 'plug'. Plugs are often seen attached to the cell membrane on one side only (Figure 6F),

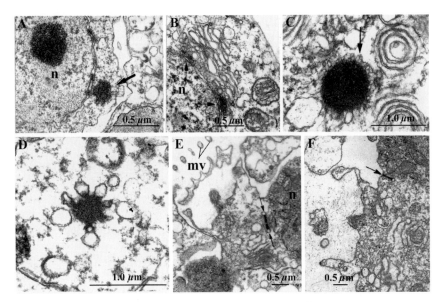

Figure 6 Formation of plugged junctions in *Rhabdocalyptus dawsoni*. A–D after Pavans de Ceccatty and Mackie (1982); E, F after Mackie and Singla (1983). (A) First appearance of plug (arrow) adjacent to pores in the nuclear membrane. (B) Early plugs (arrows) on *cis* side of Golgi complex. (C) Plug surrounded by small vesicles (arrow). (D) Plug surrounded by larger vesicles. (E) Plug still free in the cytoplasm, cut transversely, showing vesicles at the ends (arrows). (F) Plug attached to cell membrane at one end (arrow), the other still free. mv, microvilli; n, nucleus.

presumably a transitional stage preparatory to full insertion. This sequence of events is based entirely on interpretation of static images, and may be defective in some respects. For instance, if plugs eventually detach from the cell membrane and go through a recycling process some similar images might result.

3.3. Hexactinellid plugs compared with other junctions

Hexactinellid plugged junctions (Figure 7A) resemble the pit plugs of red algae and the plasmodesmata of higher plants more than they do the gap junctions of other animals. In the simplest, and presumably most primitive, algal pit plugs, the core of the plug is inserted into a narrow intercellular bridge lined by the plasma membrane (Figure 7C). In other pit plugs, caps are associated with the core on one or both sides, and cap membranes are also present, effectively isolating the core in an extracellular compartment (Pueschel, 1989). Traversing endoplasmic reticulum (ER) cisternae may be

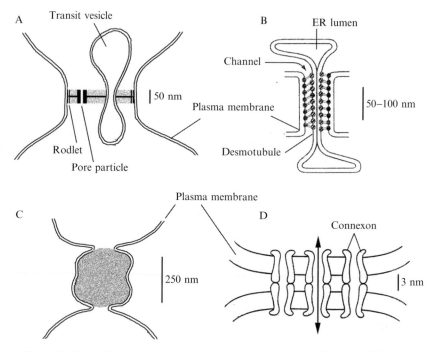

Figure 7   Intercellular junctions in animals and plants. (A) Hexactinellid plugged junction (based on *Rhabdocalyptus dawsoni*, from Mackie and Singla, 1983). (B) Plasmodesma in a higher plant (after Mezitt and Lucas, 1996, with kind permission of Springer Science and Business Media). (C) Red algal pit plug (after Pueschel, 1989). (D) Mammalian gap junction (after Goodall, 1985).

present during core deposition, but they disappear after the plug is complete. There are no reports of pores within the mature plug core and no definite information on what materials can cross, but it seems likely that the plugs allow transit of certain materials on a selective basis.

In typical plasmodesmata (Figure 7B), an ER cisterna (desmotubule) extends through a plasmalemma-lined pore. While the inner faces of the desmotubule are closely appressed, globular proteins on its outer face are loosely linked to proteins on the inner face of the adjacent plasmalemma, creating 8–10 tortuous intercellular paths (microchannels) each 2–3 nm in diameter. Thus, plasmodesmata establish both a cytoplasmic and an ER continuum that extends from the shoot apical meristem all the way down the plant axis (Mezitt and Lucas, 1996). Small metabolites, ions and signalling molecules up to about 1000 MW can pass through plasmodesmata. Viruses spread between cells by secreting movement proteins which act as carriers for the virus's nucleic acids, which in turn produce viral replicas on reaching the other side. Interestingly, the plasmodesmal microchannels undergo dilation during viral transport, showing that plasmodesmata, like gap junctions, are labile structures.

Gap junctions (Figure 7D) are formed by transmembrane proteins that form cylindrical structures (connexons) that, when aligned in apposing plasma membranes, form a continuous aqueous channel 1.5 nm in diameter connecting the two cell interiors. Numerous such channels assemble to form a single gap junction. Gap junctions differ in permeability according to the character of the component proteins (connexins, innexins) and they can close completely. In the open state, they are freely permeable to ions, allowing electrical and dye coupling. Molecules up to about 1000 MW can also cross mammalian gap junctions.

Despite profound structural differences, it is clear that gap junctions and plasmodesmata function in remarkably similar ways, and we predict that hexactinellid-plugged junctions will prove to function similarly in regulating intercellular traffic.

3.4. Trabecular syncytium

Modern understanding of the trabecular syncytium dates from Ijima (1901) who found that the entire trabecular system in *Euplectella marshalli* consisted of a network of syncytial protoplasm, containing 'free' nuclei. This ran counter to an earlier view (Schulze, 1899) that the trabeculae and other internal surfaces were covered by a flat epithelium. Ijima thought that the 'trabecular cobweb' was probably unstable during life owing to 'protoplasmic contractility'. He regarded the trabecular system as distinct from the tissue forming the walls of the flagellated chambers, and also cellular elements (archaeocytes and thesocytes). Later work by this author (Ijima, 1904, 1927) suggested that the trabecular syncytium was a feature common to all hexactinellids. Owing to difficulties in obtaining well-fixed material, little new histological work was done on the group until Reiswig (1979a) took advantage of the availability of hexactinellids within SCUBA range around Vancouver Island, British Columbia to obtain well-fixed material of *Aphrocallistes vastus* and *Chonelasma* (now *Heterochone*) *calyx*. His optical and electron microscopic study confirmed the syncytial nature of the trabecular reticulum.

The trabecular syncytium constitutes much the greater part (ca. 75%) of the soft tissues of the sponge including the dermal and atrial membranes and it provides the interface separating the living tissues from water flowing through the sponge, except for the collars of the choanocytes which project into the flagellated chambers. Hexactinellids lack pinacocytes, but the term 'pinacoderm' has been used in *Oopsacas minuta* to describe the surface layers mentioned above (Boury-Esnault and Vacelet, 1994; Perez, 1996). The trabecular tissue can take the form of flat, perforated sheets like the dermal membrane or thin trabecular strands, often <1.0 μm in diameter. The trabecular syncytium extends into the lining of the flagellated chambers (Figure 4B) providing a supporting framework (primary reticulum, R1) for the collar bodies of the branched choanocytes (Mackie and Singla, 1983). With the possible exception of *Caulophacus cyanae* (Boury-Esnault and de Vos, 1988) and *Dactylocalx pumiceus* (Reiswig, 1991) trabecular processes also extend into the lumen of the flagellated chambers forming the secondary reticulum (R2) at the mid-collar level, while in *F. occa*, centripetal projections from R2 form yet another reticular layer, the inner membrane, at the level of the flagellar tips (Reiswig and Mehl, 1991). Reiswig's comment (1979a) that the trabecular syncytium 'may constitute a single, continuous cytoplasmic network throughout an entire specimen' has been echoed by later workers and is a central concept in all discussions relating to nutrient transport and electrophysiological conduction pathways.

The trabecular syncytium contains countless small nuclei (ca. 2–3 μm), several of which may be visible in transmission electron micrographs (TEM) within the same mass of cytoplasm (Figure 8A). The cytoplasm contains mitochondria, rough ER, Golgi elements, cytoskeletal components (actin filaments and microtubules) and a wide variety of vesicular inclusions including phagosomes and residual bodies from phagocytic processing. The composition of the cytoplasm varies locally, depending on activities such as wound repair, growth, regeneration and nutrient uptake.

Ijima's suggestion (1901) that the trabecular syncytium is a labile, 'contractile' structure has been borne out by observation of thin pieces of living, intact sponge tissues ('spicule preparations') grown between glass slides and coverslips ('sandwich cultures', Leys, 1995; Wyeth *et al.*, 1996; Leys, 1998; Wyeth, 1999). Cytoplasmic streaming has been observed throughout the trabecular reticulum. It seems likely that the thicker tracts seen in fixed tissues, measuring up to 20 μm in diameter and termed 'cord syncytia' (Reiswig, 1979a), represent 'rivers' of actively streaming cytoplasm stabilised *in situ* by fixation. The velocity of particle transport in these bulk streams is ca. 0.33 μm s^{-1} (Leys, 2003b), some eight times faster than in thin areas of the syncytium, but within the range of values observed in plated aggregates (Leys, 1995). It now seems clear that the activities observed in the *in vitro* preparations fairly represent processes carried on *in vivo* within the trabecular syncytium and that the syncytium provides the main pathway for all forms of translocation within the sponge as well as for spread of electrical events. Transport in plated aggregates and electrical conduction are dealt with in later sections.

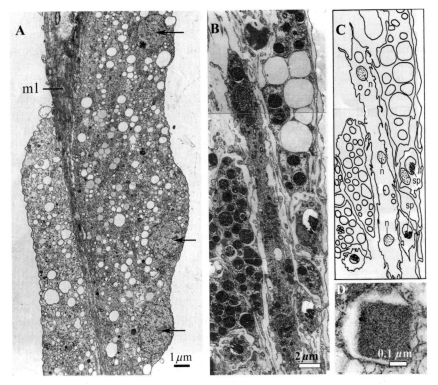

Figure 8 (A) Trabecular syncytium in *Rhabdocalyptus dawsoni* with three nuclei (arrows) in a continuous mass of cytoplasm; ml, mesolamella (from Mackie and Singla, 1983). (B, C) Sclerosyncytium in *Oopsacas minuta*. The section photographed in B passes longitudinally through a sclerosyncytium containing three nuclei in a continuous mass of cytoplasm, as drawn in C. On the right, several sclerocytes are cut transversely, showing nuclei (n) and fragments of spicules (sp) within vacuoles lined with silicolemma (after Boury-Esnault *et al.*, 1999). (D) Axial filament in a sclerocyte of *Oopsacas minuta* (from Leys, 2003a).

3.5. Sclerocytes and sclerosyncytium

In the development of *Farrea sollasii*, spicule production is described as occurring within sclerocytes, originally mononucleate but becoming multinucleate (Okada, 1928). Subsequent work on other species has confirmed that spicule production is intracellular within a syncytial structure, here termed sclerosyncytium (Figure 8B and C). Young sclerocytes in *Oopsacas minuta* embryos bear numerous, long pseudopodial extensions (Leys, 2003a). It was originally thought (Mackie and Singla, 1983) that the sclerosyncytium was completely isolated from surrounding trabecular elements, but in *Oopsacas minuta* embryos sclerocytes are connected by plugged bridges to surrounding tissues (Leys, 2003a). Spicules develop within a silica deposition space (Section 6.5.1) bounded by a membrane (silicalemma) and containing an axial filament. Silica is probably initially deposited around the axial filament. The filament is quadrangular in cross section (Figure 8D), which determines the similar form of the completed spicule (Reiswig, 1971a). After desilication, sections of hexactinellid tissues show holes where spicules were formerly located. These sometimes contain traces of organic material, but the nature of the axial filament and other organic matter present is unknown. Spicule morphology and silication is covered in Section 6.

3.6. Archaeocytes

Typically clustered in groups, or congeries (Ijima, 1901), archaeocytes are spherical or sub-spherical cells, ca. 3–5 μm in diameter in *Aphrocallistes vastus, Oopsacas minuta* and *Rhabdocalyptus dawsoni* but up to 8.8 μm in *F. occa*, with densely granular cytoplasm and numerous mitochondria (Figure 9A). The Golgi component and ER are not prominent and phago-somes are rare. Archaeocytes in congeries are frequently attached to one another and to the trabecular syncytium by plugged junctions. Their rounded form and lack of pseudopodia suggest that they are not mobile like archae-ocytes in sponges of other classes. It is extremely likely that archaeocytes in hexactinellids as in other sponges are pluripotent cells giving rise to various other cell types. Spermatogonia and oogonia arise within archaeocyte conge-ries (Section 8.1.2). The nucleated domains of the branched choanocytes closely resemble archaeocytes and are presumably derived from them, as are cystencytes according to Ijima (1901).

3.7. Cells with inclusions

Sponges have various types of cells with inclusions, some with large inclu-sions and some with small (Simpson, 1984; Harrison and de Vos, 1991). They go under a wide variety of names and their functions are poorly understood.

In hexactinellids, round or ovoid cells with large, empty-looking vacuoles were described in *Rhabdocalyptus dawsoni* as spherulous cells. They appear similar to the vacuolar cells (the preferred term) observed in other species. Thinly dispersed flocculent matter can be seen in the vacuole but the content is probably largely aqueous. In *Oopsacas minuta* and *Rossella nuda*, some vacuolar cells have several vacuoles rather than just one. This was also noted in *Rhabdocalyptus dawsoni* (Leys, 1996). Köster (1997) recognises two types of vacuolar cells in *Rossella nuda*, Type 1, large cells (<20 μm) with rounded

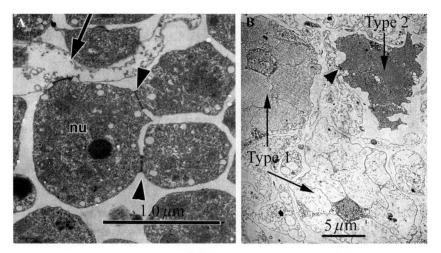

Figure 9 (A) An archaeocyte (n, nucleus) in *Rhabdocalyptus dawsoni* is attached to two other archaeocytes by plugged junctions (arrowheads) and similarly to the trabecular syncytium (arrow) (from Leys, 2003b). (B) Two types of vacuolar cell in *Rossella nuda*. Arrowhead shows a plugged junction (courtesy of E. Köster).

contours and lying within pockets in the trabecular syncytium, but not joined to them by plugged junctions, and Type 2, smaller with irregular borders and denser contents, connected to the trabecular reticulum with plugged junctions (Figure 9B). Köster's specimens were collected in late summer and vacuolar cells were very prominent in them, suggesting that they may contain nutrients stored for the Antarctic winter.

Reiswig (1979a) described thesocytes as oval or globular cells, $5 \times 7 \mu m$ to $7 \times 12 \mu m$ in dimensions, with an interior almost filled by a single homogeneous inclusion that leaves the nucleus flattened to one side in a thin peripheral layer of cytoplasm. Similar but larger ($<17 \mu m$) cells occur in *Rhabdocalyptus dawsoni*, sometimes clustered in groups. In some cases, the inclusion body contains crystals, an indication that the inclusion is proteinaceous. The cells are completely enveloped by the trabecular syncytium. Similar cells have since been described as cystencytes (the preferred term) in both *F. occa* and *Oopsacas minuta*. In *F. occa* (Figure 10A), the electron-dense inclusion contains patches of lower density crystalloid material with a regular layer spacing of 3.7 nm.

Granular cells, first observed by Schulze (1899, 1900) measure $<8 \mu m$ in *Rhabdocalyptus dawsoni* and *F. occa*, and contain numerous small ($1–2 \mu m$) electron-dense granules. Unlike cystencytes they do not lie completely within pockets of the trabecular syncytium, but are typically attached to the mesohyl and project through openings in the trabecular syncytium into the external medium (Figure 10B). In some cases, plugged junctions are seen

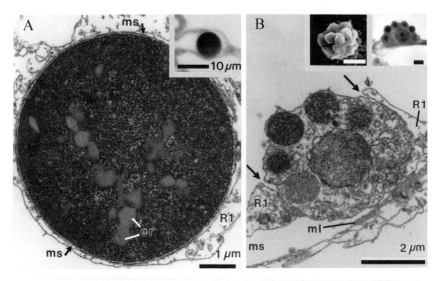

Figure 10 Cells in *Farrea occa* (from Reiswig and Mehl, 1991, with kind permission of Springer Science and Business Media). (A) Cystencyte-containing crystals (cr) lying within a thin mesohyl space (ms) bounded by primary trabecular reticulum (R1). Inset shows a cystencyte seen by light microscopy. (B) Granular cell projecting from the mesohyl space (ms) through a gap (arrows) in the primary reticulum (R1). Insets (scale bars, 2 μm) show granular cells seen by scanning electron microscopy and light microscopy.

at the trabecular interface. Granular cells are also prominent in *Rossella vanhoeffeni* (reported as *Aulorossella vanhoeffeni*, Salomon and Barthel, 1990) and in *Schaudinnia rosea* (reported as *S. arctica*, Mehl *et al.*, 1994) where they measure 10–25 μm in diameter and are densely packed along the walls of the aqueous canals. The cells show marked fluorescence with aureomycin indicating high levels of Ca^{2+} in the granules, and weaker fluorescence with calcein dyes. Mehl *et al.* (1994) suggest that the cells store lectins. These workers also note the presence of what appear to be stages in the transformation of archaeocytes into granular cells.

If vacuolar and granular cells are indeed storage sites as suggested, it would be interesting to study their abundance and distribution at different times of the year.

3.8. Choanocytes

Hexactinellids lack choanocytes of the type found in other sponges, which are compact, well-defined cells bearing a flagellum surrounded by a collar of microvilli. Instead, we find branching structures consisting of a basal nucleated domain that sends out processes bearing separate collar bodies ('choanomeres') each resembling the apical portion of a conventional choanocyte (Figure 11A). The terminology here has been especially problematic. The term choanosyncytium was used by Reiswig (1979a) to describe *all* the tissues lining the flagellated chambers. Electron microscopy (Mackie and Singla, 1983) subsequently showed that the chamber wall is a composite of trabecular

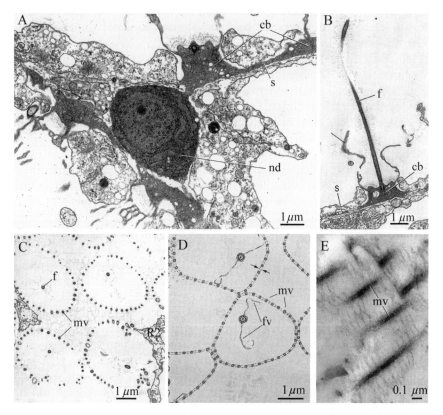

Figure 11 Structure of branched choanocytes in *Rhabdoclyptus dawsoni* (A–C, from Mackie and Singla, 1983), *Aphrocallistes vastus* (D, from Mehl and Reiswig, 1991, with kind permission of Springer Science and Business Media) and *Rossella racovitzae* (E, courtesy of E. Köster). In A, several collar bodies (cb) are shown attached by stolons (s) to a single nucleated domain (nd). This cell supplies collar bodies to two flagellated chambers. (B) Shows a collar body (cb) in a section passing vertically through the flagellum (f). Part of a stolon (s) is also visible. (C) A transverse section at the collar level shows flagellum (f) and collar microvilli (mv). (D) A similar section in *Aphrocallistes vastus* shows flagellar vanes (fv) and the mucous sheets (arrows) spanning the gaps between the microvilli (mv). (E) Mucous sheet spanning gaps between microvilli (mv) shown in a near-longitudinal section through the collar.

reticular elements and the nucleated domains ('choanoblasts') and their processes that give rise to collar bodies. The word choanosyncytium has continued in general use to refer to the nucleated cell body plus its attached collar bodies. The latter are located at the ends of cytoplasmic processes that sometimes stretch for considerable distances in the form of narrow stolons. The collar bodies may be connected to the nucleated zone by open cytoplasmic bridges, but more frequently by plugged junctions. So far as is known, only one nucleus is present in the fully formed structure although there may be three or more stolons (Figure 11A and B), each with two or three collar bodies. The exact numbers are hard to estimate, as serial reconstructions have not been carried out.

It is difficult to know what to call a complex of this sort. The term choanosyncytium is misleading, as it implies a multinucleate structure, and there is no evidence that more than one nucleus is present at any stage in the development of the complex. On the other hand, if we are dealing with a single, elaborately branched cell, it is one which has several distinct cytoplasmic domains, of which the collar bodies are segregated from the rest of the cell by plugged junctions (Figure 5A). We have earlier seen that plugs occur at intercellular junctions, but here they are clearly intracellular. Perhaps because of these anomalies, workers have continued to use the older terminology but it seems more appropriate and accurate to speak of these complex cells as *branched choanocytes*, and we will describe them as such in this account, abandoning the earlier terms choanosyncytium and choanoblast. The picture is essentially similar in *Rhabdocalyptus dawsoni*, *F. occa* and *Oopsacas minuta*, but in *Dactylocalyx pumiceus* the collar bodies appear to be completely separate structures, isolated from any nucleated structure (Reiswig, 1991).

In *Rhabdocalyptus dawsoni*, the nucleated portion of the branched choanocyte resembles a large archaeocyte having a compact, rounded form and electron-dense cytoplasm. It has a richly developed rough ER and processes run out from it leading to the collar bodies. Like archaeocytes the nucleated bodies often appear grouped in clusters, as well seen in *Aphrocallistes vastus* (Leys, 1999). The cytoplasmic strands running to the collar bodies may be short, wide, open bridges (possibly an early stage) or they may have longer process with a narrow neck containing a plugged junction. What appear to be stages in the formation of these junctions have been described, but such reconstructions are necessarily speculative (Mackie and Singla, 1983). The collar body bears a single long flagellum up to 20 μm long (Figure 11B) containing the usual $9 + 2$ array of microtubules. In most hexactinellids, the flagellum has a simple form (Figure 11C) but in *Aphrocallistes vastus*, it bears wing-shaped projections (Figure 11D, 'flagellar vanes') resembling those seen in some genera of cellular sponges (Mehl and Reiswig, 1991).

In *Schaudinnia rosea* (reported as *S. arctica*), the flagella also have vanes, 1.5 μm long, and thread-like cross section (Mehl *et al.*, 1994).

A collar of 30–50 microvilli projects from the outer surface of the collar body. In *Schaudinnia rosea* (reported as *S. arctica*), the cell membrane lying between the microvilli and the flagellum bears a 300 nm thick layer of fibrous material ('glycocalyx') that is organised into distinct proximal and distal layers. Throughout the Hexactinellida, the distal parts of the collars are inserted through pores formed within the secondary reticulum and probably fit tightly enough in life to ensure that the water entering the flagellated chambers must pass between the microvilli rather than around the outside of the collar (Figure 4B). Investing the individual microvilli and spanning the gaps between them are meshes of presumed glycoprotein. In *Rhabdocalyptus dawsoni*, the material was interpreted as a 'filter' with mesh size 50 × 200 nm (Mackie and Singla, 1983), while in *Aphrocallistes vastus*, the holes in the mesh measured ca. 22 × 70 nm (Reiswig, unpublished data). In *Rossella racovitzae*, the filter is shown as a perforated sheet (Figure 11E), the ovoid pores measuring 45 × 125 nm (Köster, 1997). Cross sections through the collar sometimes show the filter as a double structure, one sheet spanning the inner sides of the microvilli and the other the outer (Figure 11D, arrows; Figure 12). A double filter has also been seen in *Aphrocallistes vastus* (Mehl and Reiswig, 1991), *Schaudinnia rosea* (reported as *S. arctica*, Mehl *et al.*, 1994) and *F. occa* (Reiswig and Mehl, 1991) and may be a characteristic of all such filter structures in hexactinellids.

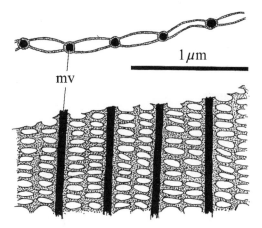

Figure 12 Mucous sheets in the collar of *Rossella racovitzae* shown diagrammatically in transverse and longitudinal section (courtesy of E. Köster).

It is clear that there is little fundamental difference between the apical structures of hexactinellid collar bodies and of choanocytes in other sponges. It is the presence of multiple collar bodies on the branches of a ramifying cell whose nucleus lies far away that makes hexactinellid choanocytes so unique.

3.9. Mesohyl

Ijima (1901) could find no evidence in *Euplectella marshalli* for anything corresponding to the mesohyl of other sponges, but a thin extracellular fibrous layer 0.05 to 0.10 μm thick was observed by Reiswig (1979a) within the trabeculae and was readily demonstrable by its acidophilic staining reaction. The fibrils composing this material showed a 45 to 50 nm periodicity. Mackie and Singla (1983) proposed that this presumably collagenous 'mesolamella' provided internal support for the trabeculae and was probably secreted by the trabecular syncytium itself, as no other cellular elements were present in many regions where it occurred. Reiswig and Mehl (1991) reach similar conclusions for *F. occa*. Dissociated cellular material from *Rhabdocalyptus dawsoni* has been found to adhere well and to spread out on surfaces coated with aqueous extracts from conspecifics (Leys and Mackie, 1994; Section 4.1). It is likely that the extracts contain extracellular matrix components deriving from the mesohyl, and that these mediate cell-substrate adhesion in the normal state of the tissue, as with jellyfish mesogloea (Schmid et al., 1991). The mesohyl is found in most parts of the trabecular syncytium (Figure 8A), but it is absent in the secondary reticulum and inner membrane. As well as forming a supportive lamella within trabeculae, the mesohyl invests various discrete cells and bacterial symbionts, but there is no evidence that it provides a substrate for movement of migratory cells as seen in cellular sponges.

4. TISSUE DYNAMICS

4.1. Reaggregation of dissociated sponge tissue

Wilson (1907) first demonstrated that if a sponge is dissociated by squeezing through cheesecloth, the cells will come together and reconstitute a new sponge. The process depends on cell adhesion molecules and is homeotypic and thus of interest in relation to self–nonself recognition. Extensive studies using reaggregation models in sponges have now shown that sponges possess an effective polymorphic immune system (see reviews by Fernandez-Busquets and Burger, 1999; Müller et al., 1999). Reaggregation experiments have mostly used demosponges, although calcareous sponges and hexactinellids also reaggregate (McClay, 1972; Pavans de Ceccatty, 1982; reviewed in Simpson, 1984). Typical aggregates in demosponges form an opaque sphere of cells; in only a few cases, however, can the cells reorganise themselves into a functional aquiferous system, and only one paper documents choanocyte chambers forming in such a sponge (Van de Vyver and Buscema, 1981).

In hexactinellid sponges, dissociated tissue also forms opaque spheres (Pavans de Ceccatty, 1982). Although much of the experimental work on hexactinellid tissue has been carried out with *Rhabdocalyptus dawsoni* because of the relative ease of collecting this species, reaggregation of dissociated tissue has also been shown in *Aphrocallistes vastus* (Leys, 1998) and *Heterochone calyx* (Leys, unpublished data). The general characteristics of aggregation are similar in all three species. If a 2 cm^2 section is cut off the sponge and squeezed through Nitex mesh into a Petri dish, the tissue dissociates into numerous spherical pieces 5–30 μm in diameter (Figure 13). The pieces may be without a nucleus or have one or several nuclei, and some may consist of several 'cells' joined by cytoplasmic bridges and plugged junctions. If left in a dish of sea water at 10°C for several hours, the pieces come together to form a large opaque sphere, up to 1 mm in diameter. Transmission electron microscopy of sections of aggregates at various times after plating has shown that the tissue is continually reorganizing the cellular and syncytial components within itself (Pavans de Ceccatty, 1982) (Figure 14). Plugged junctions are commonly encountered in the process of being produced and inserted into cytoplasmic bridges between the syncytial trabecular reticulum and a variety of cellular components (Pavans de Ceccatty and Mackie, 1982).

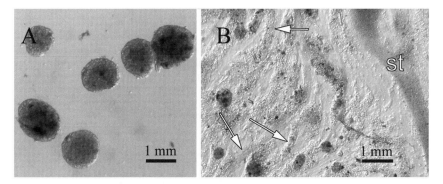

Figure 13 Live aggregates of dissociated tissue from *Rhabdocalyptus dawsoni*. (A) Opaque spherical aggregates and (B) aggregates that have adhered to a substrate coated with acellular tissue extract and in which swaths of cytoplasm (streams, st and arrows) can be seen traversing the coverslip.

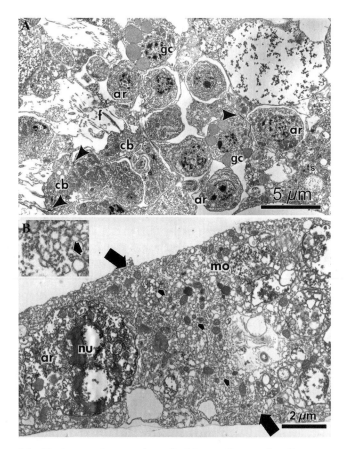

Figure 14 Electron micrographs of thin sections of adherent aggregates (*Rhabdocalyptus dawsoni*). (A) A transverse section through an adherent aggregate showing a region of stationary cytoplasm with archaeocytes (ar), collar bodies (cb) with flagella (f) and microvilli and granular cells (gc). Plugged junctions (arrowheads) connect archaeocytes to the trabecular syncytium (ts) and collar bodies to choano-blasts. (B) A cross section of an adherent aggregate shows microtubule bundles (small black arrows, and inset) form a swath from top to bottom (large black arrows) of the tissue. Microtubule bundles are always associated with numerous mitochondria and membranous organelles (mo). Other vesicles and cells such as archaeocytes (ar) with nuclei (nu) are further away from the microtubule bundles.

Although the contents of spherical aggregates can be studied by thin sectioning, this technique does not allow a view of the relationship between the multinucleate and cellular components, nor does it give a true idea of the extensiveness of the syncytial tissue.

Proof that aggregates truly form a giant cell was found by creating a thin preparation in which the tissue could be examined live by video microscopy, and which could be fixed and preserved at various stages of fusion to illustrate the extent of the cytoskeleton. The idea for this preparation came from observations made first by Stuart Arkett, a postdoctoral fellow with George Mackie, who spent some time trying to make preparations for electrophysiology. Arkett noticed that in some instances, dissociated tissue did not roll together to form opaque aggregates, but instead, adhered to the substrate and spread out to form a thin sheet. Most remarkably, the cytoplasm in the thin sheet of tissue tracked over the substrate in continually moving streams.

The result was not readily replicable. In order to obtain a consistent preparation of adherent tissue from the glass sponge, Leys (1997) resorted to making an extract of extracellular material from the sponge itself (acellular tissue extract, ATE). The extract was made following the technique of Schmid and Bally (1988) and Schmid *et al.* (1991) for causing adhesion of cells from the hydrozoan medusa *Podocoryne*. Tissue from the sponge was rinsed briefly in calcium- and magnesium-free sea water, and soaked four times for 2 hours each in 20 times the volume of distilled water at 4°C. After the final soaking, the tissue was mechanically dissociated with a glass rod in a new volume of distilled water, causing the release of a cloudy suspension. A drop of the suspension pipetted into normal sea water coagulated into a white buoyant solid. Electron microscopy of osmium- and glutaraldehyde-fixed solid extract showed it to consist of cell debris. A tenfold dilution of the extract dried onto coverslips or petri dishes caused the species-specific adherence of dissociated tissue from sponges (Leys, 1997). In subsequent experiments, not only ATE, but also the lectin, concanavalin A (100 g ml^{-1}) and poly-L-lysine (500 g ml^{-1}) also caused adhesion of the dissociated sponge tissue.

4.2. Fusion

When plated on either the ATE substrate or the commercial lectin or poly-L-lysine, dissociated tissue pieces from both *Rhabdocalyptus dawsoni* and *Aphrocallistes vastus* adhere to the substrate and spread thinly within 5–10 s (Leys, 1995, 1998). Only 20 min after plating, the pieces begin to migrate around the dish using broad lamellipodia and long filopodia. Pieces begin to encounter each other about 1 hour after plating. Lamellipodial membranes of separated pieces first overlap for 10–30 min and then, rather than forming cell–cell junctions, the two membranes fuse; very shortly afterwards, organelles can be seen moving across the bridge. The bridge rapidly expands until the two formerly separate pieces of tissue cannot be distinguished (Figure 15). Numerous pieces of tissue join in this way, and after 12–16 hours a vast amount of cytoplasm is enclosed within a single membrane.

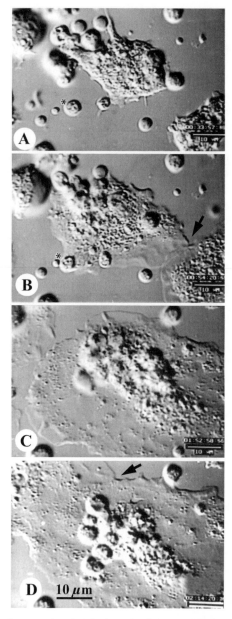

Figure 15 Video images showing fusion of adherent aggregates of *Rhabdocalyptus dawsoni* (from Leys, 1995). (A) Pieces of dissociated tissue adhere and begin to form filopodia. (B) Within 1 hour broad lamellipodia of adjacent pieces contact each other (arrow). Membranes fuse and the two pieces exchange cytoplasm, becoming a single large piece. (C) Fused pieces continue to grow and fuse with others (D, arrow). Time shows minutes after plating dissociated tissue. Asterisks indicate the same location in A and B.

Continuity of cytoplasm within a single membrane (e.g. the processes of neurons) is most frequently demonstrated by injection of cells with a fluorescent dye such as Lucifer Yellow. Unfortunately, injection proved impossible, since a stable penetration of the electrode could not be obtained, or the surface membrane adhered to and blocked the electrode. Nevertheless, by using the fluorescent acytoxymethyl esters, calcein-AM and calcein blue-AM, which 'self load' through the plasma membrane, separate lots of dissociated tissue from the same sponge were successfully filled with blue and green dyes. When two dye lots were plated together, the tissue pieces formed aggregates that were blue-green, a mixture of the two dyes. Control experiments in which dye was loaded into the cellular sponge *Haliclona* formed aggregates consisting of a mosaic of the two colours; no exchange of dye occurred (Leys, 1995) (Figure 16).

4.3. Cytoskeleton

While video microscopy showing fusion of tissue pieces clearly demonstrates that the tissues really do form a giant cell, images of the cytoskeleton of adherent aggregates illustrate the vast size of this unusual tissue. Preservation of both the actin and microtubule cytoskeleton was not a straightforward task (Leys, 1996). Using typical fixation procedures, the microtubule cytoskeleton disassembled and could not be visualised using antibodies. Lack of antibody cross-reactivity was a problem, and phalloidins did not penetrate the membrane. Techniques developed to circumvent these problems included fixation in the presence of calcium chelators (EGTA) and microfilament stabilisers (e.g. tannic acid), use of Western blots to determine a suitable anti-tubulin antibody, and briefly lysing the adherent aggregate prior to fixation to allow penetration of rhodamine phalloidin (Leys, 1996).

One-hour-old adherent aggregates are 30–100 μm in diameter. The actin cytoskeleton consists of blunt rods (Figure 17A and B) that project out from the centre of the tissue piece (Leys, 1995). In *Aphrocallistes vastus*, the actin rods are several micrometres wide and over 20 μm long (Leys, 1998). As aggregation continues and tissue pieces fuse, aggregates have large microfilament bundles, meeting at focal points around their periphery. In day-old adherent aggregates, rhodamine phalloidin-labelled bundles of microfilaments that traversed the tissue for over 500 μm, and giant 20 μm long filopodia extended from the edges of lamellipodia (Figure 17). Scanning electron micrographs of whole and lysed adherent preparations show that each filipodium contains over a dozen actin rods 60 nm wide and over 20 μm long. The microtubule bundles become equally extensive in older adherent preparations. One hour after plating dissociated tissue from either *Rhabdocalyptus dawsoni* or *Aphrocallistes vastus*, the microtubules form a network of fine

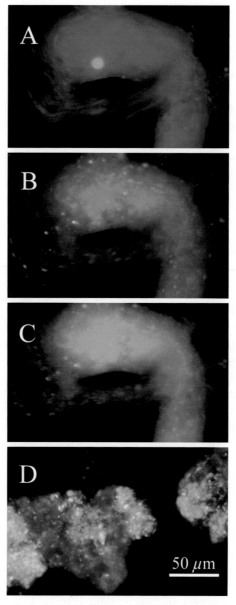

Figure 16 Dye exchange between aggregates of the glass sponge *Rhabdocalyptus dawsoni* (A–C) but not the demosponge *Haliclona permollis* (D) confirms syncytial tissues (Leys, unpublished data). Dissociated tissue from both sponges was divided into two lots, one loaded with Calcein Blue AM ester, the other with Calcein AM ester (green fluorescence). Separate lots of fluorescent tissues were mixed and allowed to continue aggregation. Mixtures of *Rhabdocalyptus* show both blue (A) and green (B)

tracks wandering throughout the 30 to 50 μm diameter aggregate (Leys, 1996). As soon as the aggregates have fused, immunofluorescence shows that multiple parallel tracks of microtubules traverse the entire preparation. Double labelling of microtubules and nuclei shows that the nuclei are randomly distributed among the microtubules; they do not form a cell soma from which the microtubules radiate. Cross sections and tangential sections of streams studied by TEM reveal large bundles of microtubules lying in parallel (Leys, 1995, 1998) (Figure 14B).

The site of nucleation of microtubules in the trabecular syncytium (or adherent syncytial tissue) remains somewhat of a mystery. Microtubule organizing centres (MTOCs, also known as centrosomes and usually distinguished by two centrioles frequently near the cell nucleus) have been found only in archaeocytes and branched choanocytes in adherent aggregates (Leys, 1996) (Figure 18). No centrioles have been found in the syncytial tissue. The

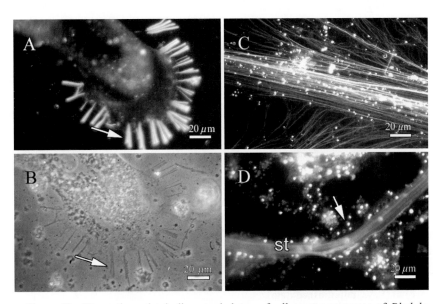

Figure 17   The actin and tubulin cytoskeleton of adherent aggregates of *Rhabdocalyptus dawsoni* (Leys, 1995). (A) Rhodamine phalloidin-labelled actin rods (arrows) at the periphery of adherent aggregates. (B) Light microscopy of the same region. (C) Microtubules labelled with anti-tubulin in an adherent preparation. (D) Nuclei labelled with Hoechst 33242 imaged in a live preparation over 30 s, shows them moving along streams (st) and in the stationary cytoplasm (arrow).

fluorescence throughout, giving the aggregate a blue-green hue (C). Mixtures of dye-loaded cells from the cellular sponge *Haliclona* (D) aggregate but do not exchange the dye, forming chimeras of blue and green aggregates.

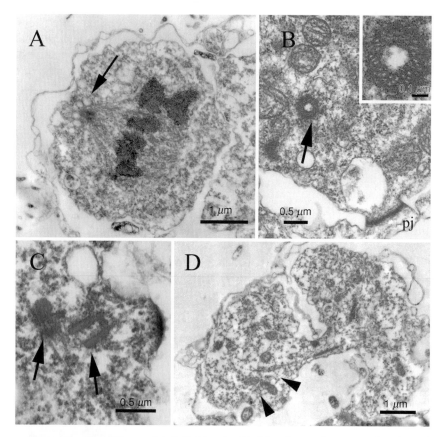

Figure 18 Centrioles and cell division in *Rhabdocalyptus dawsoni* and *Aphrocallistes vastus* (transmission electron microscopy). (A) Chromosomes on the mitotic spindle in a dividing archaeocyte in an aggregate of *Rhabdocalyptus dawsoni*. Note the vesicles clustered around the centriole visible on the left (arrow). (B) Centriole (arrow, enlarged in inset) in an archaeocyte connected to the trabecular syncytium by a plugged junction (pj). (C) Two centrioles (arrows) in an archaeocyte in *Rhabdocalyptus dawsoni*. (D) Archaeocytes dividing in an aggregate from *Rhabdocalyptus dawsoni*. Thread-like fibres (arrowheads) appear to anchor a central shaft of microtubules (arrow) (Leys, unpublished data).

fact that nuclei are transported along microtubules suggests there are no fixed regions of the cytoplasm, and thus no specific locations for nucleating microtubule formation. The freshwater foraminiferan *Reticulomyxa* also lacks centrioles and a readily identifiable MTOC. But the polarity of microtubules (inferred from the direction of microtubule assembly and organelle transport) suggests that microtubule nucleation is controlled by gamma tubulin located in the cell body (Koonce *et al.*, 1986; Kube-Granderath and Schliwa, 1995).

In other cells with extensive tracts of microtubules (e.g. neurons), nucleation occurs in the cell body and microtubules are transported as polymers to neurites (Yu *et al.*, 1993), but polymerisation of microtubules can also occur locally in neurites (Miller and Joshi, 1996).

Centrioles also play a role in organizing spindle poles during cell division. Centrioles and spindle fibres were not uncommon in TEMs of archaeocytes in aggregated tissue from both *Rhabdocalyptus dawsoni* and *Aphrocallistes vastus* (Leys, 1996) (Figure 18). Vesicles were clustered around the centrioles of dividing cells in *Rhabdocalyptus dawsoni*, and similar vesicles were found around centrioles in sections of the adult tissue from *Aphrocallistes vastus*. The cluster of vesicles is most like that found around the newly forming plugged junction (Section 3), and if so, then perhaps membrane is recruited for the daughter cells in this manner. Although one or two archaeocytes in a single aggregate may divide synchronously, by no means do all archeocytes undergo mitosis synchronously, despite connections by cytoplasmic bridges. Nuclei in the trabecular tissue were not found dividing, and it is possible that these nuclei might be terminal like the nuclei of myoblasts after fusion (Wakelam, 1988; Leys, 1996). This is a question that could readily be followed up by using markers such as Bromodeoxyuridine that incorporate into new nuclei.

4.4. Organelle transport

Almost immediately after fusion of the first membranes of tissue pieces, organelles can be seen moving across the bridge between the pieces (Leys and Mackie, 1994). Within a few hours of plating dissociated tissue, the moving cytoplasm is organised into vast channels up to 20 μm in diameter, which move steadily and unceasingly around the adherent aggregate. In live preparations broken spicule pieces can be identified moving in bulk streams, and preparations labelled with the fluorescent stain Hoechst 33242 and photographed with an extended exposure of 30 s show that the nuclei are transported within cytoplasmic streams (Leys, 1995) (Figure 17D).

The streams of cytoplasm in adherent preparations are constantly changing. They can increase in diameter by merging with other streams, run parallel but in the opposite direction to other streams and cross over other streams without any apparent interruption in volume or velocity of flow (Leys, 1995). The distance any one stream can travel is limited only by the area of the substrate. Using Nomarski optics, individual organelles, as well as bulk cytoplasm can be seen moving along the thinnest areas of the adherent aggregate. The rate of transport is fairly constant for individual organelles and bulk transport in streams: 2.15 ± 0.33 μm s^{-1} in *Rhabdocalyptus dawsoni*

(Leys, 1995) and 1.82 ± 0.15 μm s^{-1} in *Aphrocallistes vastus* (Leys, 1998). Rates were significantly faster at the edge (2.01 ± 0.44 μm s^{-1}) of a bulk stream than at the middle of the stream (1.71 ± 0.29 μm s^{-1}, $n = 31$, $p = 0.008$) (Figure 19) (Leys, 1996). Movement of organelles in lamellipodia differs, however. In these areas, individual organelles depart from the steady tracts and begin a saltatory motion, moving rapidly at times, halting and then continuing at a much slower rate (Figure 20). Rates of transport here are much slower (0.32 ± 0.11 μm s^{-1}). Pharmacological experiments suggest that the movement of bulk cytoplasm occurs along microtubules. Cytoplasmic streaming can be inhibited by nocodazole (1 μg ml^{-1}) and colcemid (10 μg ml^{-1}), both of which inhibit microtubule polymerisation, but not by cytochalasin B (10 μg ml^{-1}), which causes microfilament depolymerisation (Leys, 1995). However, not all organelles moving in a stream are directly attached to microtubules. Electron micrographs of thin sections show numerous membranous vesicles around microtubules and reaching several micrometres out into the cytoplasm (Leys, 1996). Negative stain electron microscopy of adherent pieces captures the interaction between organelles on the microtubules and those pulled along by membranous organelles, much as occurs in characean algae (Kachar and Reese, 1988) (Figure 21).

The mechanism of streaming is of interest given the basal position of the Hexactinellida within the Metazoa. Inhibition by nocodozole and colcemid clearly indicates that bulk streams are transported along microtubule tracks. However, transport within the thin lamellipodia is much slower, and has a saltatory nature, more resembling actin-based transport mechanisms (Bearer *et al.*, 1993). Unfortunately, attempts to determine which motor proteins might be involved in either bulk streaming or saltatory lamellipodial organelle transport have not been very successful so far. Most drugs that inhibit motor activity do not penetrate the plasma membrane. Hence, typical methods for identifying motor proteins involve permeabilizing the plasma membrane with a mixture of gentle detergents in a solution of ATP, and then application of a variety of inhibitors. Attempts were made to lyse and reactivate the preparation using combinations of five different buffers, five different detergents and varying the pH (Leys, 1996). Western blots showed immunoreactivity of cell lysate from *Rhabdocalyptus dawsoni* to two antibodies to cytoplasmic dynein (M74–1 and M74–2), but none to two antibodies to kinesin (SUK4 made against sea urchin kinesin and K1005 made against kinesin in the foraminiferan *Reticulomyxa*) (Leys, 1996).

Movement of individual organelles and of bulk streams of cytoplasm is also evident in pieces of tissue that have been pulled off the intact sponge and allowed to regenerate between a coverslip and slide in 10°C sea water (Leys, 1998). One might think that cytoplasmic streaming is a phenomenon of regeneration, and streaming of cytoplasm is likely the means by which the animal transports materials to repair wounds *in situ*. But transmission

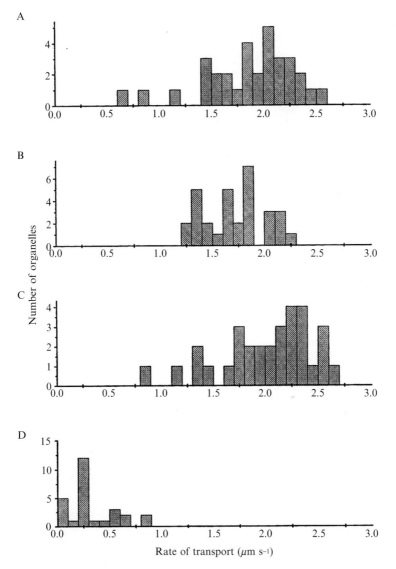

Figure 19   Rates of transport of four classes of organelles in adherent aggregates from *Rhabdocalyptus dawsoni* (Leys, 1996). (A) Single organelles moving at a constant velocity. Organelles moving at a constant velocity in the middle (B) and edge (C) of bulk streams. (D) Organelles moving by saltation in lamellipodia (32 measurements were made at each location).

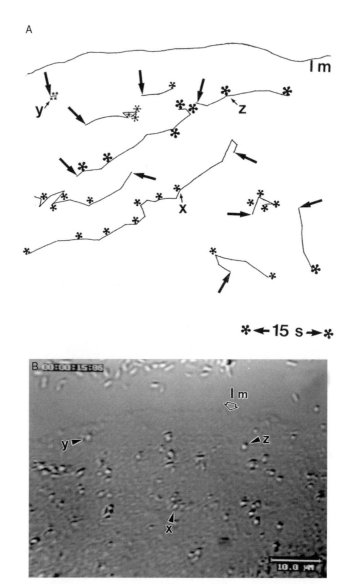

Figure 20 Saltatory movement of organelles in a broad lamellipodium (lm) of an adherent tissue culture, imaged by video microscopy (from Leys, 1996). (A) Arrows indicate the starting point of organelles at $t = 0$ and asterisks show the position of each organelle at 15-second intervals. (B) A single image showing the position of organelles x, y and z at $t = 15$ s.

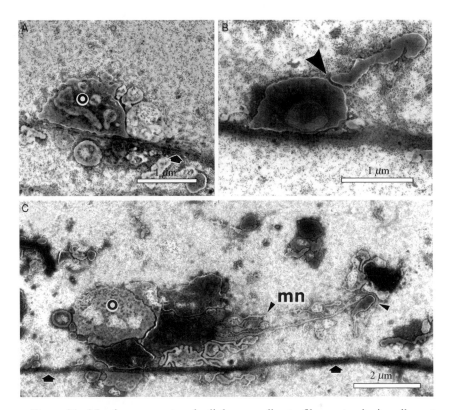

Figure 21 Membranous networks link organelles to fibrous tracks in adherent tissue of *Rhabdocalyptus dawsoni* as revealed by negative stain electron microscopy (from Leys, 1996). (A) Organelles (o) abut fibrous tracts (arrows), and contact other organelles directly (B, arrowhead), or (C) are attached to a membranous network (mn) (arrowheads) that is in contact with the fibrous tracts.

electron microscopy has shown that the content of the cytoplasm in bulk streams in adherent tissue and regenerating preparations is the same as that described as 'cord syncytia', regions of the trabecular reticulum of the whole animals (Reiswig, 1979a). Streaming is also visible in sandwich preparations of both *Rhabdocalyptus dawsoni* and *Aphrocallistes vastus* (Wyeth *et al.*, 1996; Leys, 1998; Wyeth, 1999). Thus, 'streaming' in itself is probably a normal feature of adult sponge physiology.

4.5. Comparison with cellular sponges

The role of cytoplasmic streaming in nutrient uptake in glass sponges is discussed in Section 5. This is one of the fundamental differences between the cellular and syncytial organisation. Whereas intrasyncytial transport allows nutrients to be moved over distances of many centimetres within a single cytoplasm, in a cellular system nutrients must be passed to cells that are capable of movement and which migrate through the collagenous mesohyl presumably to distribute the nutrients (Vosmaer and Pekelharing, 1898; Van Tright, 1919; Willenz and Van de Vyver, 1982, 1984; Imsiecke, 1993). Analogues of the cytoplasmic streams—nutrient transport pathways many centimetres in length—exist in cellular sponges as discrete strands of elongate cells that are embedded in collagen (Leys and Reiswig, 1998). Cells in such strands can transport nutrients, but the process occurs by the crawling of individual cells and takes up to 1–2 weeks to traverse a distance of approximately 20 cm. Movement of organelles *within* the cytoplasm (intracellular transport) occurs over much smaller distances (some 10–50 μm) in cellular sponges, but the basic mechanism of transport is much the same as in hexactinellids. In the freshwater sponge, *Spongilla lacustris*, mitochondria are transported from a centrally located nucleus towards the cell periphery and back along microtubules, as in other animal cells (Weissenfels *et al.*, 1990). Also, as in other metazoans, both microtubules and microfilaments have been implicated in the movement of the Golgi apparatus (Wachtmann *et al.*, 1990; Weissenfels *et al.*, 1990; Wachtmann and Stockem, 1992a,b). There is one report of cytoplasmic streaming seen during the reaggregation of cells of *Microciona prolifera* (Reed *et al.*, 1976), but because the aggregates were demonstrated to be cellular using scanning electron microscopy, it has been supposed that Reed *et al.* (1976) were referring to the intracellular movement of organelles within each cell (Leys, 1996).

4.6. Immune response

A calcium-dependent cell adhesion molecule was isolated from *Aphrocallistes vastus* and shown to agglutinate preserved cells and membranes in a non-species-specific manner (Müller *et al.*, 1984). Subsequently, two C-type lectins with molecular weight of around 22 kDa have been purified from the isolate (Gundacker *et al.*, 2001). It is suggested that the lectins bind to the cell membrane by the hydrophobic segment and interact with carbohydrate units on the surface of other cells and syncytia. The finding that the aggregation factor from *Aphrocallistes* works in a non-species-specific manner implies that glass sponges differ from cellular sponges in lacking either individual or species-specific immunoreactivity. Two experiments suggest this is not the case. Dissociated tissue from a selection of cellular sponges does not adhere to the ATE made from *Rhabdocalyptus dawsoni* (Leys, 1997), as would otherwise be expected if species un-specific binding were the case. Evidence for recognition of individuals of the same species comes from the unusual 'graft' preparation that was developed to allow extracellular recording of electrical events in the sponge (Leys and Mackie, 1997) (Section 5). If aggregates are made from one individual and placed on a slab of the body wall of a different sponge, no fusion between aggregate and host occurs. In fact, the cytoplasm from the host sponge 'thickens' directly under the aggregate as though the animal was transporting to the site of interaction material that causes the rejection of the non-host tissue. The only successful grafts are those that are made between aggregate and host from the same individual. Thus, glass sponges do possess molecules on their membranes that are capable of recognizing and rejecting tissue from other individuals of the same species.

5. PHYSIOLOGY

5.1. Hexactinellids as experimental animals

Most hexactinellids are adapted to life in deep-water habitats (Tabachnik, 1991) or in submarine caves (Vacelet et al., 1994) and subsist on a meagre supply of organic matter originating in surface waters far from their natural habitats, or possibly in a few cases benefiting indirectly from primary production at hydrothermal vents (Boury-Esnault and de Vos, 1988). Most are probably unaffected by diurnal or even seasonal changes and few are built to withstand the mechanical stresses associated with high-energy shallow water environments. Collecting and transporting specimens to the laboratory has to be done with extreme care, avoiding temperature increases, mechanical stresses and introduction of air bubbles into their interiors. Once in the laboratory, they must be kept cool and disturbed as little as possible. The critical role of temperature is illustrated by the findings that in *Rhabdocalyptus dawsoni* pumping is abolished below $7°C$ and arrests, which are normally exhibited by healthy specimens, are unusual above $12°C$ (Leys and Meech, 2006).

To date, the only place in the world where specimens have been kept alive in good physiological condition for long periods is the Bamfield Marine Sciences Centre in Barkley Sound, British Columbia, where a continuous supply of high-salinity, cold sea water is pumped into the laboratory from 35 m depth, not far from where *Rhabdocalyptus dawsoni* occurs naturally. As a species adapted to relatively shallow water (>25 m), *Rhabdocalyptus dawsoni* is rugged enough for laboratory research. Specimens have been maintained in the Bamfield sea water system at $10°C$ with only the food brought in with the laboratory water supply. Gaseous supersaturation of the water proves fatal, as bubbles rapidly fill up the water passages in the sponge. Movement to aquaria in other laboratories has been attempted with little success. Transported to the University of Washington Laboratories at Friday Harbor, *Rhabdocalyptus dawsoni* rapidly deteriorated and became moribund. The water there, though drawn freshly from the sea, is warmer and generally of lower salinity than that at Bamfield.

Although useful results have been obtained on specimens of *Rhabdocalyptus dawsoni* and *Aphrocallistes vastus* kept in aquaria at the University of Victoria, conditions there are far from ideal. In Wyeth's (1996) sandwich culture work (see below), the preparations deteriorated after 2 weeks. Attempts to improve longevity by providing additional food, artificially circulating the water or cooling it below 10°C were unsuccessful. The flagellated chambers of *Rhabdocalyptus dawsoni* kept in this system all became compartmentalised by inner membranes (Leys, 1999). Viable aggregates could be obtained from dissociated sponge tissue (Leys, 1995), but only very rarely did the aggregates mature into recognisable sponges with oscula. The most obvious sign of physiological deterioration, however, was the loss of the sponges' ability to arrest their feeding currents, which rendered them useless for electrophysiological experiments on the conduction system.

Oopsacas minuta, collected from a submarine cave near La Ciotat and transported under controlled temperature conditions to sea water tanks at the Station Marine d'Endoume, Marseille, were kept at 13°C and proved useful for short-term studies of particle retention (Perez, 1996), but it appears that no solution has yet been found to the problem of long-term maintenance in this species.

Until the water conditions necessary for long-term culture of hexactinellids are better understood, workers on these animals will probably have to go to marine stations located close to the sponges' natural habitat and offering appropriate culture facilities.

5.2. Food and wastes

It has always been assumed that hexactinellids like other sponges are filter feeders, removing particulate matter from the incurrent water stream, but until recently there has been little exact information on what is taken up. In a study comparing the composition of exhalent water from the atrial openings with ambient (inhalent) water, Reiswig (1990) showed that *Aphrocallistes vastus* retained particulate material, including bacteria, as its primary organic carbon source (89% of the total) and made relatively little use of dissolved organic carbon (11%). In a study of *Aphrocallistes vastus* and

Rhabdocalyptus dawsoni in laboratory tanks, Yahel *et al.* (2007b) found that bacteria and protists accounted for the entire uptake of organic carbon. The sponges showed surprising evidence of size-independent selectivity (Figure 22). Small, non-photosynthetic bacteria (<0.4 μm) and eukaryotic algae (3–5 μm) were removed with almost equal efficiency, but the retention of intermediate-sized bacteria varied seasonally and was sometimes much less

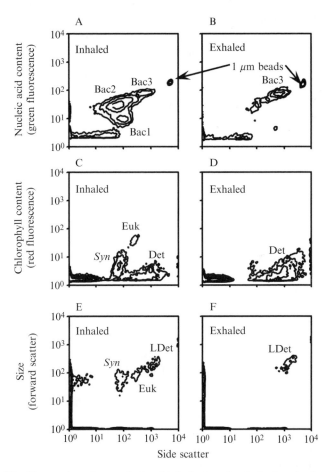

Figure 22   Flow cytometry analysis of inhalant and exhalant water in *Aphrocallistes vastus*. (A, B) Of three populations of bacteria the smaller ones (Bac1, 2) were selectively retained in preference to larger bacteria with higher nucleic acid content (Bac3); 1 μm beads are for reference. (C, E) Both eukaryotic algae (Euk) and a cyanobacterium *Synechococcus* (Syn) were present in the inhalant water, along with detritus (LDet), but only detritus was present in the exhalent water (E, F) (courtesy of G. Yahel).

efficient. An intermediate-sized cyanobacteria, *Synechococcus* (1.1–1.5 μm), was also retained less efficiently.

While the sponges in these experiments efficiently removed up to 99% of the smallest and most abundant bacteria and up to 94% of the eukaryotic algae, the amount of detritus present in the exhalent water was not reduced, indeed it was greater than in the ambient water. Clearly, the sponges have some way of getting rid of the detritus as fast as it is brought in with the feeding current. How they do so is unclear, but two possibilities can be considered: (1) indigestible particles are somehow recognised in their passage through the inhalant water passages and allowed to pass directly into the exhalent stream around the edges of the collars or by special 'bypass routes', possibly equivalent to those described on the basis of latex injection preparations (Bavestrello *et al.*, 2003) and (2) indigestible particles are taken up by endocytosis along with digestible, and are then transported through the cytoplasm and egested into the exhalent stream. The first possibility is hard to reconcile with the extraordinarily high retention rate for the smallest category of bacteria. If bypass routes existed, it would be expected that large numbers of bacteria would slip through them along with the detritus. The second proposition is therefore more plausible. It implies that the sponges endocytose all particles indiscriminately but then sort them out intracellularly, targeting the vacuoles containing detritus along exit routes, while breaking down the digestible material in phagosomes. Indigestible residues from intracellular digestion would also be egested, which would explain the fact that the detritus fraction was larger in the exhalent water than in the inhalent. A way of testing these two propositions might be to feed sponges on readily identifiable indigestible particles (e.g. polystyrene beads) and determine how long it takes for the particles to appear in the exhalent water. If they appeared within a few seconds it would support proposition (1) above. If it took a few minutes, (2) would appear more likely.

Microscopic observations on *Rhabdocalyptus dawsoni* (Wyeth *et al.*, 1996) showed that the sponge phagocytised not only *Escherichia coli* and *Isochrysis galbana* but also latex beads. Similarly, *Oopsacas minuta* took up both phototropic sulphur bacteria (1.0–6.0 μm) and latex beads (0.5–1.0 μm), although tentative estimates based on colorimetry suggested that the sponge 'preferred' bacteria to latex beads (Perez, 1996). Beads smaller than 0.1 μm diameter were not retained at all. These observations are consistent with the idea that inorganic particles are phagocytised along with organic and with the evidence for selectivity covered above.

In a study on *Aphrocallistes vastus* and *Rhabdocalyptus dawsoni* carried out with inhalant and exhalent water collected *in situ* from the sponges at 120 to 160 m depth using a remotely operated vehicle, Yahel *et al.* (2007a) were able to assess removal of bacteria and protists along with excretion of nitrogenous waste in the natural environment. The water at this depth contained a high inorganic sediment load and little phytoplankton compared with the water from nearer the surface used in the tank study. Both sponge species collected bacteria, removing up to 95% of them (median efficiency 79% in both cases). Heterotrophic protists (4–10 μm) were also efficiently removed and contributed ca. 30% of the total organic carbon uptake. Neither in tank experiments nor in the *in situ* study was evidence found for uptake of dissolved organic carbon and the entire organic carbon uptake along with excretion of ammonium could be accounted for on the basis of organic particulates. Surprisingly, given that silica constitutes nearly 80% of the dry weight of a hexactinellid such as *Aphrocallistes vastus*, silica uptake was below detection levels in the study by Yahel *et al.* (2007a), but this can be attributed to the slow growth rate of the sponges (Section 7.4).

Ectosymbiotic diatoms have been suggested as a source of nutrition in the case of *Rhabdocalyptus racovitzae*, where Cattaneo-Vietti *et al.* (1996) propose that long spicules act as optical fibres collecting and delivering light to diatoms in the sponge's interior. Whether they actually function in this capacity (which seems doubtful, Section 7.7), a study of the basalia of *Euplectella marshalli* shows that they have the structural properties of optical fibres and are capable of acting as light pipes (Sundar *et al.*, 2003; Aizenberg *et al.*, 2005). In *Hyalonema sieboldi*, spectral transmission studies suggest that the stalk spicules filter out wavelengths below 615 nm (Müller *et al.*, 2006). As it is these shorter wave lengths that penetrate deepest in sea water, it seems unlikely that the spicules are adapted for detection of light coming from the surface. Equally, the reduction of transmission below 615 nm argues against spicular transmission of bioluminescent emissions from deep-sea organisms, which typically peak in the range 460–490 nm (Nicol, 1967).

The upper size limit for particles entering the sponge is presumably set by the pores (ostia) in the dermal membrane, the thin, flat sheet covering the outer surface (Figure 23A). In *Rhabdocalyptus dawsoni*, the pores, as seen in living and osmium-fixed preparations, are 4–20 μm in diameter (Mackie and Singla, 1983). Once inside the sponge, there would be nothing to impede passage through the trabecular strands, which are thin and far apart (Figure 23B). Access to the flagellated chambers would require passage through prosopyles (Figure 23C) which have an average diameter of 4.5 μm in *Rhabdocalyptus dawsoni* (Leys, 1999). The dermal pores and prosopyles are not associated with circular bundles of contractile filaments and there is no evidence from observations on living material that hexactinellids can open or close these structures, or not in the short term. Particles too large to enter the flagellated chambers are presumably phagocytised by the trabecular syncytium.

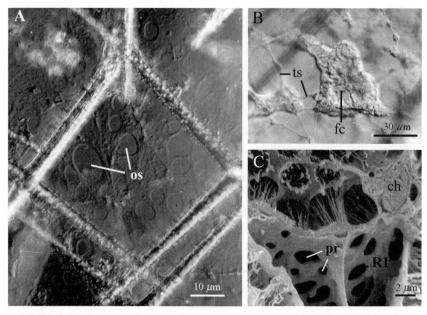

Figure 23   Pores in the inhalant pathway (*Rhabdocalyptus dawsoni*). (A) Ostia (os) in the dermal membrane as seen in a living preparation by Nomarski interference microscopy (from Mackie and Singla, 1983). (B) Wide channels are seen *in vivo* between the trabecular strands (ts) surrounding a flagellate chamber (fc) (from Wyeth *et al.*, 1996). (C) Prosopyles (pr) in the primary reticulum (R1) shown by scanning electron microscopy; ch, choanocyte cell body (from Leys, 1999).

5.2.1. Particle endocytosis

Uptake could in theory occur at any point in the pathway through the sponge, which includes all the water channels bounded by the trabecular syncytium. In *Rhabdocalyptus dawsoni*, phagocytosis of latex beads has been observed in the syncytial strands of plated aggregates, which can be taken as a model of the trabecular syncytium (Leys, 1996). Beads began to change position within 15 min of uptake and after 30 min many were moving in cytoplasmic streams. In sandwich cultures of the same species however, uptake of latex beads, as determined using differential interference contrast and fluorescence microscopy, appeared to be concentrated almost entirely in the vicinity of the flagellated chambers (Wyeth *et al.*, 1996). Only very rarely were beads seen moving in streams far from flagellated chambers. Electron microscopy confirmed the presence of internalised beads in vesicles within the trabecular syncytium close to flagellated chambers.

Uptake in the flagellated chambers has been studied in detail in *Oopsacas minuta* by Perez (1996) (Figure 24) and in *Rhabdocalyptus dawsoni* by

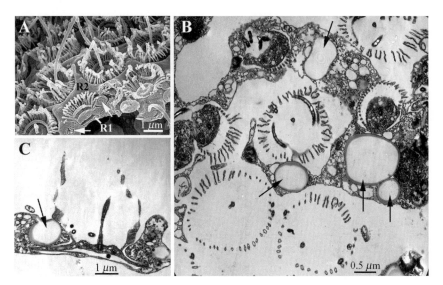

Figure 24 Uptake of red sulphur bacteria (Chromotiaceae) and latex beads in *Oopsacas minuta* (courtesy of T. Perez). (A) Scanning electron micrograph showing bacteria (white arrows) in the primary reticulum, one of which is in an advanced stage of intracellular digestion. A symbiotic bacterium is seen attached to the collar. The picture shows how closely the choanocyte collars fit within holes in the secondary reticulum (R2). (B) Numerous latex beads (black arrows) shown in a tangential section passing through several collars and reticular processes. (C) Vertical section of a collar body showing a latex bead in the adjacent primary reticulum.

Wyeth (1999). Both workers found that particle uptake occurred within components of the trabecular syncytium adjacent to collar bodies rather than within the collar bodies themselves. Both primary and secondary reticula can take up particles. While collar bodies might take up very small particles (0.05 μm), their main role appears to be simply one of providing the propulsive force for water movement by means of the beating of their flagella. This contrasts with the situation in demosponges where choanocytes are primary sites of food endocytosis (Simpson, 1984).

In Wyeth's proposed sequence of events (1999) (Figure 25), particles pass through prosopyles and enter the space between the primary and secondary reticula. Reiswig (1979a) suggested that one function of the secondary reticulum was to occlude the space between adjacent collar bodies, forcing water through the microvilli. The collars fit neatly into the pores in the secondary reticulum (Figure 24A) leaving little space around their perimeters for particles to escape into the effluent pathway (Boury-Esnault and Vacelet, 1994; Perez, 1996; Leys, 1999). Particles will tend to lodge against the sides of the collars and will be taken up by the primary or secondary

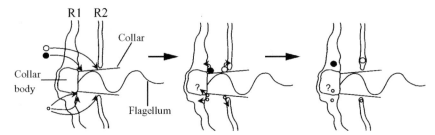

Figure 25 Proposed sequence of events during particle feeding in *Rhabdocalyptus dawsoni*, from experiments using 1 μm latex beads (after Wyeth, 1999, with kind permission of Springer Science and Business Media). A single collar body is shown lying in the primary reticulum (R1) and with its microvillar collar fitting tightly within a pore in the secondary reticulum (R2). Water flows through the spaces between the microvilli and enters the flagellated chamber but particles are trapped within a space between primary and secondary reticula, where phagocytic uptake occurs. Filled circles show the main uptake pathway (into the primary reticulum). Open circles denote particles following other routes. Phagocytosis of smaller (<1 μm) particles is not excluded and could occur in the reticula or collar bodies.

reticulum. In two species that apparently lack a secondary reticulum, *Caulophacus cyanae* (Boury-Esnault and de Vos, 1988) and *Dactylocalx pumiceus* (Reiswig, 1991), uptake would presumably be restricted to the primary reticulum. Yahel *et al.* (2007b) calculate that a typical flagellated chamber in *Aphrocallistes vastus* contains 2750 collar bodies and given the known volumes of water filtered and density of particles in typical samples, each collar body may be expected to trap one bacterium and possibly 10 indigestible particles per day.

A meshwork or sheet of glycoprotein-like filaments has been observed in several species spanning the gaps between the microvilli of collar bodies, as described in Section 3.8 and Figures 11 and 12. Water drawn into the collar would have to pass through the pores in this double layer which are variously estimated at 20–50 nm in breadth. The presence of the pores probably causes larger particles to remain outside the collar, circulating within a space largely lined by primary and secondary trabecular reticula, and increasing the probability of capture and uptake. The significance of the double filter is unknown but it seems possible that the filter is a single, continuously secreted sheet that is conveyed up the collar on the inside and down the outside, on the principle of the continuous mucous filter seen in tunicate branchial sacs and elsewhere (Werner, 1959) but, if so, it is not clear what the propulsive force for movement of the filter might be.

If large particles entering the prosopyles are unable to escape into the flagellated chamber around the edges of the collars because of the snugly fitting secondary reticulum, and if they are too large to go through the pores in the mucous filter, they should be retained and ingested. Given the dimensions of the pores, all objects larger than about 0.1 μm ought to be retained and ingested. As noted above, however, *Oopsacas minuta* was unable to retain 0.1 μm objects (Perez, 1996). It is clear that more precise data on the mesh structure and dynamics will be needed, along with more experimental data on particle retention, before the role of the mucous net can be properly understood.

On passing into the flagellated chamber, incurrent water is usually free to escape through apopyles to the atrial cavity and thence to the exterior. In some species however, including *F. occa* (Reiswig and Mehl, 1991), *Aphrocallistes vastus* and *Rhabdocalyptus dawsoni* (Leys, 1999), an inner membrane system is present that might further regulate or impede passage of water, or be a site of phagocytosis. In *Rhabdocalyptus dawsoni*, material freshly collected from the natural habitat, inner membranes were found in only 1–10% of the flagellated chambers. The membranes were no more prevalent in one season than another, but in specimens kept in aquaria in the laboratory for 3 weeks the membranes appeared in all the flagellated chambers, forming an elaborate system of internal partitions. Leys (1999) suggests that the membranes are 'amoeboid extensions of the trabecular reticulum that absorb non-functioning flagellated chambers'. It seems unlikely that they are significant in the normal feeding process.

5.2.2. Translocation

While a role for the collar bodies in nutrient uptake cannot be excluded, the bulk of the evidence points to the trabecular syncytium, in particular the primary and secondary reticula of the flagellated chambers, as the primary site where phagocytosis takes place (Perez, 1996; Wyeth *et al.*, 1996; Wyeth, 1999). The trabecular syncytium is also clearly the transport route for digested nutrients en route to other regions. Vesicles of all sizes can be seen moving in cytoplasmic streams in trabecular strands in pieces of living tissue sandwiched between a slide and a coverslip (Figure 26). Phagosomes containing internalised latex beads have been observed moving slowly in the trabecular syncytium close to collar bodies (Wyeth, 1999) but these movements did not resemble the faster, directional movements of particles moving in cytoplasmic streams, which may mean that primary phagocytic breakdown is largely accomplished locally in the vicinity of the flagellated chambers. In plated aggregates, particles move in streams at a rate of $2.15\ \mu m\ s^{-1} \pm 0.33\ \mu m\ s^{-1}$ (Leys, 1996, 2003b), as also in intact tissue.

Nutrient transport in hexactinellids thus appears to resemble the 'symplastic' transport seen in plants, where cells remote from sites of nutrient

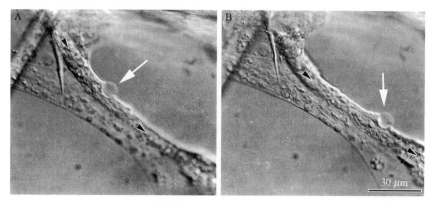

Figure 26 Cytoplasmic streaming in the trabecular syncytium observed in a sandwich culture of living tissue from *Rhabdocalyptus dawsoni* (from Wyeth *et al.*, 1996). A vesicle (white arrow) is seen being transported within a cytoplasmic stream moving from left to right as indicated by black arrowheads. The two pictures were taken 20 seconds apart. Velocity of transport was 2.25 μm s^{-1}.

uptake are supplied by flow of materials from cell to cell through aqueous intracellular channels. There is no evidence that archaeocytes play a role in hexactinellid nutrient transport by migrating through the mesohyl as described for other sponges (extracellular or 'apoplastic' transport). Mackie and Singla (1983) in making the symplastic/apoplastic distinction point to the rounded form and lack of pseudopodial extensions in hexactinellid archaeocytes as evidence that they do not migrate, and to the thinness of the mesohyl as evidence of its unsuitability as a migration pathway. Transfer of nutrients to archaeocytes, cystencytes and other cells would require passage across plugged junctions. Much remains to be determined about the composition of these junctions and what materials can pass through them, but the junction is not a membrane barrier (Section 3.2) and there are structures interpreted as pores in the material of the plug (Mackie, 1981; Pavans de Ceccatty and Mackie, 1982; Mackie and Singla, 1983; Köster, 1997). The central channel through the pore particles is given at ca. 7 nm in diameter in *Rhabdocalyptus dawsoni* and 11 nm in *Rossella racovitzae*. In addition to the pores, 'transit vesicles' are frequently seen, apparently caught in the act of crossing plugged junctions, for instance in *F. occa* (Reiswig and Mehl, 1991) and *Chonelasma choanoides* (Reiswig and Mehl, 1994). Both pores and transit vesicles are reported for *Aphrocallistes vastus* (Leys, 1998). Despite the lack of physiological evidence regarding transloca- tion of materials across plugged junctions, it appears reasonable to assume that materials can cross them either directly through the pores or within membrane-enclosed transit vesicles.

5.2.3. Autophagocytosis

Electron micrographs of *Rhabdocalyptus dawsoni* tissues kept in sandwich cultures (Wyeth, 1996) show phagosomes containing what appear to be collar bodies in various stages of breakdown. This, along with other evidence, suggests a process of remodelling whereby collar bodies and possibly other components are internalised and digested. Whether such processes are characteristic of healthy, normal animals is not clear. Observations by Leys (1999) on specimens of *Rhabdocalyptus dawsoni* kept in laboratory aquaria suggest that the inner membranes that appear within the flagellated chambers are amoeboid processes involved in absorption of non-functioning tissues. This species undergoes seasonal regression (Leys and Lauzon, 1998), but it is not known if the tissues undergo autophagocytosis during these periods. Cell death as such has not been investigated in hexactinellids, but it should be noted that homologues of mammalian genes involved in apoptosis have been identified in other sponges (Wiens and Müller, 2006).

5.2.4. Comparison with other sponges

Contrary to the situation in cellular sponges (Simpson, 1984), there is no evidence that hexactinellids can phagocytise particles in significant amounts in the incurrent canals or in any location other than the primary and secondary reticula of the flagellated chambers. If collar bodies phagocytise particles at all, it is likely that only very small particles are taken up. The arrangement seen in some hexactinellids whereby most or all of the water entering the flagellated chambers is apparently forced to pass through the microvillar mucus net by the presence of tightly fitting rings of secondary reticulum at mid-collar level is apparently unique to this group and has no counterpart in cellular sponges. The inner membranes seen in some hexactinellids may have counterparts in the central cells of certain cellular sponges but the significance of these structures is unclear. In hexactinellids, intracellular digestion occurs in the primary and secondary reticula and transport of phagosomes and food breakdown products almost certainly occurs symplastically by cytoplasmic streaming within the trabecular syncytium and does not involve apoplastic transport by migratory cells. Bypass routes allowing water to pass directly from incurrent to excurrent canals avoiding the flagellated chambers have been reported in some hexactinellids (Leys, 1999; Bavestrello *et al.*, 2003) and have also been described in demosponges (Bavestrello *et al.*, 1988), but experimental evidence of their functioning in this capacity is still lacking and recent evidence (Section 5.2) suggests *Aphrocallistes vastus* and *Rhabdocalyptus dawsoni* lack such routes.

5.3. Production and control of feeding currents

As viewed by Bidder (1923), hexactinellids were essentially 'inefficient', passive feeders, their bodies interposed like filters in the path of slowly moving water masses in the 'eternal abyss'. He saw flagellar beating as having only local significance in assisting movement of water through the meshes in the choanosome. It is now clear that hexactinellids, like other sponges, can pump water efficiently through the entire body by the beating of the flagella in their flagellated chambers and in the absence of external water currents. This is not to say that passive ventilation may not also play a role. Currents generated by the sponge not only bring in food but also serve for respiratory exchange and removal of metabolic wastes.

5.3.1. Spontaneous and evoked arrests: sensitivity to environmental variables

Flow meter recordings from *Rhabdocalyptus dawsoni* in its natural habitat (Lawn *et al.*, 1981) showed that the sponge usually pumped water at a steady rate but that pumping was occasionally interrupted by spontaneous arrests of variable duration. Arrests were frequently seen when divers were working in the vicinity of the sponge (G. Silver, personal communication), suggesting that seemingly spontaneous arrests may have been due to extrinsic factors, such as sediment in the water. Flow meter recordings from sponges taken into tanks in the laboratory showed a similar pattern of steady pumping interrupted by arrests, occurring both 'spontaneously' and in response to mechanical disturbance (Mackie, 1979; Lawn *et al.*, 1981). Leys (1996) and Leys *et al.* (1999) showed that arrests could also be evoked by sediments introduced into the inhalent water stream. Leys and Tompkins (2004) tested sponges on filtered (25 μm) and unfiltered sediments whose main ingredients were organic debris, clay and silica from diatom frustules. The minimum concentration of sediment causing arrests was 10 mg litre^{-1}. Arrests in *Rhabdocalyptus dawsoni* were typically isolated events (Figure 27). When the sponge stopped pumping the flow rate declined sharply and then started to recover after about a minute, returning to the resting level in another minute. A second arrest occurring before the sponge had recovered from the first resulted in a further downward deflection. Stepped deflections of this sort were seen both at the start of an arrest sequence and at the end, as pumping resumed, as if the sponge was testing the water. In *Aphrocallistes vastus*, such stepped deflections and recoveries were much more common than in *Rhabdocalyptus dawsoni* (Figure 27), and also appeared to be more tolerant of increased sediment loads than the latter.

In *Aphrocallistes vastus*, with continued application of filtered sediment, the resting level baseline, representing maximal pumping, gradually declined,

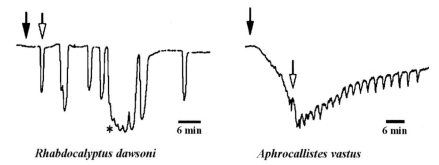

Rhabdocalyptus dawsoni   **Aphrocallistes vastus**

Figure 27 Arrests (downward deflections) induced by addition of 25 μm filtered sediment to the water. Black and white arrows indicate the start and finish of sediment addition. In *Rhabdocalyptus dawsoni*, single arrests are all-or-none events lasting about 2 min. Arrests occurring in sequence at short intervals (asterisk) depress the baseline further, as water flowing through the sponge progressively loses momentum, resulting in a stepped pattern. Arrests in *Aphrocallistes vastus* tend to occur in rapid succession, leading to gradual depression of the baseline, and recovery is often also stepped, as if the sponge was testing the water (courtesy of S. P. Leys and G. J. Tompkins).

presumably because of clogging of the canals. When addition of sediment ceased, pumping gradually returned to normal levels. It is not clear how hexactinellids rid themselves of sediments, but it seems probable that the particles are endocytosed and transported along cytoplasmic exit routes and released into the effluent stream, as discussed in Section 5.2.

Arrests can readily be evoked by mechanical stimuli such as tapping the walls of the tank, pinching a corner of the sponge with forceps or even twanging a single spicule. Electrical stimuli are also effective (Lawn *et al.*, 1981; Mackie *et al.*, 1983; Leys *et al.*, 1999). In each case, arrests spread through the whole sponge; they are never restricted to the vicinity of the stimulus.

There is no clear evidence that hexactinellids are sensitive to changes in light intensity. Sensitivity or susceptibility to ultraviolet (UV) light may seem unlikely in a predominantly deep-water group of animals, but a gene has been isolated from *Aphrocallistes vastus* that shows a high degree of sequence similarity to genes encoding invertebrate (6–4) photolyase (Schröder *et al.*, 2003a). This is a DNA repair gene, and its presence may mean that the sponge is adapted to repair damage caused by UV irradiation. *Aphrocallistes vastus* occurs at <20-m depth in British Columbia waters, where significant amounts of UV penetrate. The gene is expressed chiefly in the upper parts of the body where irradiation damage would be most likely to occur. The photolyase photo-reactivating system may have evolved in hexactinellids early in their evolution when exposure to UV light exceeded present levels and when hexactinellids lived in shallower water. Whether the sponge can detect UV or light of any sort in the behavioural sense is unknown, and there is no evidence of diurnal rhythmicity in the pattern of spontaneous arrests.

5.3.2. Nature of arrest events

All the evidence available at the present time suggests that water current arrests are simply due to cessation of flagellar beating, but this has yet to be confirmed by direct visual observation. Tissues maintained in sandwich cultures should in theory allow visualisation of flagellar arrests, but pieces of *Rhabdocalyptus dawsoni* prepared in this way in tanks at the University of Victoria (Wyeth *et al.*, 1996; Wyeth, 1999) did not show the normal pattern of evoked and spontaneous arrests. In fact, they, like intact sponges in the same water system, lost the ability to arrest altogether and pumped continuously. Even in optimal conditions at Bamfield, sponges often failed to show arrests or started doing so only after many hours or days of continuous pumping. The arrest system is clearly very susceptible to disturbances in the environment. The mechanical trauma involved in setting up a sandwich culture would be relatively severe.

There is no reason to suppose that arrests are brought about by closure of pores such as those in the dermal membrane (ostia) or the prosopyles leading into the flagellated chambers. Observations of ostia in living sponges showed no changes, and electron microscopy showed no aggregations of microfilaments around any of these pores that might function as sphincters. Arrests take place rapidly (within 20 seconds), suggesting flagellar arrest as much the most likely mechanism. Ciliary or flagellar arrests as a means of regulating water flow are well documented in tunicates, bivalve molluscs and other invertebrates (Aiello, 1974).

5.3.3. Spread of arrests

Arrests were found to propagate through the sponge on the all-or-none principle at velocities of 0.26 cm s^{-1} (Lawn *et al.*, 1981; Mackie *et al.*, 1983). They can spread along zig-zag paths created by cuts and along thin strips cut vertically and horizontally through the body wall (Figure 28). In specimens bearing asexual buds, arrests evoked in the sponge spread to the bud and *vice versa*.

Lawn *et al.* (1981) proposed that the trabecular reticulum was the histological substrate for conduction of signals causing arrests. As a syncytium, it could 'act as a single neuron', conducting electrical impulses throughout the sponge. Attachment of recording electrodes was made difficult, however,

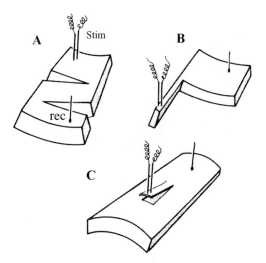

Figure 28 Spread of arrests in *Rhabdocalyptus dawsoni* recorded with a flow meter (rec) following electrical stimuli delivered at distant points (stim) (from Mackie *et al.*, 1983). Conduction can take place along zig-zag pathways (A), in narrow slices through the sponge wall (B) and in flaps of dermal membrane (C).

by the porous structure of the sponge and all attempts to record the suspected electrical signals failed until an approach using aggregates of dissociated sponge tissue grafted on to the sponge was devised (Leys and Mackie, 1997; Leys *et al.*, 1999). The aggregates were placed on the dermal or atrial surfaces of the same sponge from which the tissue was taken, and became attached, evidently fusing with the trabecular syncytium of the host. These provided substantial, non-porous attachments for recording electrodes (Figures 29–31). Impulses were found to propagate into and out from these autografts. With two such grafts, it was possible to record the passage of an impulse directly over a measured distance between recording electrodes (Figure 30). In the example shown, impulse conduction velocity was 0.28 cm s^{-1} at 10°C. This agrees with values for the velocity of spread of arrests as measured on whole sponges with a flow meter. Preparations in which current arrests and electrical impulses were recorded simultaneously showed a perfect correspondence between spread of impulses and spread of arrests (Figure 31). Repeated arrests and long-duration arrests were shown to be triggered by sequences of electrical impulses. An important observation was that conduction could still occur even when the feeding currents were completely arrested (Figure 31C and D). This shows that the conduction and effector systems are independent. It seems very clear that the conduction system is indeed the trabecular

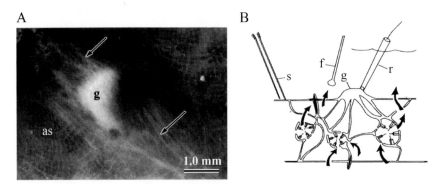

Figure 29 Setup for recording from attached autografts in *Rhabdocalyptus dawsoni* (from Leys and Mackie, 1997). (A) Shows a graft (g) attached to the atrial surface (as) that has fused with the trabecular syncytium. Streams of cytoplasm (arrows) run out from the graft into the sponge tissue. (B) Diagram of a section through the body wall bearing a graft (g). A stimulating electrode on the atrial surface (s) evokes an impulse that propagates to the graft, where it is recorded electrically (r). Arrows show water flow through the sponge, changes in flow rate being recorded by a flow meter (f).

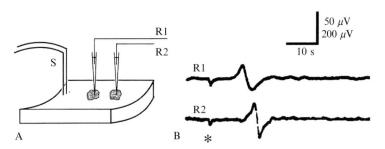

Figure 30 Recording from grafts in *Rhabdocalyptus dawsoni* (from Leys *et al.*, 1999). (A) Stimulus (*) delivered as S evokes an electrical impulse that is picked up in linear sequence by recording electrodes (R1, R2) in grafts (stippled) attached to the atrial surface of a piece of sponge. (B) Shows the time relationships as the impulse passes from R1 to R2, a distance of 1.55 cm, giving a conduction velocity of 0.28 cm s^{-1} in this example.

reticulum, including the flattened dermal and atrial layers, and that impulses spread everywhere on the all-or-none principle through this system.

The effectors are presumably the flagella of the choanocytes. To enter these cells impulses must cross plugged junctions. As noted in Section 3.2, hexactinellid plugs are not membrane barriers, and probably therefore offer little resistance to the forward flow of action currents.

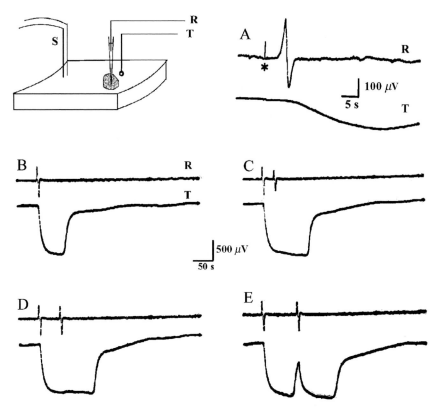

Figure 31   The diagram shows a recording setup where electrical stimuli delivered at S evoked impulses that propagated to a recording electrode (R) in a graft, while an adjacent flow meter (T) recorded arrests (*Rhabdocalyptus dawsoni*). A single evoked impulse and accompanying arrest are shown on an expanded time scale in A (asterisk indicates moment of shock). Responses to single (B) and paired shocks (C–E) are shown on a compressed time scale. In C–E, shocks were delivered 35, 57 and 93 seconds apart, respectively. The second shock prolonged the arrest (C, D) or evoked a second, longer arrest (E), where the sponge had started to pump again when the second shock was delivered (from Leys *et al.*, 1999).

The system shows the typical characteristics of an electrical conduction system such as fatigue, chronaxie and refractoriness, but compared with excitable tissues in other animals the absolute and relative refractory periods are very long (29 and 150 seconds, respectively). The conduction velocity of 0.27 cm s^{-1} is also slow compared with conduction in excitable epithelia, muscles and nerves in animals, though it falls within the same range as action potential propagation in some plant systems, for example 0.1–0.4 cm s^{-1} in tomato seedlings (Wildon *et al.*, 1992). Impulses propagating through the sponge probably do not travel in straight lines 'as the crow flies' owing to the branching, reticulate character of the substrate and the actual conduction velocity within individual strands is probably considerably higher. The system is unusually temperature sensitive. Conduction velocity varied with temperature showing a peak at 10°C (Figure 32), a temperature which corresponded well with the temperature at the collection site. Above 12.5°C propagation was usually lost, although flagellar beating continued. Below 7°C flagellar beating ceased, but this does not mean that impulses necessarily ceased to propagate. Between 7 and 10°C the Q_{10} was about 3. In later tests, some individuals were found that were able to function at considerably higher temperatures, but the same Q_{10} value applied to the pooled sample used in these tests. The sensitivity of the system to temperature is probably a reflection of the sensitivity of calcium channels on which the action potential depends. Action potentials based on sodium are much less temperature sensitive. These results with *Rhabdocalyptus dawsoni* suggest that temperature may well be a limiting factor affecting hexactinellid distribution in other habitats (Leys and Meech, 2006).

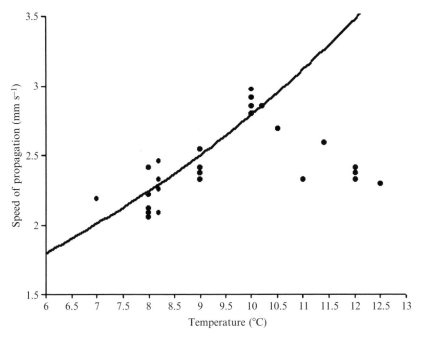

Figure 32 Effect of ambient temperature on the speed of propagation in *Rhabdocalyptus dawsoni*. The line through the points shows the slope of the relationship expected for a Q_{10} of 3 (courtesy of R. W. Meech).

5.3.4. The action potential

Experiments with drugs and ions affecting membrane channels strongly suggest that the electrical impulses are calcium spikes (Leys *et al.*, 1999). Sodium-deficient solutions had little effect on the wave form of the action potential, but propagation was blocked in solutions containing elevated levels of cobalt and manganese ions known to block calcium channels (Figure 33A and B). Nimodipine, a calcium channel antagonist, blocked conduction in relatively low concentrations. Cessation of flagellar beating—the effector response triggered by impulses arriving in the choanocytes—is

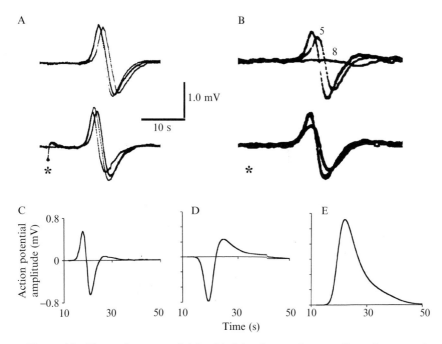

Figure 33 The action potential in *Rhabdocalyptus dawsoni* (from Leys *et al.*, 1999). (A) Lack of sodium dependency. The picture shows the effect of a 50% reduction (upper) and 75% reduction (lower) in $[Na^+]$. The superimposed traces show the reduced $[Na^+]$ trace along with sea water controls made before and after the experiment. The amplitude of the potentials was not significantly decreased. (B) Addition of 10 mmol of Co^{2+} to the sea water. In the upper set, the first trace is a sea water control. The second shows the reduced amplitude of the potential after 5 min, the third after 8 min (complete block). In the lower set, addition of 5 mmol Co^{2+} reduced but did not abolish the event. (C–E) Integration of the externally recorded action potential (C) produced the curve shown in D, and a further integration produced E, which approximates to the shape of the spike as it would appear in an intracellular recording.

presumably due to influx of calcium ions accompanying spiking, as in other ciliated cells. Beating would start again when the normal intracellular calcium levels were restored by ion pumping.

Intracellular and patch recordings have not yet been attempted on this preparation, so interpretation of the underlying events is based on evidence from extracellular recordings alone. Integration of the bipolar wave form of the impulse recorded with a suction pipette revealed the diphasic external current associated with propagation. A further integration produced a monophasic signal which may be taken as an approximation to the action potential as it would appear in an intracellular recording (Figure 33C–E).

5.3.5. Functional significance of arrests

The ability to turn off the feeding current is assumed to be adaptive in terms of preventing entry of sediments with the incurrent water stream. Sediments could be stirred up by any natural disturbance in the locality, as by the activities of interlopers such as crustaceans moving on or near the sponge. The response finds a close counterpart in the behaviour of ascidians, which arrest the feeding current when disturbed in any way, including when there is excessive sediment in the water (MacGinitie, 1939; Takahashi *et al.*, 1973).

5.3.6. Comparison with cellular sponges

An ability to arrest the feeding current is reported for some tropical demo-sponges (Reiswig, 1971b). In some cases contraction of exhalent water channels is held to be responsible while in others cessation of flagellar beating is thought to take place along with contractile movements. Other workers have described spreading waves of contraction in various sponges but such responses are typically local, though in some cases (e.g. *Tethya lyncurium*, described by Pavans de Ceccatty *et al.*, 1960) they affect the whole sponge. They are also usually very slow compared with *Rhabdocalyptus dawsoni*. In *Ephydatia muelleri*, for example, contraction waves run up the osculum at 30 to 350 μm s^{-1}, while those spreading through the canals in the choanosome travel at 4 μm s^{-1}. An exception is *Phorbas amaranthus*. This demosponge drops flaps over its ostial fields when stimulated, with a latency of <1 second (Reiswig, 1979b). The mechanism is unknown. All these phenomena and the possible mechanisms underlying them are discussed by Leys and Meech (2006). Cellular sponges lack not only nerves (Pavans de Ceccatty, 1989) but also gap junctions (Green and Bergquist, 1982), and in the absence of these or any other known aqueous channels interconnecting the cells, it is hard to see how electrical signals could propagate through their tissues. Indeed, all attempts to record electrical correlates of the contractions have failed. What distinguishes hexactinellids such as *Rhabdocalyptus dawsoni* is that purely local stimuli such as twanging a single spicule can send an electrical signal through the entire animal that causes an immediate, all-or-none effector response affecting all parts of the body within a few seconds. This is possible because the trabecular syncytium, lacking internal membrane barriers, can act as a conduction pathway. No counterpart to this tissue exists in cellular sponges.

6. THE SILICEOUS SKELETON

6.1. Discrete spicules

The taxonomy of Hexactinellida has historically been, and still is, based on their siliceous skeletons. Indeed, distinction between the two primary sub-classes, Amphidiscophora and Hexasterophora, is on the form of their smaller spicules, amphidiscs versus hexasters. The skeleton supporting the thin network of living tissues is a delicate scaffold of siliceous spicules, some of which may be fused together by secondary silica deposition to form a rigid framework. Silica is deposited in the form of amorphous opal, $SiO_2 \cdot nH_2O$, in production of a variety of distinct types of structures, one, some or all of which may be present in different species at various stages of growth: (1) discrete spicules that remain as loose elements throughout life of the sponge; (2) rigid siliceous networks formed by fusion of the main supporting spicules and extending through parts or all of the main body tissue mass (choanosomal or parenchymal position) and (3) rigid networks or thin lacework plates of fused minor spicules and synapticulae, formed commonly at contacts with hard foreign objects (basidictyonal plates) and more rarely on free exposed outer surfaces (surface crusts or external capsules).

As a group, hexactinellids secrete an amazing array of spicules of various shapes and sizes, but any one specimen or species produces repetitive copies of only a few (2–12) types of spicules. Here we introduce only those names of spicule types needed to follow general discussion. The first naming system for sponge spicules developed by Bowerbank (1858) was unwieldy, little used and supplanted for hexactinellids by a more comprehensible system, based on spicule shapes, developed by Schulze (1887). Since then, as new species and spicule shapes were discovered, new names for spicule types, often without definitions, were added to hexactinellid nomenclature. Hexactinellid taxonomists, not immune from occasional failure of logic, have proposed some unfortunate names for spicule groupings that should be allowed to disappear by disuse. There is no single complete and authoritative illustrated list of hexactinellid spicule names available. Several sources provide names and illustrations of part of the variety (Koltun, 1967; Hooper and Wiedenmayer, 1994; Boury-Esnault and Ruetzler, 1997), and Tabachnick and Reiswig (2002) provide text definitions of spicule names but without illustration. The complete range of hexactinellid spicule forms and their nomenclature can be surveyed by perusal of the Hexactinellida section of Systema Porifera (Hooper and Van Soest, 2002, pp. 1201–1509).

The spicule types produced by any single species are generally consistent across a wide range of specimen sizes, compelling evidence that spicule form and size are genetically controlled. They are thus used as primary characters for species descriptions. During juvenile development, however, expression of the different spicule types occurs as a succession of stages and the first spicules of a type may be smaller than the corresponding spicules in mature adult specimens. In addition, while the fidelity of shape and size of any spicule type formed by a specimen is impressive, considerable variation is always present within a sample of every spicule type. Furthermore, there are always some spicules produced that are obvious aberrancies or mistranslations of a shape/form programme. These are usually ignored in descriptions unless they occur at a frequency that assures they will be found by an observant investigator.

6.2. Megascleres and microscleres

6.2.1. Size

Spicules of Hexactinellida, like those of Demospongiae, are generally divided into megascleres and microscleres based on the features of size, shape and function. Megascleres, as larger components, usually with largest dimension 0.2–30 mm, obviously include the major supporting skeletal elements. The largest megascleres are those that project from the body surface and serve as protective lateral spines or basal attachment roots— the single basal spicule of *Monorhaphis* can attain a size of about 3 m in length and 1 cm in diameter (Schulze, 1904). Microscleres are usually less than 0.1 mm in diameter and are commonly astral in form, here meaning having more than six terminal rays. They are clearly accessory skeletal elements by their small size, but some astral microscleres attain diameters of 1 mm, overlapping significantly with megascleres in size. When encountering unusually large spicules of a type that is elsewhere small in size, astral in form and assigned to the microsclere fraction, they are likewise considered to be simply large microscleres. In practice, distinction between megascleres and microscleres is subjective for intermediate spicule types, has changed over time, and is best decided by concensus of practicing specialists. Microscleres are usually not fused to skeletal elements except for their rare and apparently accidental incorporation in frameworks and the common insertion of small oxyhexactins throughout the frameworks of many Hexactinosida. Reiswig's recognition (1992) of medium-sized, surface-associated and non-supporting spicules as 'mesoscleres' has not been consistently applied by that author nor followed by other recent authors.

6.2.2. Form

All spicules of hexactinellids, both megascleres and microscleres, have either the form of a regular hexactin or can be considered derivations of that basic form (Figure 34A). The regular hexactin has six rays (actins) of equal length and shape, intersecting at a point where each ray is normal to the centre of a circumscribing cube (cubic symmetry), and each ray intersects four others perpendicularly and is in line with the remaining fifth ray extending out from the opposite side of the intersection point. Since the spicule consists of three axes or axons (two rays per axis), the spicules are also called 'triaxons' and the class has been referred to in the past as the 'Triaxonia', a junior synonym of Hexactinellida. Every hexactinellid spicule has a system of internal organic axial filaments developed to varying degrees. Each ray consists of an organic axial filament surrounded by a cone of silica, sometimes as a single unit but conspicuously concentrically layered in larger spicules. Early suggestions that organic material is either incorporated in each layer or a thin sheet of spiculin is deposited between layers of silica have not been supported by spicule analysis (Sandford, 2003). Collagen fibrils have been demonstrated in the layered matrix of *Hyalonema sieboldi* root spicules (Ehrlich *et al.*, 2005), but whether this organic matrix material is a general feature of siliceous spicules remains as yet unknown. In the regular hexactin, the six axial filaments, one running axially in each ray, intersect perpendicularly at the centre of the spicule in cubic symmetry. The intersection of the axial filaments is known as the 'axial cross' (Figure 34A) and that point is considered the 'spicule centre'. Those spicules where the axial filaments typically extend distally to the tip of each ray are known as 'holactins' (Kirkpatrick, 1910)—antonym 'heteractin' (new term) with some distal rays lacking axial filaments (see below). If the filament remains open and exposed at the tip, lengthening of the ray can continue. When the filament is closed off at the ray tip by silica deposition, extension of that ray by primary growth cannot continue.

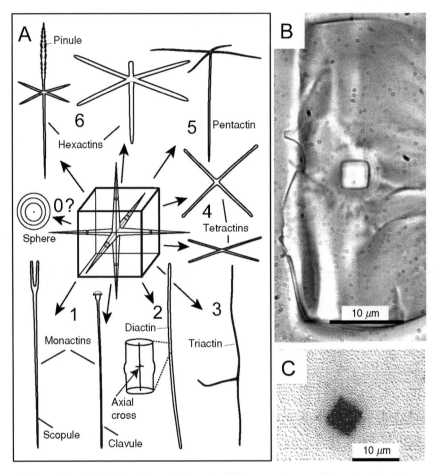

Figure 34 (A) Derivation of hexactinellid megascleres with undeveloped rays from a basic hexactin of cubic symmetry. (B) Cross section of a pentactin of *Crateromorpha* sp. with square axial canal (from Reiswig, 1971). (C) Atomic force microscope image of the square axial filament in a polished cross section of a spicule of *Rossella racovitzae* (from Sarikaya *et al.*, 2001).

There has not yet been a systematic study of open versus closed spicule tips within or between hexactinellid species or even spicule types.

The organic axial filament of Hexactinellida is square in cross section (Figure 34B and C) while that of Demospongiae is either hexagonal or triangular. The square sectional shape is retained in uneroded spicules when the axial filaments are artificially removed by treatment with nitric acid, bleach or hydrogen peroxide. The exposed space is called the axial canal. Walls of axial canals of spicules from non-living sources in nature (sediments, spicule pads or mats, dead fused skeletons and so on) erode more rapidly than the rest of the spicule and the canal is thus enlarged and rounded in section. This occurs in spicules of museum specimens stored for extended periods in acidic fluids. The axial filament consists mainly of a mixture of proteins previously known as 'spiculin' as well as a small fraction of inorganic minerals. The main proteins in axial filaments of both hexactinellids and demosponges have been sequenced; they have been more specifically renamed as 'silicateins' indicating their presumed enzymatic activity in silication (see silication below).

The impressive diversity of spicule forms in Hexactinellida can be ascribed to variation of only a few features of the basic regular hexactin. Derivations from the regular hexactin involve: (1) modification of one or two of the six rays by elongation, shortening, formation of special spination and so on to form a variety of irregular hexactins (Figure 34A); (2) complete reduction of one or more rays (Figure 34A, 1–5) and (3) addition of secondary rays on the tips of one, some or all of the developed rays or on the undeveloped spicule centre (Figure 35). In addition to these major sources of form variation, all spicules may be modified by ray curvature (rare in hexactinellids) or formation of small to large spines or thorns in various patterns on some or all ray surfaces. Modification of one or both rays of a hexactin axon are common in surface-related spicules (Figure 34A, 1), producing spicules known as pinules with a single long bushy ray projecting from the sponge surface, or reduction of the projecting ray from slight to nearly complete. Total ray reduction from the basic hexactin results in five-rayed pentactins, four-rayed tetractins (called stauractins if all developed rays lie in two axes and one plane), three-rayed triactins (called tauactins if all developed rays lie in two axes and one plane), two-rayed diactins (rhabdodiactins if both rays lie in one axis) and single-rayed monactins (Figure 34A).

Virtually all of the irregular variants that can be imagined for each general form occur as well. Although these general spicule forms are listed and typified as discrete, continuity exists between forms such that only arbitrary distinction can be applied to some spicules, for example hexactins with one very short ray and pentactins with a short nub as a rudiment of a sixth ray. With some exceptions, the axial filament system of these derived spicules contains an axial cross with the normal six filaments emanating from it, but those of the undeveloped rays are only rudiments, a few micrometres long. This is considered strong evidence that these forms are indeed modifications of the basic hexactine spicule. Some pentactins and most stauractins, however, do not develop even the rudiments of the undeveloped rays, the significance of which is unclear. Since examination of the axial system requires tedious, careful microscopy, a survey of rudiment patterns has not been made for the various spicule types across taxonomic groups. The exceptional spicules that characteristically lack an axial cross are:

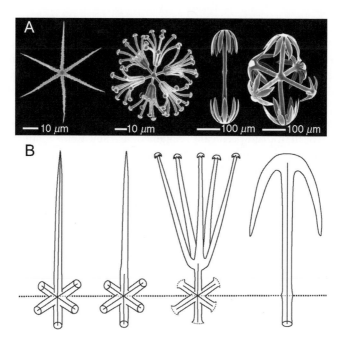

Figure 35   (A) Hexactinellid microscleres, from left to right: oxyhexactin from *Chonelasma lamella*, discohexaster from *Euryplegma auriculare*, macramphidisc and hexadisc from *Hyalonema populiferum* (SEM). (B) Transparency drawings of microscleres (centres and one of the primaries with terminal rays) showing axial filament system, left to right: holactine oxyhexactin from Amphidiscophora, heteractine oxyhexactin and discohexaster from Hexasterophora and amphidisc from Amphidiscophora (redrawn from Kirkpatrick, 1910).

(1) the common thorned diactin known as the 'uncinate', with an axial filament extending tip to tip; (2) the amphidisc of Amphidiscophora in which a slight swelling of both the axial filament and first layer of silica gives clear indication of the spicule centre and (3) the rare non-actine sphere or pearl with only a formless central granule as a presumed axial rudiment (Figure 34A). The likelihood that the amphidisc is indeed derived from a hexactin is strongly supported by rare reports of a visible axial cross (Mehl, 1992, Plate 5, Figure 1) and occurrence of hexactine variants, the hexadisc (Figure 35A, right), among typical amphidisc spicules.

Secondary or terminal rays ('pseudoactines' of Reid, 2003) lacking detectable axial filaments are commonly added to microscleres but rarely to megascleres. Microscleres have secondary rays appended to the distal ends of all six or only some primary rays, resulting in the attractive complexity and symmetry of these hexactinellid spicules. These structures, sometimes numbering only one per primary ray (more usually three to five), or occasionally as many as one hundred, are identical or proportionately similar to silica extensions arrayed in definite patterns on simple or variously inflated primary ray tips. Additional variation in form and nomenclature stems from the moderate variety of shape of the secondary rays—straight, sigmoid, hooked and so on—and shape of the terminal tip—sharp-pointed, digitate, serrated disc, group of thorns and so on. Secondary structures carry no detectable axial filament and are referred to as anaxial rays by Reid (2003). The overall form of microscleres bearing more than one secondary ray per primary ray is suggestive of ray branching, but indeed the primary rays and their axial filaments never branch. Such anaxial secondary rays are not appended to the ends of regular megasclere rays, but similar anaxial structures borne on the central knob of monactine megascleres (tines of scopules, Figure 34A, 1 left) can be considered equivalent to secondary rays of microscleres. Spicules bearing both primary axial rays (all spicules) and secondary anaxial rays may be called 'heteractins' to distinguish them from holactins that consist only of primary axial rays. The term 'astral' proposed by Kirkpatrick (1910) as antonym of holactin is inappropriate for inclusion of scopules and clavules. Although it is, in practice, extremely difficult to ascertain presence of primary versus secondary rays in microscleres with only six pointed rays (microxyhexactins), this distinction has major taxonomic and phylogenetic importance. All such spicules in Amphidiscophora are holactins, while all such spicules in Hexasterophora are heteractins (Figure 35B).

Other structures such as the whorls of anaxial spines or teeth on the ends of the diactine amphidisc microscleres characteristic of the subclass Amphidiscophora are generally not considered equivalent to secondary rays, since they do not exhibit complexity of tip form, and are equivalent to the tip ornaments (spines) of secondary rays of Hexasterophora. An extensive analysis of anaxial ornamentation structures with the aim of determining their equivalence or homology has not yet been made. Likewise, it is not entirely certain that axial filaments are totally absent from secondary rays, spines, thorns and other structures simply because they cannot be perceived with standard light microscopy. As has been shown by Drum (1968), some demosponge microscleres have submicroscopic axial filaments in secondary structures, and for hexactinellids, such a possibility remains unexplored. A general 'rule' for hexactinellid axial filament patterns is that, although these structures intersect at the spicular centre of most spicules, they never branch as occurs in dichotriaenes of some demosponges. A corollary of this restriction is that axial rays can never be added to a spicule after silication has begun and no hexactinellid spicule can have more than six axial rays unless such occurs through an accident during its early development.

6.3. Spicule locations

Different spicule types often have precise locations in hexactinellids, attesting to specificity of function, although this is not always apparent. The original positional designations made by Schulze (1887) are still used, with a few modifications, for original descriptions of spicule locations and diagnoses of patterns for higher taxa (Figure 36). There are basically three types of location recognised—spicules associated with surface membranes (dermalia, atrialia, canalaria), spicules distributed through the main body wall between the surfaces (parenchymalia), and spicules partly projecting from the outer body surface (prostalia). Dermal spicules supporting the outer living tissue layer are generally distinct types and sizes of pentactins, hexactins or sometimes tetractins and diactins. When an additional larger spicule type supports these immediately below the surface, the dermalia are termed autodermalia and the larger supporting category, usually large pentactins but sometimes diactins, are called hypodermalia. In Recent hexactinellids, surface-related spicules almost always remain as loose skeletal elements, presumably to allow movement of surface tissues during expansion, by hydrostatic inflation, and collapse, by elastic tissue rebound, of fluid spaces circumscribed by the trabecular syncytium. They are rarely interconnected by silica fusion to form networks, although in many fossil forms they were fused to the main framework (Mehl, 1992). A similar nomenclature is applied to equivalent spicules lining the atrial cavity and larger canals.

Parenchymalia include several distinct types of spicules with different distributions and roles. The major spicules providing primary physical support for the sponge are the principalia, usually large, robust hexactins or diactins (rhabdodiactins), sometimes pentactins, tetractins or triactins. They are often very intimately entwined by much thinner diactin spicules, the comitalia, which presumably provide fabric-like resiliency to the combined network. All other spicules, both megascleres and microscleres, distributed throughout the parenchyme without detectable pattern (so far), are collectively called intermedia. Unlike dermalia, the parenchymal principalia and comitalia are sometimes fused by secondary silication to form a rigid skeleton. In lyssacine Hexactinosa, such fusion of parenchymal megascleres is an informal process, but in the dictyonine Hexactinosa, hexactin principalia are often (not always) regularly arranged at fusion to form intricate lattices—the principalia then known as dictyonalia.

The remaining positional designation, prostalia, includes three groups of megascleres usually in the form of monactins, pentactins or sometime diactins. The most well known are the basalia extending from the lower body, for example the basal spicule tuft of *Hyalonema*, that provide support of the body over the benthic surface but also enable anchorage in soft muds and grapnel attachment to irregular hard substrates. Basalia are typically

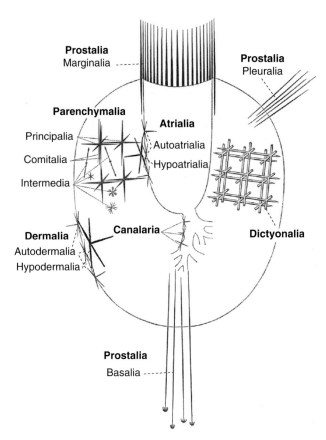

Figure 36 Nomenclature of spicule locations in Hexactinellida (redrawn and modified from Schulze, 1887).

monactins or diactins with specialised toothed anchors or pentactins or rough diactins serving as very effective grapnels. Prostalia which project from the lateral surface, known as pleuralia, are mainly diactins or monactins, but some hypodermalia are erected from their positions of origin below the dermal surface to extend as a veil several millimetres over the outer surface. Pleuralia presumably provide protection of the lateral surface from large predators and support that surface above the substrate should the sponge fall over (prevent occlusion of ostia and potential burial). Special prostalia around the oscular margin, the marginalia, likewise protect that delicate growing edge and prevent easy entrance to the osculum by larger crawling invertebrates. The distal projecting parts of prostalia listed above are bare siliceous structures freely colonised by epibionts (Boyd, 1981; Beaulieu, 2001a,b). Large uncinates, the barbed diactins or monactins present in many groups, may project from lateral surfaces and are then considered by most practicing hexactinellid specialists as prostal megascleres, although they always remain totally covered by a thin film of living tissues. We find no support, neither in size nor form, for their historical assignment as inter-medial microscleres. Likewise, the special surface-related monactin group called sceptrules (clavules and scopules and related forms) were historically regarded as microscleres but by size and form they are clearly megascleres. By position, they are best considered comitalia of dermalia and gastralia.

6.4. Fused silica networks

Spicules of hexactinellids can be physically joined (fused or ankylosed) by deposition of secondary silica in several ways: by simple spot-soldering with a very thin siliceous film at points of contact between spicules, by formation of anaxial bridges or synapticula (silica filaments of Reid, 2003) between surfaces of spicules not in direct contact or by enclosure of rays or entire spicules in a continuous layer of secondary silica. These are best-considered stages of a continuous series. Spot-soldering is differentiated from synap-ticular bridging only by visible gaps between the original spicule surfaces adjacent to the connection. Synapticula may also extend from spicule sur-faces as free-ended filaments; these may be unconnected distally or join to other such filaments, occasionally forming anaxial networks. Although small, completely anaxial networks have been reported in a few cases, it is uncertain whether they originate free of other skeletal elements or are out-growths of simple synapticula that have broken free of their original location on axial spicule surfaces.

Hexactinellids that form fused skeletons are much more likely than those with only loose spicules to be found as fossils. While spicule fusion is common among hexactinellids, it is equally important to appreciate that this process is completely unknown among the subclass Amphidiscophora where all spicules remain loose throughout life. Even in those groups where fusion is well known or ubiquitous (orders Hexactinosa and Lychniscosa of Hexasterophora), specific types of spicules (microscleres, sceptrules, un-cinates and so on) always remain loose or are only occasionally incorporated by apparent accident.

6.4.1. *Lyssacine parenchymal networks*

While some members of all three families of Recent Lyssacinosida may form fused basal skeletons, fusion of parenchymal megascleres is known to occur in only two families, Euplectellidae and Rossellidae; it remains unknown in

Leucopsacidae. In Euplectellidae, the often regularly arranged principalia and the more irregularly interlaced comitalia and intermedia are typically found to be entirely loose spicules, and in many species parenchymal spicule fusion is unknown. However, in some species of many genera, for example *Euplectella* and *Corbitella*, spicule fusion appears to begin from the base and spread upwards, ultimately involving the entire suite of parenchymal megascleres, including those of the sieve plate. Some specimens have only patches of spot-soldering between megascleres of the lower body, but fusion in other specimens is extended to include synapticular bridges and, in some, expanded to full enclosure of the entire parenchymal skeleton into a rigid glassy network. The fused skeletons are quite irregular (Figure 37A) since the connected spicules are of mixed types, diactins to hexactins, and remain in their original orientations relative to one another. The fusion process appears to be informal in the sense that it does not seem to be a predictable result of an early developmental programme. It may be related to aging since specimens with completely fused skeletons are unable to grow in length or width, but wall thickening remains possible. A detailed study relating the extent of spicule fusion with size, a surrogate for age, has not yet been carried out for any species of this family. Note that fusion of parenchymalia remains unknown for many species of Euplectellidae.

6.4.2. Dictyonal parenchymal frameworks

Four orders of Recent hexactinellids, Hexactinosida, Aulocalycoida, Fieldingida and Lychniscosida, have fused siliceous parenchymal frameworks composed primarily or exclusively of hexactins connected by secondary silica enclosure. Framework forms are fairly regular and predictable in all of these except Fieldingida where the irregular structural pattern remains unresolved. The component hexactins are known as dictyonalia and the frameworks as dictyonal. Dictyonalia of the first three orders are simple hexactins (Figure 37B), but those of the Lychniscosida have 12 synapticular struts developed equidistantly and symmetrically around each dictyonal centre, outlining a regular octahedron (Figure 37C). Such dictyonalia are known as lychniscs. Interconnection of hexactins in all these orders is a normal and necessary part of growth and occurs at or just behind the free margin, resulting in longitudinal extension of the fused framework. Hexactinellid specialists have long known that basic differences exist between dictyonal frameworks of various groups, but to date there has been little progress in understanding and communicating those differences. Only four types of differences are now appreciated: presence and absence of various types of skeletal channelisation (canals of palaeontologists), how adjacent

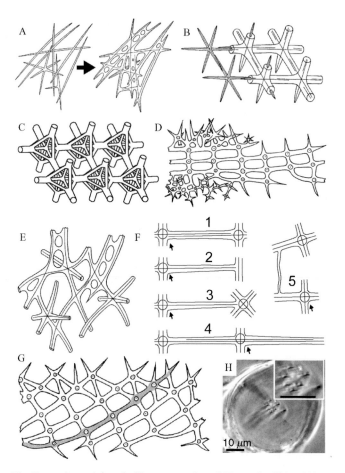

Figure 37 Parenchymal-fused silica networks of Hexactinellida. (A) Loose (left) and fused (right) networks of parenchymalia in lyssacine hexactinellids (from Reid, 1958). (B) Part of a dictyonal framework of hexactinosan showing classic, but erroneous, terminal addition of hexactin dictyonalia (from Reid 1958). (C) Part of a dictyonal framework of a lychniscosan with characteristic lychnisc nodes. (D) Longitudinal section through the wall of a hexactinosan with cortices developed on both dermal (upper) and atrial (lower) surfaces of the primary framework which alone extends as a series of longitudinal strands to the growing margin (right) (from Reid, 2003). (E) Portion of the irregular parenchymal dictyonal framework of an aulocalycoid hexactinellid (from Reid, 2003). (F) Transparency views of five methods of fusion of dictyonalia including relationships of axial filaments; new appended dictyonalia indicated by arrow. (G) Longitudinal section through hexactinosan dictyonal framework showing longitudinal dictyonal strands (one emphasised in grey) curving to dermal surface (from Reid, 2003). (H) Oblique section of longitudinal strand of *Farrea occa* with central bundle of six axial canals plus one from the most recently added dictyonalium.

dictyonalia are connected, primary and secondary layers in frameworks and dictyonal strands.

Skeletal channel types were first described by Ijima (1927) and later refined by Reid (1958) when only two groups of Recent dictyonal Hexactinellida, the Hexactinosa and Lychniscosa, were known. These consisted of simple tubular channels, in the size range of one-half to a few millimetres in diameter, termed epirhyses when passing vertically but shallowly into the body wall from the dermal (inhalant) surface, aporhyses from the atrial exhalant surface, diplorhyses when the two channel types deeply overlapped through the middle of the wall, diarhyses when passing completely through the wall, schizorhyses when interconnected to form a single labyrinthic system of channels within the wall and a special form termed amararhyses when channels branch in slits on the atrial side and open on projections on the dermal side. These macroscopically obvious gaps in framework structure are still used as major characters in defining families of Hexactinosa. They are, however, negative characters in the sense that they indicate patterns where addition of dictyonalia is suppressed, presumably to enhance water flow to deeper wall layers, and thus add little or nothing to show how the different frameworks are actually constructed. These different types of skeletal channelisation may, however, be attained convergently in different families.

Distinction of primary and secondary components of dictyonal frameworks, informally recognised by Ijima (1927), was formalised by Reid (1958). The dictyonal framework formed at the growth margin is considered the primary framework, and can be traced back to earlier stages through the entire skeleton. Dictyonalia may be added onto the outer and inner surfaces behind the growth margin (radial accretion), resulting in body wall thickening. The patterns of connection and mesh geometry of these added superficial dictyonal layers are usually strikingly different from those of the primary framework, and such layers are called secondary or cortical layers (Reid, 1958) (Figure 37D). All dictyonal frameworks thus have a primary framework, with or without cortical layers on dermal and atrial surfaces. This distinction of layers allowed recognition that channelisation may be restricted to only the cortical layers or involve the primary layer as well, adding another method of distinguishing pattern in skeletons of different taxa. Empowered with this recognition of dictyonal layers, Reid (1958) recognised and defined farreoid (two-dimensional sheet), euretoid (three-dimensional sheet) and aulocalycoid (irregular, Figure 37E) types of primary frameworks among Hexactinosa.

Details of how dictyonalia are individually inserted onto or into an existing framework remain elusive for all Hexactinellida. Nonetheless, five basic patterns of dictyonal fusion have been identified so far (Figure 37F): (1) parallel ray fusion, (2) tip-to-ray fusion, (3) tip-to-centre fusion, (4) axon-to-ray fusion and (5) synapticular bridging. The first pattern appears to be the most common for connection of dictyonalia in primary frameworks and was regarded by Ijima (1927) as the original and basic process of dictyonal framework formation, as seen in *Farrea*. Here the resulting internodal beams have two axial filaments, one from the component rays of each of the two hexactins. In tip-to-ray and tip-to-centre fusions, beams have a single axial filament. Mixtures of these indicate a haphazard attachment process, perhaps best interpreted as a lack of pattern. In axon-to-ray fusion, the axial centre and both axial filaments of one axon are appended to another ray so the two beams in question have two or more axial filaments, one of which is continuous throughout the two beams (Reiswig, 2002a). Synapticular bridging is rare but is an important method of connection in Aulocalycoid skeletons (Reiswig and Tsurumi, 1996).

Longitudinal dictyonal strands were undoubtedly recognised by early hexactinellid specialists, but their existence was first formulated by Reid (1958). Here the longitudinal rays of dictyonalia are aligned in a continuous series to form major longitudinal structural components of the primary framework. In basic farreoid and euretoid frameworks, strands are more or less straight (Figure 37B and D), but in groups with thicker body walls, strands curve smoothly to dermal or atrial or both surfaces (Figure 37G). Reid (1958) presumably accepted Ijima's statement (1927) that all beams in the basic two-dimensional *Farrea* framework resulted from parallel ray fusion and thus carried two axial filaments. Hence, the relatively straight path of strands was thought to be due only to the sequential step-by-step alignment of each dictyonalium in series. No one has questioned Ijima's statement for 85 years. One of us (H.M.R.) has recently inspected the axial filaments in beams of *F. occa*, the type species of *Farrea*, which was the basis of Ijima's claim, and found transverse beams to have the expected two axial filaments but longitudinal beams contain numerous (five to eight or more) closely bunched axial filaments passing entirely by the axial centre of all added dictyonalia (Figure 37H). It is clear that dictyonalia are added here in axon-to-ray fusion, not parallel ray fusion, and linearity of longitudinal strands here, and likely in all taxa with such structures, is due to dictyonalia being attached to the side of an existing continuous silicified axis composed of a bundle of rays. This finding suggests that present hypotheses about how dictyonal frameworks with longitudinal strands are constructed must be completely rethought and casts doubt on the distinctness of the order Aulocalycoida. Several groups of hexactinellids lack dictyonal strands and their methods of framework construction also need detailed examination in light of this discovery. Unfortunately, determination of axial filament patterns requires very careful examination of framework elements with high magnification oil immersion objective, a necessarily destructive process.

6.4.3. Basidictyonal frameworks

Special fused basal skeletons have long been known in a few dictyonine hexactinellids (Schmidt, 1880; Weltner, 1882), but recognition of their general occurrence in all basiphytous (attached to hard substrate) dictyonine and lyssacine forms was delayed until many species had been surveyed. Ijima (1901) introduced the term 'basidictyonalia' for frameworks composed of small, thick-rayed hexactins (Figure 38A and B) in the basal regions of lyssacines. Since its earliest recognition, the basidictyonal framework was known to consist of two distinct components—the main three-dimensional mass composed of fused hexactins and the thin lattice network of a quite different structure apposed to the substrate (Figure 38C). A broad range of names has been applied inconsistently to the entire framework and its component parts. We suggest 'basidictyonal mass' for the hexactine skeleton, 'limiting basal plate' for the superficial surface membrane and 'basidictyonal framework' for the entire structure; the term 'basidictyonal plate' often used for the structure is often physically inappropriate and is easily confused with the macroscopic 'basal plate' used to describe a spreading body part at the attachment site.

The basidictyonal mass is mainly or totally composed of simple, stubby hexactins, sometimes pentactins, of ray length 30–120 μm, fused together directly at contact points or by synapticula. Even in Lychniscosa, the basidictyonalia never have lantern nodes. Junctions are tip-to-tip unlike those of most true dictyonalia (see above). This structure may be a single layer or may be several millimetres in thickness. In most lyssacines, the main parenchymal spicules (diactins) may insert loosely into this lattice, but fusion between parenchymals and basidictyonalia occurs only in those forms that have extended secondary fusion. Details of the relationship between the basidictyonal mass and the main fused parenchymal framework of dictyonal hexactines is poorly known but they are probably joined in early development. Basiphytous hexactinellids settle on two basic types of substrate relative to the structural components of the limiting basal plate—flat surfaces of rocks, shells and so on, and cylindric surfaces of prostal spicules and dictyonal frameworks of dead hexactinellids. On flat substrates, the limiting basal plate, where known in detail, consists of stauractins (basidictyonalia?) joined by synapticula to form a thin porous silica sheet with 20–40 μm pores (Figure 38D). On curved surfaces, stauractine, pentactine and/or hexactine dictyonalia appressed to the surface form the building elements for encompassing substrate curvature by synapticular junctions and threads (Figure 38E). The plate is not in direct contact with the substrate, as is the case with calcareous or organic spongin deposition in other sponges, but attachment is attained by encompassing (grasping around) substrate

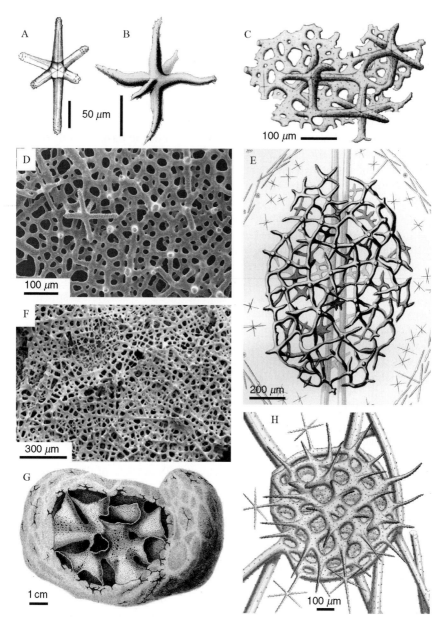

Figure 38 Non-parenchymal siliceous networks of Hexactinellida. (A, B) Stout, small basidictyonalia from lyssacinosans, (A) *Regadrella okinoseana* (from Ijima, 1901) and (B) *Rhabdocalyptus mirabilis* (from Schulze, 1899). (C) Basidictyonal skeleton of *Rhabdocalyptus unguiculatus* from sponge side showing limiting basal plate with fused hexactins of the basidictyonal mass above them (from Ijima, 1904).

irregularities with a closely conforming rigid plate. The limiting basal plate is fused to the network of basidictyonal hexactins above by direct fusion and synapticula. Although the basidictyonal skeleton is characteristically formed at the site of larval attachment, indistinguishably similar structures are apparently formed at sites where adult tissues contact foreign objects and form secondary attachments. In the two known series of post-larval development, basidictyonal elements in the dictyonine, *F. sollasii*, have begun formation in the earliest stage encountered (0.53-mm body length; 67-nl volume) (Okada, 1928), but develop later in the lyssacine, *Leucopsacas scoliodocus*, where they are first seen in juveniles (1.85-mm body length; 1650-nl volume) (Reiswig, 2004). In both cases, parenchymal megascleres appear later than the basidictyonal elements.

6.4.4. Surface networks

Networks of fused spicules and synapticula, variously referred to as cover layers, cortices, tunics, capsules, rinds, Hüllschichte and so on, were characteristic of several groups of fossil dictyonine hexactinellids (Reid, 1958) but are known in only three Recent dictyonine genera: *Lychnocystis* and *Neoaulocystis* of the order Lychniscosida and *Fieldingia*, the only Recent genus in the new order Fieldingida (Tabachnick and Janussen, 2004). These structures are composed of stauractin and pentactin spicules, with a few hexactins, interconnected by an extensive network of anaxial synapticula, forming finely porous lamina, always fused to the underlying dictyonal frameworks. In the two lychniscid genera, the spicules of the networks differ from normal dictyonalia in always having simple nodes. In *Lychnocystis* and *Fieldingia*, the surface networks are rigid, up to 1 mm thick, and multilayered, containing stacks of laminae joined by occasional fusions with the unpaired ray of constituent pentactins stretching between them (Figure 38F). In these genera, every specimen so far encountered has well-developed surface networks; hence the process of development is unknown. In *Neoaulocystis*, most specimens lack any surface networks, but some have early plates extending laterally from the outer margins

---

(D) Margin of a flat basidictyonal framework of *Rhabdocalyptus dawsoni* from sponge side showing limiting basal plate with only a few rays of constituent pentactins projecting up a sites for future hexactin attachment. (E) Basidictyonal framework of a *Rhabdocalyptus mirabilis* juvenile settled on two prostal spicules (vertical rods) of an adult; note the lack of direct basidictyonal cohesion with the prostal spicules (from Schulze, 1899). (F) Surface network of *Fieldingia valentini* from internal side; compare to basidictyonal limiting plate, D (from Tabachnick and Janussen, 2004). (G) Body of *Neoaulocystis zitteli* partially enclosed in surface network (from Schulze, 1887). (H) Weltner body from *Lychnocystis superstes* (from Schulze, 1887).

of tube walls (Figure 38G). The soft, flexible polygonal plates meet adjacent plates, circumscribing stellate clefts which are eventually closed to cover the entire outer surface excepting large oscular openings in a continuous 0.3 mm thick membrane of porous silica. Multilayering has not been noted in *Neoaulocystis*. Without clear definitions, the few data on mesh and beam dimensions are incomparable, but pore sizes generally range from 2 to 45 μm. While porosity of the structures seems adequate for inhalent water flow through these rigid structures, formation of an unexpandable surface shell seems mal-adaptive in blocking further growth. However, it is not known whether these structures are permeated or covered with living tissue and generally entire body forms are unknown for the species bearing these structures. Most workers (Schmidt, 1880; Schulze, 1887; Ijima, 1904) have commented on the similarity of surface networks with basidictyonal frameworks, especially those formed at new contacts with foreign objects, but a direct comparison has not yet been carried out.

6.4.5. Weltner bodies

Sharply delineated spherical knots, skeletal nodes or spheres known as Weltner bodies in recognition of Weltner's detailed descriptions (1882) of them are known sporadically from many dictyonines, but occur predictably in only a few species, *Lychnocystis* (earlier *Cystispongia*) *superstes*, *Neoaulocystis zitteli* and both known species of *Fieldingia*. These tight condensations of the regular internal dictyonal framework, liberally covered and sometimes filled with fused small spiny hexactins (Figure 38H), are commonly 0.7–2.0 mm diameter and visible to the naked eye. They are regularly spaced and probably provide stress focal points in the very loose skeleton of *Fieldingia* species, but it is doubtful that they contribute to strengthening the sturdy lychnisc skeletons of the other species. They have been considered products of local stimulus (irritation) by foreign organisms (Schulze, 1900) and possible basidictyonal skeletons of internal buds that have been later integrated into the parental skeleton (Ijima, 1927). These and a variety of other less well-described networks of silica deserve detailed study, but lack of availability of specimens is a continuing deterrent.

6.5. Silication

6.5.1. Basic process and forms of siliceous structures

It has long been assumed that initial silica deposition in hexactinellids occurs as spicule formation as in demosponges by polymerisation of silicic acid on an organic axial filament deposited within an intracellular vacuole in a cell, generally termed a scleroblast, and specifically called a silicoblast. The specialised limiting membrane of such vacuoles is known as a silicalemma, but the space in which silica deposition occurs remains unnamed. We propose 'silica deposition space' (sds) for this, avoiding "vacuole" since in some or all cases it may be confluent with extracellular spaces. Silica deposition has not been directly observed in hexactinellids, but morphological details of preserved hexactinellids and direct observation of the process in living freshwater demosponges (Weissenfels and Landschoff, 1977) offer strong support that initial silication is similar in both sponge groups. In hexactinellid megascleres, additional layers of silica are added either over the entire surface of the primary spicule or only onto the ends of the longer ray in very large prostalia. Such layers are easily seen due to presumed variations in refractive index as 'ghosts' of earlier spicule surfaces, in lateral view as complete outlines or serially stacked elongate cones and in cross sections as concentric rings surrounding the primary spicule. The initial siliceous cylinder immediately around the axial filament is known as the adaxial zone (Hartman, 1983) or axial cylinder (Claus, 1868) and the added secondary layers are collectively referred to as the peripheral zone. Since basic morphology of both the initial spicule and added layers appears to be determined by the geometric form of the axial filament system, these can best be termed primary and secondary stages of actinal silication (Reid, 2003).

Ornamentation added either to megascleres as spines, hooks, anchor teeth of basalia or long tines of scopules, or to microscleres as unbranched or extensively branched terminal rays, are often considered results of a distinct silication process ('anactinal' in Reid, 2003) since such structures bear no axial filaments resolvable by light microscopy. At our present stage of understanding silica deposition, there seems to be no basic difference 'in kind' between inconsistent variable development of thorns on rays of pentactine megascleres and consistent, channelised formation of terminal rays on hexaster microscleres. Spine formation on megascleres clearly occurs as part of secondary layering of the primary spicule, but whether layer addition occurs on microscleres has not been determined. In both cases, ornamentation does not seem ascribable to the templating function of the axial filament system since that element is quickly enclosed in an impermeable (?) silica sheath.

Additional forms of silica deposition such as spot-soldering of spicules, enclosure of part or entire spicules by silica layering in dictyonal and lyssacine frameworks, formation of synapticular bridges and networks between spicules, all appear to be aspects of anaxial silication since silica deposition here cannot be patterned directly by axial filament geometry. Deposition here appears to differ only qualitatively from the simplest layer addition on primary spicules. Thus, the process of silica deposition in fusion of spicules and synapticula formation, though generally known as 'secondary silication' is unlikely to differ from that involved in layer addition to basic axial spicule structures. In addition to the rules of hexactinellid spicule form noted earlier (axial filaments never branch and new rays cannot be added on to pre-existing spicules), two additional features of silica deposition require consistency testing: silica deposition is irreversible (notwithstanding discovery of a silicase gene in a demosponge, Schröder *et al.*, 2003b) and silica can only be deposited directly on autochthonous axial filaments or on autochthonous siliceous surfaces deposited on autochthonous axial filaments.

6.5.2. *Silica constituents and characteristics*

Spicules of hexactinellids are nearly indistinguishable from those of demosponges in most measured characteristics. They have the general formula of opaline silica, $SiO_2 \cdot nH_2O$, where *n* varies from 0.33 in *Monorhaphis* to 0.25 in *Rossella* and most other hexactinellids, reflecting the bound water (water of hydration) of 7–12% dry weight driven from 'dry' spicules when heated to 900°C (Minchin, 1909; Sandford, 2003); some water continues to be evolved at least to 1400°C (Sandford, 2003). Sandford (2003) found that bound water bonds, as indicated by Fourier-transform infrared spectra, are generally similar for silica gel and spicules of both hexactinellids and demosponges. He found an anomalous endothermic event in glass transition temperatures at 425–500°C in *Hyalonema* (amphidiscophoran) spicules that was absent in hexasterophoran spicules. In addition to bound water, hexactinellid spicules consist of 83.8–88.3% SiO_2 and 1.7–4.2% of other minerals, including sulphur, aluminium, potassium, calcium and sodium as major elements (Sandford, 2003). The amount of organic matter in hexactinellid spicules remains unknown and can only be inferred as less than (part of) the weight loss on combustion, given as 2.9% of dry weight in *Poliopogon amadou* by E. Fischer (Schulze, 1904). The refractive index is reported as 1.49 in *Rossella racovitzae* (Sarikaya *et al.*, 2001), 1.467–1.47 in four hexactinellid species (Sandford, 2003, with reservation on accuracy) and 1.425–1.48 in basal spicules of *Euplectella aspergillum* (Aizenberg *et al.*, 2004), values slightly higher than those of demosponges. The long held notion that the microscopically visible layering in spicules is due to variations in refractive index caused by differences in bound water (Schwab and Shore, 1971 in demosponges) were rejected for spicules of *Rossella racovitzae* by measurements of Sarikaya *et al.* (2001), but substantiated by interferometric measurements of spicules of *Euplectella aspergillum* by Aizenberg *et al.* (2004). Density of hexactinellid spicules is 2.03–2.13 (Sandford, 2003), encompassing most values reported from demosponges.

Sarikaya *et al.* (2001) provided additional measurements on spicules of *Rossella racovitzae* for which comparative data on demosponge spicules is unavailable. Hardness is 3.22 ± 0.33 GPa and elastic modulus is 38 ± 3 GPa, both measured by atomic microforce indentor and both about one-half the values for glass fibres and silica rods. Hexactinellid silica is thus softer and less elastic than glass. In three-point bends, these authors found hexactinellid spicules to be exceptionally tough and flexible, with fracture strength of 900 mPa, fracture toughness of 2–5 mPa (both five times the values for glass rods) and fracture energy 15 times that of glass rods, ascribed to energy storage in spicule layers prior to general rupture of the entire spicule. Mechanical properties of individual spicules and the entire skeleton of *Euplectella aspergillum* were elegantly explored in terms of engineering design by Aizenberg *et al.* (2005). They concluded that details of construction at seven levels of hierarchy provided a supporting framework of outstanding mechanical rigidity and stability for the basically brittle silica-building material.

6.5.3. Microscopy of silica deposition

Observations of sections of fixed hexactinellid tissues have led to a series of conclusions on spicule formation. Ijima (1901, 1904) combined observations on adult tissues of many hexasterophorans and concluded that microscleres were formed in dense, multinucleate scleroblast masses which he interpreted not as discrete cells as in demosponges, but as local condensations of the general trabecular syncytium. The earliest stages had only the six basal rays of the spicule and no definite number of associated nuclei. He discerned a repeating pattern of expansion, condensation and differentiation of the scleroblast mass with microsclere maturation, with its eventual disappearance or transformation to a diffuse tissue network identical to, and continuous with, the general trabeculum. Ijima's conclusions were entirely supported by studies of Woodland (1908). Okada (1928), in his description of hexaster development in larvae and juveniles of *F. sollasii*, reported prospective scleroblasts to be discrete mononucleate cells of the embryonic dermal epithelium that move below the outer layer and become multinucleate scleroblast masses when the first hexactine stages of the siliceous primordium are encountered. Nuclear numbers of this tissue, which we now recognise as a sclerosyncytium, increase as spicules mature and scleroblast nuclei differ from regular nuclei of the trabeculum in being larger, more spherical, packed with chromatin granules, often possessing nucleoli and staining more deeply with borax carmine. Okada concluded that the scleroblast mass (sclerosyncytium) became part of the general trabeculum when spicule formation was complete.

Development of larval stauractins (megascleres?) has been reported by Okada (1928) in *F. sollasii* and by Leys (2003a) in *Oopsacas minuta*. In *F. sollasii*, stauractin scleroblasts are specific columnar epithelial cells that move under the outer epithelium and begin spicule formation, but remain discrete mononucleate cells throughout spiculogenesis. Using transmission and scanning electron microscopes, Leys (2003a) concluded that the 14 larval stauractins in *Oopsacas minuta* begin development by formation of typical hexactinellid axial filaments in intracellular sds of discrete sclerocytes and deposition of silica on these axial structures. The sclerocytes are first mononucleate cells rich in mitochondria, clear vesicles and pseudopodia and connected to other tissues by plugged junctions. As spicule maturation proceeds, the sclerocytes become multinucleate and then known as sclerosyncytia, but after completion of spicule formation, the ultimate fate of these sclerosyncytia remains unknown.

Few observations on development of megascleres in adult tissues or their relationship to living tissue have been published. In *F. sollasii*, Okada (1928) described formation of clavules and hexactine dictyonalia in multinucleate scleroblast masses that are distinct from the general trabeculum only by nuclear form. Mackie and Singla (1983) summarise observations on tissues of adult *Rhabdocalyptus dawsoni*, without making distinction between conditions in microscleres and megascleres, that spicules are formed intracellularly in multinucleate giant sclerocytes (sclerosyncytia) which are the only large tissue components differentiated from and connected to the trabecular syncytium by plugged cytoplasmic bridges while silica is being deposited. These authors also found mature spicules within the general trabecular syncytium, but whether they are formed there remains an open question. They also report electron-dense materials remaining in/on the walls of the cavity left when spicules are dissolved from tissues by hydrofluoric acid.

Several questions on tissue–spicule relationships remain incompletely resolved. Does silication take place only in sds lined by a special silicalemma membrane? Many authors have taken this to be a general rule for all siliceous sponges because it has been repeatedly verified (or unfalsified) in freshwater demosponges where spicule deposition has been monitored in special living microscopic preparations. Extracellular secretion of spicular silica has been reported only by Uriz *et al.* (2003) in the demosponge, *Crambe crambe*. Although it may be extremely difficult to demonstrate continuity between silicalemma and plasmalemma in ultra-thin sections and transmission electron microscopy, it should be obvious that, in hexactinellids, silica secretion and often axial filament extension must occur at sites which are bordered by membranes that are continuous with the plasmalemma of either the scleroblast or the general trabecular syncytium. These include all projecting prostalia, including the conspicuous basal anchors of Hyalonematidae, which are mostly emergent from the living tissue but remain embedded in trabecular syncytium at one end where axial filament and silication continues. Extension of the axial filament and silication must take place here in spaces that are continuous with the external milieu, but that are presumably functionally isolated by close apposition of membranes to spicule surfaces. The same situation must hold for dictyonal skeletons in general where living tissues have retracted from any part of the skeleton, most commonly the basal attachment. Initial formation of axial filaments and silica secretion may take place within closed intracellular sds, but lack of continuity with the plasmalemma would be difficult to prove. Many spicules may be enclosed in intracellular sds during their entire development, but some, as just noted, cannot remain so, even during their development, which may continue as long as the sponge lives and never be completed.

Do spicules that have completed development remain within intra-syncytial sds or are they exocytosed and thus freely exposed to the exterior milieu? Again, as noted, some spicules and dictyonal structures are clearly exposed at least in part. Observations by several workers suggest most spicules remain entirely coated (enveloped within) the trabecular syncytium (Woodland, 1908; Okada, 1928; Wyeth *et al.*, 1996), but others have suggested that most megaslceres are attached to or enclosed by strands of the trabecular syncytium only at discontinuous points (Mackie and Singla, 1983).

Do spicules have an organic spicular sheath or organic matrix? The fibrillar organic material found by Travis *et al.* (1967) on/in *Euplectella* spicules may be part of the collagenous mesolammella appended to outer spicule surfaces when in extra-syncytial position, but the electron-dense lining remaining on cavities of *Rhabdocalyptus dawsoni* spicules in intracellular locations (Mackie and Singla, 1983) might represent an organic sheath involved in the process of silica deposition. The undissolved amorphous electron-dense matter remaining after spicule dissolution with hydrofluoric acid (HF) (Travis *et al.*, 1967; Mackie and Singla, 1983; Leys, 2003a) may be the remains of matrix molecules, but this material has yet to be completely characterised. Aizenberg *et al.* (2004) interpreted gaps formed between outer layers of basalia of *Euplectella aspergillum* after NaOCl exposure, but other interpretations for this effect were not explored. An energy dispersive X-ray analysis (EDX) carbon map of a spicule section, likewise, interpreted to document organic matter within the spicule matrix was not convincingly portrayed. This may be the same fibrillar collagen demonstrated by Ehrlich *et al.* (2005) in root spicules of *Hyalonema sieboldi* using a novel method of silica extraction.

6.5.4. Axial filaments

Early history of observations and nomenclature of hexactinellid spicules and their parts is summarised by Schulze (1904), but he omits his own earlier

fanciful interpretation (Schulze, 1887) of axial filaments as channels to
supply nourishment to the living distal parts of root spicules of *Hyalonema*
and *Euplectella*. While most studies of axial filaments have been carried out
on demosponges, conclusions obtained there can be generally assumed to
apply also to those of hexactinellids pending their falsification in that group.
Axial filaments of hexactinellids vary in size (width of side) from 0.3 to
0.5 μm in larval stauractins of *Oopsacas minuta* (Leys, 2003a) to 7 μm in
pentactins of *Rossella racovitzae* (Sarikaya *et al.*, 2001, Figure 34C). Bütschli
(1901) concluded from staining characteristics that axial filaments of demosponges were mainly protein. Dendy (1926) proposed that axial filaments
were a group of microbes, his 'sclerococci', which induced host sponges to
encapsulate them in silica, thereby forming their skeletons. This hypothesis
attracted no support by other biologists. The nature of filaments of the
demosponge *Haliclona rosea* was convincingly settled by Garrone (1969)
through use of electron microscopy, X-ray diffraction and pepsin digestion.
He concluded that these filaments were composed primarily of a hexagonally
crystallised protein. Subsequent studies on demosponges showed that filaments undergo submicroscopic branching, extending fine cores as small as
0.05 μm diameter into spines and thorns of megascleres and microscleres
(Garrone *et al.*, 1981; Simpson *et al.*, 1983, 1985). The filaments were also
shown to be more heterogeneous, both within a single spicule and between
species, than previously thought, with occasional repeating surface beading,
less dense centres and refractile cores (perhaps containing silicates) or voids.
The crystalline protein figured earlier by Garrone (1969) was considered
only a small part of the mature filament.

Shimizu *et al.* (1998) purified three main structural protein subunits from
axial filaments of the demosponge *Tethya aurantium*, cloned and sequenced
the cDNA of the most abundant one and determined that these were similar
to members of the cathepsin L and papain family of proteases. Since axial
filaments were already suspected of serving as templates for silica deposition
(Imsiecke *et al.*, 1995), and Shimizu *et al.* (1998) found that the purified
protein filaments and subunits promoted polymerisation of silica and silicones from corresponding silicon alkoxide precursors at room temperature
and neutral pH, they named the three protein subunits as silicatein α, β and γ
to emphasise their enzymatic properties. Although axial filaments and silicateins of hexactinellids and demosponges are no doubt derived from a
common ancestral source, Croce *et al.* (2003) found significant differences
between them in small angle X-ray diffraction studies. Filaments of the
hexactinellid *Scolymastra joubini* have two different two-dimensional lattices
due to protein units repeating at $+50°$ and $-50°$ relative to the spicule axis
and consistent with a hexagonal packing of spirally oriented cylindrical
protein units elongated along the filament axis (Figure 39B and C). This
contrasts to the hexagonal protein lattice of the demosponge *Geodia*

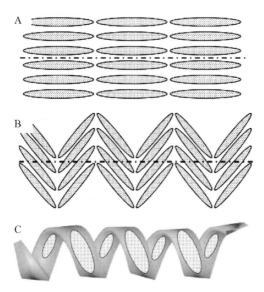

Figure 39   Comparison of two-dimensional structural models of protein (silaca-tein) units in axial filaments of the demosponge *Geodia cydonium* (A) and the hexactinellid *Scolymastra joubini* (B), with longitudinal spicule axes represented by dot-dashed lines. (C) Three-dimensional representation of the model of spirally oriented cylindrical protein units along the axial filament axis of *Scolymastra joubin* (from Croce *et al.*, 2003).

cydonium with units oriented parallel (apparently erroneously reported as perpendicular) to the spicule axis and packed more densely than in hexacti-nellid filaments (Figure 39A). Spacing between units also differs in the two filaments, 8.4 nm in the hexactinellid and 5.8 nm in the demosponge. A more detailed comparison of the silicateins of Hexactinellida and Demospongiae awaits sequence determination of hexactinellid silicateins.

6.5.5. Molecular processes and silication model

Results of experiments on silication in demosponges, summarised here, have led to proposed models that probably apply also to hexactinellids. Silicat-ion rate *in vivo* is influenced by silica concentration, temperature, particulate organic matter concentration and Fe^{3+} concentration (Uriz *et al.*, 2003). Silicateins are not entirely restricted in their distribution to the axial filament but have also been shown by immunofluorescent labelling to occur on the outer surfaces of developing spicules (Cha *et al.*, 1999) and may be the same material seen lining outer spicule surfaces after spicule extraction

(Mackie and Singla, 1983; Leys, 2003a). Spicules dissociated with HF expose axial filament silicateins which remain active in polymerisation of silicon alkoxides but lose their templating function (Shimizu *et al.*, 1998). Silicatein *in vitro* catalyses deposition of 70 nm diameter particles (nanospheres) on axial filaments several layers thick (Cha *et al.*, 1999). A similar process probably occurs *in vivo* since atomic force microscopy has shown that HF-etched spicules of the demosponge *Tethya aurantia* have nanoparticles of 50 to 80 nm diameter in alternating layers about one particle thick (Weaver and Morse, 2003). Similar nanospheres 50 to 200 nm in diameter have been shown to be the basic form of building material in the spicules of *Euplectella* sp. (Aizenberg *et al.*, 2004, 2005). Nanospheres as small as 17 nm diameter have been detected in demosponge spicule formation *in vivo*; these apparently fuse to form particles 100 to 1200 nm in diameter and intervening spaces are then smoothly filled with amorphous silica (Uriz *et al.*, 2003). Layering in spicules may indicate pauses in nanosphere deposition and smoothing (filling in) steps. High silica concentrations have been found in the cytoplasm near developing spicules (to 50%) and in the spicule space between spicule and bounding silicalemma (50–65%) (Uriz *et al.*, 2003).

The following model of silication in sponges is derived from these observations and propositions put forth by Croce *et al.* (2003), Müller *et al.* (2003), Weaver and Morse (2003) and Uriz *et al.* (2003). Sclerocytes secrete axial filaments composed mainly of silicateins but other organic molecules are included. Silicatein acts as a structural template for formation of the highly organised mesoporous axial filament. Inorganic silicate diffuses or is actively taken up into the sclerocyte and complexed with specific proteins to form an as yet unidentified organic-silica substrate. The organo-silica complex is transported across silicalemma to the sds. Silica from organo-silica substrate is polymerised on the outer surface of the mesoporous axial filaments as nanospheres, perhaps at serine and histidine catalytic centre sites, and the complexing protein is released and possibly recycled to the sclerocyte cytoplasm. Growth of the axial filament continues at spicule tips, providing primary patterning of the spicule, while at the same time transport and deposition of silica continue on lateral spicule surfaces between the tips. Once the first few layers of silica nanospheres encase the axial filament, enzymatic activity can no longer be expressed by silicateins of the axial filament. Continuing deposition of silica on the outer surface of spicules is facilitated and controlled by silicatein at active sites on the silicalemma, resulting in production of specific patterned morphologies of spicules.

Several elements of this model remain unverified. Other unknowns include the roles of the minor non-silicatein organic compounds incorporated in axial filaments, whether silicateins associated with the silicalemma are the same forms as those in the axial filaments and how the final high-fidelity submicroscopic external patterning on spicule surfaces is genetically controlled.

7. ECOLOGY

7.1. Habitats: distribution and abundance

Glass sponges are found in deep water (greater than 500 m) in all oceans of the world, but only inhabit shallow waters (up to 20 m) in four known locations: Antarctica, southern New Zealand, submarine caves in the Mediterranean and coastal waters of the North Pacific. Their restriction to deep waters is shared with only one other sponge group—lithistid demosponges—which, like glass sponges, have a heavily fortified glass spicule skeleton (Pisera and Lévi, 2002). It has been speculated that silica limitation in shallow waters forms the upper limit for both of these groups (Austin, 1983, 1999; Maldonado et al., 1999), but light, temperature, food availability and turbulence (current or wave-driven) may also restrict the distribution of glass sponges in particular to deeper waters (Leys et al., 2004). Currently, there is no experimental evidence to support or reject any one of these hypotheses.

Although in many regions glass sponges appear sparsely distributed, in some localities, both deep and shallow, glass sponge populations are very dense (Figure 40A and B). High densities (1.5 m^{-2}) of Pheronema carpenteri occur within a narrow depth range (1000–3000 m) in the Porcupine Seabight, southwest of Ireland (Bett and Rice, 1992). Stalks of Hyalonema can be found every 5 m ($\sim$2–3 per 10 m^2) at 4100 m in a bathyl basin off California (Beaulieu, 2001a,b). The benthos at 100–300 m in the Weddell Sea, Antarctica is heavily occupied by sponges (Barthel and Gutt, 1992; Barthel and Tendal, 1994). Although demosponges are most abundant (up to 200 per 10 m^{-2}), some sites are dominated by seven species of hexactinellids; of these Rossella racovitzae was the most abundant with up to 23 individuals in 10 m^2. Some hexactinellids like Scolymastra joubini can obtain substantial sizes of over a metre in height and breadth (Figure 40C).

The tiny 3 to 7 cm long cave sponge Oopsacas minuta occurs at a high density of up to 100 m^{-2} (Vacelet et al., 1994). But the most abundant glass sponges in terms of number and biomass are found in NE pacific fjords and continental shelf waters where dictyonine forms (those with a fused skeleton) each up to a metre in breadth, reach individual abundances of up to 250 individuals per 10 m^2 on vertical walls and form solid reefs up to 160 km^2 on continental shelf habitats (Conway et al., 1991, 2001; Krautter et al., 2001; Conway et al., 2004; Leys et al., 2004) (Section 7.3). Glass sponges tend to have a patchy distribution in all regions studied which could suggest that they either have very particular ecological requirements or larval dispersal is only local (Barthel and Gutt, 1992; Bett and Rice, 1992; Leys and Lauzon, 1998).

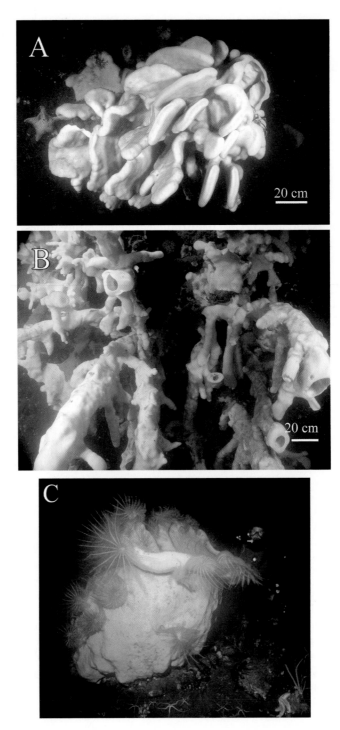

7.2. Succession: glass sponge skeletons as substrates

Except for a few species such as the minute cave sponge *Oopsacas minuta*, most glass sponges tend to be large, conspicuous inhabitants of the benthos. When these large animals die, their spicule skeletons, whether loose (lyssacine) or fused, form a firm mat or scaffold which dramatically alters the landscape for other animals.

In a study of sponge associations at Kapp Norvegia, Antarctica (Barthel and Gutt, 1992), glass sponge skeletons of lysaccine forms were found to be intertwined in such a way that when the tissue disappeared the skeleton was still resistant to tearing (Barthel, 1992). As few demosponges contribute similar distinct spicule masses, Barthel (1992) suggests that the spicule mats of hexactinellids restructure soft substrates, which then allows colonisation by demosponges and a host of other invertebrates (Figure 41).

At the Porcupine Seabight, spicule mats were found to cover one-third of the sea floor (Bett and Rice, 1992). Principal component analysis of abundance of organisms in box core samples suggests that most variation in abundance of organisms is related to the presence of spicules. Specifically, of the 10 most abundant taxa found in the cores, 7 increased in abundance with increased spicule mat density. What remains unclear is how these soft sediment substrates are initially colonised. Possibly, disturbance caused by other conditions, such as anchor ice (Dayton, 1989), can cause sufficient variation in sedimentation and food availability to allow the first hexactinellid larvae to settle.

Hyalonema is able to colonise soft substrates. Every individual of this remarkable genus lives suspended on a tall shaft of spicules anchored in 20–40 cm of sediment (Beaulieu, 2001a,b). The bulk of the body forms a thick disc or cone whose upper surface is oriented away from the prevailing current. The spicule stalk of live specimens may have some tissue coating, but once grown, the adult sponge withdraws to its high-rise habitat, thus leaving the spicules free for settlement of other animals (Beaulieu, 2001a). When the sponge dies, the disc disappears leaving an upright stem of long glass shards (Figure 42). Of 2105 *Hyalonema* stalks studied photographically in an abyssal plain at 4100 m near California, only 14% supported a living sponge on top. The remaining 1810 stalks were inhabited by a huge array of predominantly filter feeders (Beaulieu, 2001b). Common colonisers on the stalk are zoanthids (which also inhabit 30% of live stalks), but other

---

Figure 40 Abundance and distribution of glass sponges. (A, B) Dictyonine sponges, *Aphrocallistes vastus* and possibly *Heterchone calyx*, at 160 m in the fjords of British Columbia, Canada (Leys, unpublished data). (C) *Scolymastra joubini* with crinoids attached and a nearby diver at 50 m at McMurdo Sound, Antarctica (courtesy of G. Bavestrello).

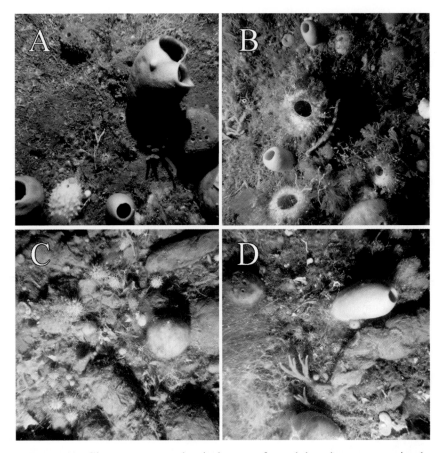

Figure 41 Glass sponges and spicule mats formed by glass sponges in the Weddell Sea (A, B are from 71°06,211′S11°39,032′W, 194-m depth; C, D are from 71°38,2′S12°09,5′W, 112 m. All photos represent 0.56 m². Courtesy of J. Gutt, copyright: J. Gutt, AWI). (A) *Rossella nuda*, a tube-shaped, smooth-sided, cream-brown sponge (top right and bottom left and middle); *Rossella racovitzae*, the white sponge bottom left; two demosponges, probably *Cinachyra barbata*, right below *Rossella nuda*. (B) Several specimens of *Rossella racovitzae* (white with projecting spicules) and *Rossella nuda* (cream with smooth exterior) on a mat of bryozoan skeletons. (C) A dense glass sponge spicule mat supporting the growth of several demosponges, including the small round white sponge *Cinachyra antarctica*. (D) *Rossella nuda* (right) and *Cinachyra barbata* (left middle) on a mat of glass sponge spicules.

inhabitants included tunicates, ophiuroids, anemones, bryozoans, other sponges (as well as other hexactinellid sponges) and even a benthic cteno-phore. So many organisms inhabit these stalks that within the slim 20 to 50 cm vertical habitat all the ecological interactions of competition for space,

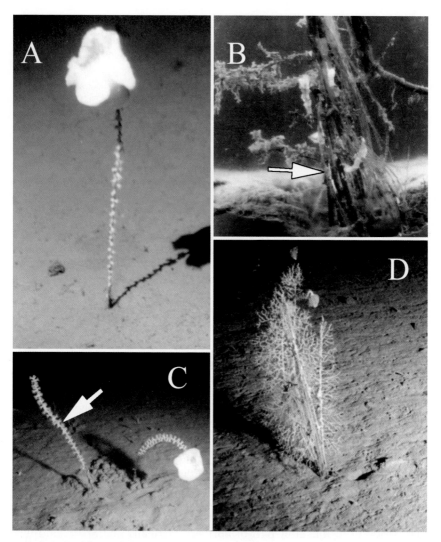

Figure 42   Live and dead stalks of *Hyalonema* at 4100-m depth in the NE Pacific (A, from Beaulieu, 2001a, with kind permission of Springer Science and Business Media; B–D, Beaulieu, unpublished data). (A, B) The sponge forms a disc of living tissue on the end of a shaft 25 to 50 cm long of spicules that are twisted together (B, arrow). In time, tissue covering the stalk recedes, allowing other animals colonise it (C, arrow). (C, D) Eventually, the disc-shaped body also dies, but the stalk remains for many years as a substrate for colonisation by other invertebrates.

flow and food, predation and succession can be found. Clearly, even the smallest collection of glass sponge spicules can structure the benthos of abyssal and shallow water habitats.

7.3. Reefs or bioherms

7.3.1. Structure of the reefs

Vast sponge reefs are at the other extreme of sponge skeleton habitats. Reefs and mounds are formed by the accumulation of generations of glass sponge skeletons. Massive mounds of spicules (biostromes) are found in locations of high densities of glass sponges, such as Antarctica, and have a structuring effect on the other biota as described above. But sponge reefs (bioherms) are formed by only three species of hexactinellid whose skeleton is fused or dictyonine. The rigid scaffold formed by the dictyonine skeleton remains after the tissue has died allowing juveniles to settle on the exposed skeletons of adults, but to form the reef the base of the skeletons must also become locked into a rigid structure by the accumulation of sediment (Conway *et al.*, 2001; Krautter *et al.*, 2001; Conway *et al.*, 2005b). It is presumed that the sponges need to grow just fast enough to keep ahead of sedimentation.

During the Jurassic, dictyonine sponges thrived and left a record that can be found throughout the world in outcrops of weather resistant rock. The first sponge reefs are known from the Atlas Mountains of Morocco, a region that formed the southern margin of the Tethys Sea. In the Middle Jurassic, bioherms are known from India, Spain, France and Hungary (Mehl and Fürsich, 1997; Pisera, 1997). The reefs reached their maximum in the Late Jurassic, when siliceous sponges formed a discontinuous deep-water reef belt over 7000-km long spanning the northern margin of the Tethys Sea, a region that now covers much of Europe (Ghiold, 1991; Leinfelder *et al.*, 1994; Krautter *et al.*, 2001). Although the reefs declined during the Cretaceous, sponge mounds with a high diversity of hexactinellids persisted throughout that period in regions now part of northern Germany and Spain, and then disappeared entirely world-wide. No modern analogues were thought to exist until the discovery in 1987–1988 of several bioherms and biostromes in Queen Charlotte Sound and Hecate Strait on Canada's western continental shelf (Conway *et al.*, 1991). Continued surveying of Canadian waters including southern locations using multibeam and side-scan sonar have now revealed at least seven reefs. In the north there are four discrete reefs, the largest nearly 160 km^2, which together form a discontinuous band covering some 425 km^2 at 165–240 m. In the Strait of Georgia three more reefs have been identified, one very close to the city of Vancouver, directly under the outflow of the Fraser River (Conway *et al.*, 2004) (Figure 43).

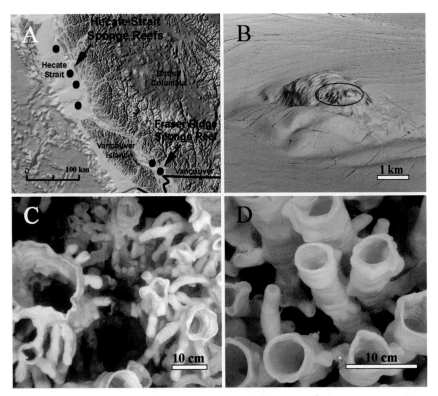

Figure 43 Sponge reefs. (A) The location of sponge reefs on the continental shelf of the Northeast Pacific. (B) A multibeam image showing the southernmost 1 km long reef (black oval) on the Fraser Ridge near Vancouver, Canada (courtesy of K. Conway). (C, D) The 5000-year-old Fraser Ridge Reef supports the growth of *Aphrocallistes vastus* and *Heterochone calyx* (courtesy of V. Tunnicliffe and VENUS).

It is quite remarkable that the presence of the reefs was not known earlier, especially considering the proximity of the Fraser Ridge to Vancouver. Although grab sampling from much earlier surveys did reveal abundances of glass sponges in a region of the Strait of Georgia 40 km northwest and southeast of Nanaimo (Fraser, 1932), the region under the Fraser River plume was not sampled. Development of modern benthic survey technology was certainly key to their discovery and description and multibeam data sets combining elevation and backscatter data can precisely delimit reef areas (Conway *et al.*, 2005a), but by far the best way to view the reefs is by submersible or remote operated vehicle. As shown in videos by these researchers, the reefs loom out of the dark waters as a forest of yellow and orange bushes. Some species have vast, gaping oscula (e.g. *Heterochone calyx*) while others have undulating and billowing palm-like extensions

(*Aphrocallistes vastus*), yet others are a mass of snow white frills (*F. occa*); four lyssaccine sponges are also found on the northern reefs (Krautter *et al.*, 2001). Detailed information on the fauna of the southern reefs is not yet available.

The living sponges on the reef surface are only about 1 to 2 m high, but cores have shown that the average depth of reef mounds is 5–8 m high, but they can be as tall as 19 m, an estimated 6000–9000 years old (Conway *et al.*, 2001), which suggests that initial colonisation occurred shortly after the retreat of glaciers on Canada's West Coast. Reefs range from symmetrical and circular in plan view biohermal forms to steep-sided elongate ridges. The first sponge larvae to settle encountered a surface of glacially derived boulders concentrated and piled high at the edges of iceberg scours (Figure 44) (Krautter *et al.*, 2001). Bottom currents are thought to keep surfaces of these mounds sediment free for the continued attachment of new sponges (Krautter *et al.*, 2001).

7.3.2. Organisms associated with the reefs

Common macrofauna on the reefs are several species of rockfishes, crustaceans and echinoderms that live in and around the sponges (Krautter *et al.*, 2001). Although a variety of annelid worms (terebellids and serpulids) and bryozoans encrust the dead sponge skeletons, endobenthic and semi-infaunal organisms are rare. The presence of two bivalve species that are adapted for low oxygen and for reducing conditions (*Thyasira fouldi* and *Thyasira flexuosa*) in cores of the reef sediment suggests this is not a thriving infaunal habitat (Cook, 2005). On the other hand, the reef supports a remarkable diversity of foraminifera (Krautter *et al.*, 2001; Guilbault *et al.*, 2006).

7.3.3. Fisheries and conservation of the reefs

Side-scan sonar data collected in July 1999 revealed numerous sets of parallel tracks traversing many mounds within several of the British Columbia sponge reefs (Conway *et al.*, 2001; Krautter *et al.*, 2001). The tracks—usually 70–100 m apart—are suspected to be carved by the heavy (2 tons) doors of the otter trawl, which fishers use to indiscriminately harvest fish and invertebrates. Trawl marks were not evident in the same area in 1988. Observations by submersible in 1999 showed piles of sponge skeletal debris (Conway *et al.*, 2001). A startling summary of observations from the British Columbia groundfish bottom trawl fishery has shown that between 1996 and 2002 about 253 tons of corals and sponges were harvested as bycatch; however,

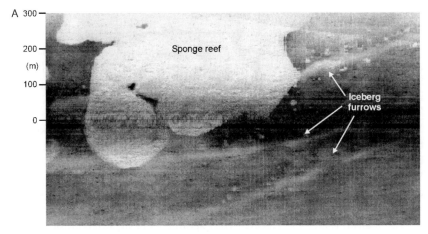

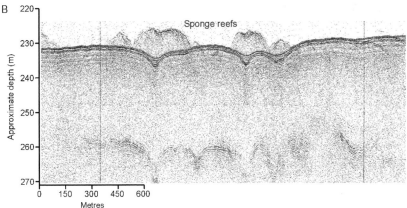

Figure 44 Side scan (A) and bathymetry (B) show profiles of one of the sponge reefs discovered in Hecate Strait in 1990 (courtesy of K. Conway). The 9000-year-old reef grows on the edges of iceberg scours at approximately 200-m depth.

because these are non-commercial species it is thought that many additional observations were unreported (Jeff Ardon, personal communication). An analysis of the regions in which most bycatch occurred suggests there are 12 patches in which 97% of all coral/sponge bycatch was observed, many adjacent to and on the sponge reefs. Voluntary avoidance of the reefs was agreed on by groundfish trawlers in 1999, but although fishing was reduced, landings continued within these areas and over the reefs (Jamieson and Chew, 2002). In 2002, the areas directly over reefs were officially closed to fishing by the Department of Fisheries and Oceans, Canada, and the long-term goal is to implement Marine Protected Areas covering an additional 9-km buffer zone around each reef (Jamieson and Chew, 2002).

7.3.4. Chemical oceanography of the reefs

The uniqueness of the Canadian sponge reefs in the modern world has prompted attempts to define the key oceanographic characteristics that delimit the distribution of glass sponges in general and of reef-building sponges in particular (Maldonado *et al.*, 1999; Leys *et al.*, 2004; Whitney *et al.*, 2005). The principal factor thought to be critical for glass sponge existence is silica, because of the massive amounts of the nutrient they require for skeleton formation (Austin, 1983, 1999; Maldonado *et al.*, 1999; Krasko *et al.*, 2000), but high levels of biogenic materials must also be necessary. The Canadian West coast sponge reefs occur in a high silicate (43–75 μm) and low oxygen environment at 140–240 m (Whitney *et al.*, 2005). Oceanographic surveys of bottom currents, nutrient and oxygen levels and particulate materials in waters near and over the Hecate Strait sponge reefs suggest that both silica and nutrients are enhanced in waters surrounding the reefs during summer months (Whitney *et al.*, 2005). As deep ocean water crosses the shelf during weak summer upwelling, silica levels increase and oxygen levels are depleted due to remineralisation of waters. Furthermore, the bottom currents around the canyons in which the reefs occur trap particulates keeping them within the reef for up to 6 days at a time. The particulates are suggested to provide nutrients to the sponges as well as enhance reef construction (Yahel *et al.*, 2007).

7.4. Growth rates and seasonal regression

Growth rates of glass sponges have been difficult to determine because of their preferred deep sea habitat and because of the unusual shapes that some species have. Dayton (1979) carried out the first long-term growth study on three species in McMurdo Sound, Antarctica, using underwater photography of marked specimens at 60 m depth. He found that while two species, *Scolymastra joubini* and *Rossella nuda*, showed only slight growth over 10 years—one specimen increasing in diameter from 75 to 77 cm and the other from 34 to 37 cm—others showed no measurable change in size. A third species, *Rossella racovitzae*, grew considerably. Because this species regularly forms buds, joins with other individuals and normally lives within the mat of spicules of previous generations of sponges, growth rates were particularly difficult to determine from photographs. Linear growth rates of portions of 13 sponges, caged and uncaged, ranged from 11 to 16 cm in 10 years (Figure 45). Small individuals, however, showed a massively faster growth rate. Of 40 individuals caged in 1974 for 3 years, 15 sponges increased their volume by 100–300%, 12 increased in volume by up to 100%, 5 died and 5 showed negligible (1–10%) growth.

Figure 45 During the 1960s and 1970s, P. K. Dayton and colleagues used cages to prevent predation on sponges by asteroids at 54 m depth in McMurdo Sound, Antarctica. (A) P. K. Dayton carrying a cage. (B) Cages had no inhibitory effect on growth of the demosponge *Mycale* which grew right through the mesh (courtesy of P. K. Dayton).

The trend to slower growth in larger animals appears to be also true for the NE Pacific species, *Rhabdocalyptus dawsoni*. Growth rates in these tube-shaped sponges were measured by underwater photography for 3 years (Leys and Lauzon, 1998). The average linear rate of growth of sponges 5–95 cm in length was nearly 2 cm year^{-1}, and the average increase in volume was 167 ml year^{-1}, but larger sponges grew marginally less during the observation period than smaller sponges. Since much larger sponges occur in other inlets on the British Columbia coast, it is suggested that the maximum size obtainable depends on local environmental conditions and that in general these sponges can be considered to show indeterminate growth as do corals (Sebens, 1987).

Attempts have been made to quantify growth rates in the aptly named 'cloud' sponge *Aphrocallistes vastus*, a principal component of the massive sponge reefs in the NE Pacific. Austin (cited in Krautter *et al.*, 2001) reported many individuals of *Aphrocallistes vastus* growing on a cable that had been submerged in the Strait of Georgia 10 years earlier. Surveys in 1991 revealed no sponges; thus animals that were 67 cm in length in 2000 may have grown 3–7 cm year^{-1} or even more depending on when they settled. Recent *in situ* measurements of small specimens of *Aphrocallistes vastus* at Senanus Island in Saanich Inlet suggest linear growth rates of 1–3 cm year^{-1} (Austin, 2003). Preliminary results from an *in situ* study of multiple photographs of marked animals on fjord walls taken with the remotely operated vehicle ROPOS (ropos.com) also suggest linear growth rates of 1–3 cm year^{-1} in small specimens. However, these images also illustrate the massive shape changes that accompany growth: increased width of the base, width of the apex and remodelling of the 'flanges' or projections that will subsequently form the characteristic 'mittens' or palm-shaped projections of this species (Figure 46). Throughout the remodelling process the osculum changes location.

7.5. Predation, mortality and regeneration

Two studies suggest that the skeleton accounts for almost 90% of the animal by dry weight; only 10% being organic material (Barthel, 1995; Whitney *et al.*, 2005). Barthel's calculations of the calorific values of glass sponge tissue suggest there is not much of a meal to be had in a glass sponge, but that does not seem to deter either asteroids or nudibranchs, the typical sponge predators.

The most significant study of predation on glass sponges is by Dayton and colleagues (Dayton, 1979, 1989; Dayton *et al.*, 1974). The asteroids *Odontaster meridionalis*, *Acodontaster conspicuus*, *A. hodgsoni*, *Perknaster fuscus antarcticus*, and the nudibranch *Austrodoris mcmurdensis* were the main predators of *Rossella racovitzae* and *Rossella nuda* (the 'volcano' sponge), but the extent of predation (by predators on different species) is highly variable. On one hand, up to 18 *Austrodoris* nudibranchs can feed on a

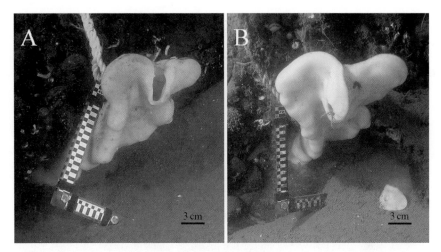

Figure 46   Growth rates in glass sponges are difficult to determine accurately because of the unusual shapes of the animals. These two images taken 1 year apart show the increase in breadth and depth of a 20-cm tall individual of *Aphrocallistes vastus*. Scale markings are 0.5 cm. The linear growth rate of this specimen is approximately 1 cm year^{-1} (Leys, unpublished data).

single *Rossella nuda* without significant damage to the sponge, yet it takes only one to four *Acodontaster* asteroids to kill a large *Rossella nuda*. Attacks by asteroids are not always fatal. Field observations showed that all large individuals of *Scolymastra joubini* had depressions and grooves indicative of browsing by nudibranchs and asteroids (Dayton, 1979). By caging freshly wounded individuals of *Scolymastra joubini*, Dayton's team was able to show that sponges could regenerate if no more than 10% of the sponge volume was wounded (Dayton, 1979); a larger wound usually caused the sponge to die. They concluded that large sponges could tolerate occasional attacks over several years, but survival was unlikely if several asteroids converged on the same individual (Figure 47).

In the NE Pacific, glass sponge populations observations of predation are scarce. Few asteroids are ever seen directly on the hexactinosan (reef-building) sponges, but three species of asteroid—*Pteraster tesselatus*, *Henricia* sp. and *Mediaster* sp.—are typically found on *Rhabdocalyptus dawsoni*. *Henricia* is thought to feed on bacteria and tiny particles that it captures in mucous (Morris *et al.*, 1980), thus it may feed on detritus in the shaggy 'spicule jungle' of *Rhabdocalyptus dawsoni*. *Pteraster* is certainly after the sponge tissue, however, since individuals pulled off the sponge come away with stomach everted and spicules and tissue attached (Leys, unpublished observations; Figure 47C).

The ability of *Rhabdocalyptus dawsoni* to rapidly recover from small wounds appears to be similar to that of *Rossella nuda*. Experimental wounds

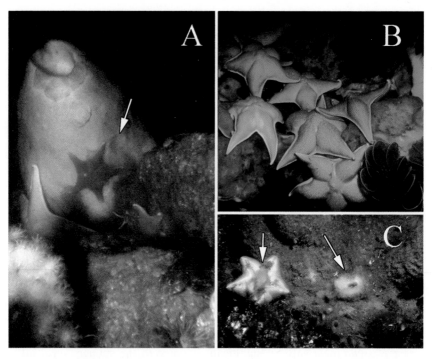

Figure 47   Asteroid predators of glass sponges (A, B, courtesy of G. Bavestrello; C, Leys, unpublished data). (A, B) The asteroid *Acodontaster* on a specimen of *Scolymastra joubini* in Antarctica. Depressions in the outer surface of the sponge (A, arrow) can be seen where the asteroid was previously feeding. (C) *Pteraster tesselatus* was pulled off the outer wall of *Rhabdocalyptus dawsoni* inhabiting North-east Pacific fjords. Spicules are still attached to the everted gut (arrows) of the asteroid.

inflicted by removing a 5 cm^2 core from 33 sponges at 20 m depth in Barkley Sound, British Columbia, regenerated at 0.05 ± 0.03 cm^2 day^{-1}, filling in the open wound within 5 months (Leys and Lauzon, 1998). These rates are in close agreement with those from an earlier study also on *Rhabdocalyptus dawsoni* by Boyd (1981), who found the oscular lip region regenerated at an average of 0.08 cm^2 day^{-1}. Several interesting observations merit note: first, several of the wounds increased in size before beginning to regenerate; second, 4 of the 33 sponges wounded died; third, regeneration of tissue targeted the atrial side of the wound first so that a thin cover was formed over the opening before the wall of the sponge was thickened. As with the Antarctic sponges observed by Dayton (1979), in most cases the site of the wound was visible as a scar for several years afterwards. The process of tissue remobilisation is presumed to involve cytoplasmic streaming (Section 4.4), rather than the crawling of individual cells as occurs in cellular sponges, but histological studies are lacking.

Hexactinosan sponges have also been found to readily regenerate soft tissue when kept in flow through sea water tanks at the Bamfield Marine Sciences Centre. Because rates of regeneration for *Rhabdocalyptus dawsoni* in laboratory tanks were identical to rates achieved *in situ* (Leys, unpublished data), it is likely that hexactinosan tissues do regenerate easily in the field, although it is not known how long it takes to reconstruct the fused skeleton once damaged.

An interesting phenomenon that Dayton noticed during his studies in Antarctica was the mortality of very large and presumably very old sponges over relatively brief periods (Dayton, 1979). Similar mortality of *Rhabdocalyptus dawsoni* has been observed in British Columbia (Leys, unpublished observations). No concrete cause of the mortality could be determined in either case, but for *Rhabdocalyptus dawsoni* it is possible that some sponges are not able to recover from the observed seasonal sloughing or shedding of the outer tissue coat that may be caused by lack of food during winter months (Leys and Lauzon, 1998).

7.6. Recruitment

Knowledge of recruitment of glass sponges varies greatly depending on what is known of the reproductive periods (Section 8), population dynamics and size class distributions of different species. To find the smallest juveniles a microscopic survey of hard substrates is necessary, but because juveniles can often be found attached to the skeleton of adults, surveys of the adult skeleton have produced the bulk of information on recruitment.

For both *Oopsacas minuta* and *F. occa* which reproduce year round (Okada, 1928; Boury-Esnault and Vacelet, 1994), juveniles are readily found. In *Oopsacas,* juveniles as small as 200 μm have been found attached to rock that was chiseled from the wall of the 3PP cave near Marseille, France. Sponges smaller than 1 cm long are already reproductive and the bulk of the population in the cave is approximately 3 cm long. Since the largest adults can be 7 cm long, the population (after several years of disturbance and collecting by divers) is considered to be young.

In *Farrea*, juveniles were found attached to the base of adults in all the collections Okada (1928) made (Figure 48), but nothing is known about population dynamics. In Antarctica, numerous small specimens of *Rossella racovitzae* were found by Dayton's team (1974), but as this species regularly formed buds, recruitment was considered to be from asexual reproduction. A few juveniles (<1 cm) of *Rhabdocalyptus dawsoni* can be found among the

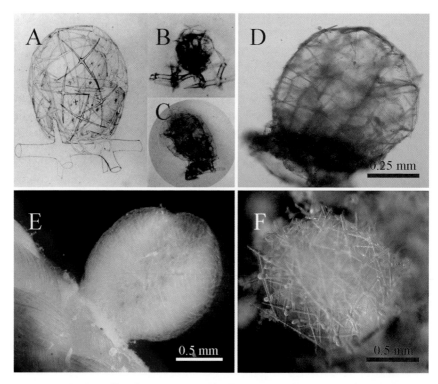

Figure 48 Juvenile glass sponges. (A–C) *Farrea sollasii* juveniles (A: Plate 2, Figure 14; B: Plate 2, Figure 10; C: Plate 3, Figure 1; from Okada, 1928). Dimensions of the juveniles are: (A, 0.6×0.7 mm^2; B, 0.53×0.49 mm^2; C, 1.02×0.76 mm^2. (D) A juvenile of *Oopsacas minuta* approximately 2 weeks old chipped from the rock wall of the 3PP cave in Marseille, France. (E) A juvenile of *Oopsacas minuta* attached to rope in the cave. (F) A juvenile dictyonine sponge found on the skeleton of a dead sponge collected from 160 m depth in Barkley Sound, Canada (D–F, S. P. Leys, unpublished data).

spicule coat of adults (Leys and Lauzon, 1998), and Reiswig (personal observation) has also found them on the hairy tunicate *Halocynthia*.

Young individuals, slightly more than 2 weeks old of *Oopsacas minuta*, can be readily found on the walls and on plastic ropes suspended from the roof of the 3PP cave near Marseille (Figure 48D and E). Juveniles of hexactinosan dictyonine sponges also settle on the skeleton of adults, thus forming the reef framework. Krautter *et al.* (2001) found 1 mm diameter 'globular' skeletons of young sponges attached to the larger skeleton of adults. The juvenile appears to attach initially with very fine, tendril-like spicules (some 10–50 μm in diameter), which wrap around the adult's spicule (Neuweiler, 2000; Krautter *et al.*, 2001). In the smallest specimens there is a space between the spicules of the new sponge and the substrate; as the sponge gets larger the space is filled in with what appear to be 'pillar-like' attachment structures. Larger juveniles (2–3 mm) also have hexactin spicules, which are joined to the previous skeleton by fine siliceous tendrils. Live juvenile glass sponges (possibly *Heterochone calyx*), 1–2 mm in diameter, were found on the skeletons of dead sponges (also *Aphrocallistes vastus*) collected from fjord walls in Barkley Sound, British Columbia in early July 2003 (Leys *et al.*, 2004).

7.7. Symbioses: animal–plant associations

It appears that glass sponges lack the kinds of secondary metabolites found in so many demosponges and which have generated interest among bioprospectors (P. Anderson, University of British Columbia, personal communication). It is unknown whether the absence of chemical deterrents makes glass sponges more vulnerable to predation or less able to resist infection after wounding (Section 7.5). Glass sponges do form associations with bacteria and other invertebrates and even algae, but while some associations appear positive, others cause the degeneration of the sponge.

Although many organisms are associated with the dead spicule skeleton of glass sponges (Sections 7.2 and 7.3.2), the live tissue of most species is pristine. In fact, considering that many of these species live in regions of high productivity, remarkably little marine detritus remains on the surface tissues. While the surfaces of dictyonine hexactinosan sponges are particularly 'clean', the long projecting spicules common to lyssacine species can trap some of the detritus, but usually in a very minor way, such that the sponges appear for the most part to be 'clean'. One NE Pacific species differs. *Rhabdocalyptus dawsoni*, the shallowest and thus most accessible of the British Colombia glass sponges, is completely covered with 'sediment' and a veritable 'jungle' of invertebrates (Figure 49A and B). In his study of the associates inhabiting this 'spicule jungle', Boyd (1981) found 56 different species from six phyla; he counted a total of 2163 associated invertebrates on 13 sponges. The most abundant epifauna by far were the polychaete worms, *Syllis* sp., *Harmathoe multisetosa* and *Harmathoe extenuata*, and the terrebellids *Eupolymnia heterobranchiata* and *Polycirrus* sp. which together composed 58% of the inhabitants, but brittlestars were also very numerous, with *Amphipholis* sp. alone counting for 15% of the associates. Some of the species were ubiquitous to all sponges (e.g. *Harmothoe* sp., *Syllis* sp., *Polycirrus* sp., *Eupolymnia heterobranchiata*, *Pugettia richii* and *Eualus pusiolus*), and yet other combinations were correlated with the depth that the sponge was collected from. Boyd determined that the majority of the macrofauna were

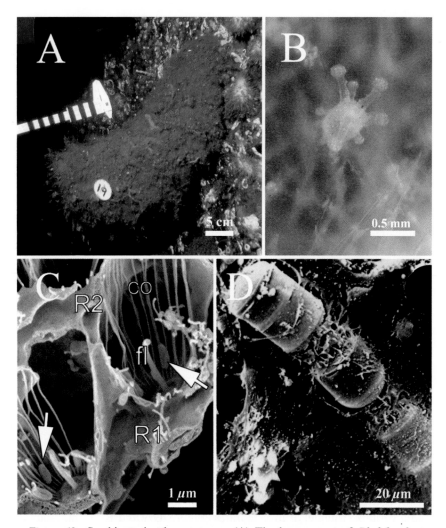

Figure 49 Symbionts in glass sponges. (A) The brown coat of *Rhabdocalyptus dawsoni* hosts a great diversity of invertebrate species. (B) A colonial hydroid lives entwined in the atrial tissue of *Rhabdocalyptus dawsoni* (light micrograph). (C) Symbiotic bacteria live along the inner surface of the collar (co) microvilli in *Aphrocallistes vastus*. Primary and secondary reticula, R1, R2; flagellum, fl. (D) Scanning electron micrograph of a diatom embedded in the tissue of *Scolymastra joubini* (D. Courtesy of G. Bavestrello).

inquilines, simply using the sponge as a substrate. Several species that were 'built in' to the sponge spicule habitat (e.g. terebellid worms and ascidians) were considered facultative commensals, based on their ability to benefit from the sponge's feeding current. Apparently, Boyd did not find any juvenile glass sponges in his survey of the spicule jungle. He did, however, find a number of juvenile echinoderms, polychaetes, gastropods and bivalves, and suggested that the presence of a 'tangled web of sharp glass fibres combined with the continual accumulation of organic matter create a nutrient-rich habitat free from the threat of predation (aside from other associates)'.

Whereas autotrophs are common associates of demosponges and are considered to contribute to the sponges' nutrition (Sarà et al., 1998), sections of glass sponges rarely reveal bacteria or algae among the tissues, with only a few notable exceptions. In *Oopsacas minuta*, a rod-shaped bacterium is harboured within the circle formed by the collar microvilli and within the mesohyl of the adult sponge (Section 5, Figure 23) (Boury-Esnault and Vacelet, 1994; Vacelet, 1996). Similar bacteria lie along the inside of the collar microvilli in *Aphrocallistes vastus* (Figure 49C). Although it is unclear what beneficial exchange may occur between the two organisms, the bacterium does not appear to be detrimental to the sponge tissue.

Diatoms are found in or on the atrial surface of some individuals of lyssacine sponges in Antarctica and have been suggested to have formed a symbiosis with the sponge by benefiting from the light-transmitting properties of the glass spicules (Cattaneo-Vietti et al., 1996). A similar alga–sponge symbiosis has been observed in a few individuals with upward facing oscula in shallow (<20 m) water in British Columbia. In British Colombia waters, however, less than 20% of the surface light reaches depths of 15 m at any time in the year (Sancetta, 1989). Given the direction of the rays of light, very little luminance would strike a spicule protruding from the dermal side and even less would be transmitted through to the atrial side of the sponge. The situation with respect to light is not much different in Antarctic waters, which suggests that the alga found on *Rossella racovitzae* is more likely to receive light that directly hits the surface of tissue facing upwards.

Diatoms have also been found embedded in the tissue of the Antarctic sponge *Scolymastra joubini*. Four of 15 sponges studied by Cerrano et al. (2000) had irregular green-brown spots $0.5–1.5$ cm^2 in size. In an extreme case, green areas covered up to 40% of the sponge. Scanning electron microscopy of the tissues revealed chains of the diatom *Melosira* partially embedded in the dermal tissue of the sponge (Figure 49D). Although the dermal tissue appeared normal, vast regions of the choanosome were absent where dense aggregates of the diatoms were wedged in among the spicule skeleton.

Colonial hydroids are common associates of demosponges, but associations with hexactinellids are less documented (Puce et al., 2005). Schuchert and Reiswig (2006) described a new species of hydroid *Brinckmannia hexactinellidophila* from *Heterochone calyx*. The hydroid lives entwined within the tissues of the sponge at densities of up to 10 individuals per square millimetre. A different, yet unidentified species lives in *Rhabdocalyptus dawsoni.* The hydroid tentacles and mouths are visible among the spicules and tissue of the atrial surface, while their stolons are completely entwined among the other tissues of the sponge (Figure 49B). The species inhabiting *Rhabdocalyptus dawsoni* is quite distinct from that in *Heterochone* and curiously, hydroid symbionts are absent from all individuals of *Aphrocallistes vastus* studied to date (H.M.R. and S.P.L., unpublished data) even though the sponges inhabit exactly the same environment.

8. REPRODUCTION

8.1. Sexual reproduction

8.1.1. Reproductive periods

Reproductive periods for hexactinellids vary depending on the species and possibly even the population. Records are scant and most stem from early collection expeditions in which scientists were focusing specifically on morphology and thus carried fixatives to preserve tissues and larvae (Schulze, 1880, 1887, 1899; Ijima, 1901, 1904; Okada, 1928). There are three recent studies (Boury-Esnault and Vacelet, 1994; Leys and Lauzon, 1998; Leys *et al.*, 2006).

Ijima (1901) made a special search at different seasons for reproductive cells in *Euplectella marshalli* but found neither developmental stages nor larvae. In subsequent collections of rossellid hexactinellids (Ijima, 1904), he still only found developmental stages and larvae in very few specimens. In *Vitrollula fertilis*, he found larvae in April, larvae and various developmental stages in July but only archaeocyte congeries in November, and from this presumed that the main active reproductive period is in early summer.

Okada (1928) probably carried out the most extensive survey of reproduction in a single population by collecting specimens of *F. sollasii* (Farreidae, Hexactinosida) monthly from the Nakanoyodomi (about 600 m deep) in the Sagami Sea. After Ijima's work, he was surprised to find reproductive specimens every month, and suggested that the breeding season was likely yearlong because of the fairly constant temperature and uniform environmental conditions that exist at that depth. Other surveys have not had the same luck as Okada, however. Although deep sea expeditions do occasionally report finding a reproductive glass sponge, a survey of tissue collected monthly from the NE Pacific rossellid species *Rhabdocalyptus dawsoni* failed to turn up anything but archaeocyte congeries (Leys and Lauzon, 1998). However, a study of samples of *Heterochone calyx* collected in October 1982 by one of us (H.M.R.) has revealed one specimen with numerous spermatocysts (accounting for approximately half of the congeries), some of which appeared to have emptied their contents, thus possibly having spawned, and many early embryos in various stages of cleavage. Of the many specimens of the NE Pacific reef-building sponge *Aphrocallistes vastus* that have been collected since the early 1980s, developing embryos were only found in one specimen collected in November 1995. Despite extensive work on glass sponges in Antarctic waters, there are no reports of sexually reproductive individuals. Thus, it is likely that shallower populations (those in the NE Pacific and Antarctica) are affected by seasonality of the surface waters more than deep-water populations. The hint from the two *Heterochone* and *Aphrocallistes* individuals is that reproductive period occurs in the autumn.

8.1.2. Gametogenesis

The bulk of our knowledge on reproduction in glass sponges stems from studies on two animals: *F. sollasii* (Okada, 1928) and *Oopsacas minuta* (Boury-Esnault and Vacelet, 1994; Boury-Esnault *et al.*, 1999; Leys *et al.*, 2006). This section of development will draw heavily from these accounts and from that information provided in Ijima's description (1904) of larvae in *V. fertilis.* The genera used below refer to the species used in these studies.

Gametes—both sperm and eggs—arise within archaeocyte congeries that are suspended within the trabecular reticulum between flagellated chambers (Okada, 1928; Boury-Esnault *et al.*, 1999). Spermatocysts are first identifiable as dense groupings of archaeocytes up to 30 μm by 23 μm in *Oopsacas minuta* surrounded by a thin (0.5 μm) layer of the trabecular reticulum (Figure 50A and B). In young spermatocysts, cells are larger in the centre than at the periphery of the cyst, ranging from 2.7 to 5.3 μm in diameter with a nucleus 1.6–2.6 μm and a nucleolus 0.5–1 μm (Boury-Esnault *et al.*, 1999). Each cell has a flagellum that coils around the cell. All spermatocytes cells are connected by plugged cytoplasmic bridges, and cells at the periphery of the cyst are connected to the surrounding trabecular envelope by plugged cytoplasmic bridges. At this point, the characteristics of free sperm remain unknown.

Oogenesis also occurs within archaeocyte congeries. The first oocyte is identifiable as a large cell (10 μm in diameter) within the congerie that has begun to accumulate yolk and lipid inclusions (Okada, 1928; Boury-Esnault *et al.*, 1999) (Figure 50C and D). Archaeocytes are connected to one-another by plugged cytoplasmic bridges and are suggested to act as nurse cells providing the lipid and yolk to the developing oocyte. But at some point the oocyte presumably breaks this connection because the mature oocyte is a

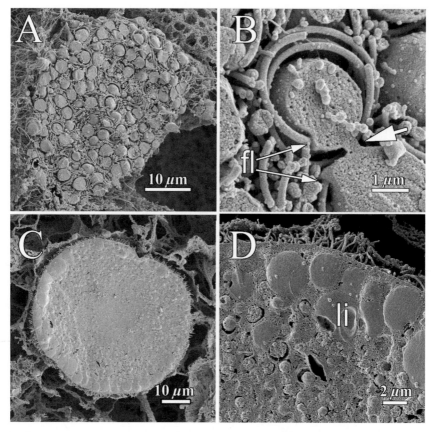

Figure 50 Gametogenesis. (A) Scanning electron micrograph of a spermatocyst in the adult tissue of *Oopsacas minuta*. (B) Higher magnification of two spermatids connected by a cytoplasmic bridge (arrow) in the spermatocyst shown in A. The flagellum (fl) is coiled around each cell body. (C) Scanning electron micrograph of an oocyte fractured in half. (D) Higher magnification of the oocyte in C showing lipid-dense inclusions occupy the periphery and the surface has numerous microvilli (Leys, unpublished images).

completely independent ovoid cell 100–120 μm in diameter in *Oopsacas minuta* (Boury-Esnault *et al.*, 1999) and 70–130 μm in *F. sollasii* (Okada, 1928), with large lipid inclusions (3.2- to 6.7-μm diameter) at the periphery and membrane-bound yolk inclusions (1.3–2.7 μm) and very small vacuoles more centrally. Boury-Esnault *et al.* (1999) describe a 45 to 50 μm diameter nucleus with a 10 μm nucleolus within the oocyte, but although a subsequent study (Leys *et al.*, 2006) found identifiable nuclear regions at the light microscope level, a nuclear membrane was not visible in any thin section studied by transmission electron microscopy. This may be what Okada referred to as a 'vesicular' nucleus (Okada, 1928, p. 3). The presumed nuclear material forms a dense osmiophilic region that occupies the centre of the cell and radiates out into the peripheral regions of the cytoplasm, not unlike the chromatin in the nucleoids of some bacteria (Robinow and Kellenberger, 1994). The surface of the oocyte has numerous short pseudopodia. Okada found most oocytes at the outer trabecular layer of *F. sollasii* and suggested that after fertilisation they migrate in to the inner trabecular layer to lie beside a flagellated chamber. In *Oopsacas minuta*, however, oocytes and developing embryos can be found throughout the body wall, from just under the dermal membrane to just under the atrial membrane, where they lie adjacent to flagellated chambers.

8.1.3. Embryogenesis

Cleavage is total and equal, but asynchronous, for the first five cycles until the embryo has approximately 32 cells (Figures 51 and 54). The embryo remains of the same size during these divisions, partitioning cytoplasm, yolk and lipid into daughter blastomeres. Early blastomeres retain all the characteristics of the oocyte, with pseudopodia extending from their surface, a dense nuclear region, large lipid inclusions at the periphery and membrane-bound yolk inclusions more centrally. It is not until the 32-cell stage (blastula) that a distinct nucleus can be seen in individual blastomeres (Leys *et al.*, 2006). This feature may reflect the fact that divisions are rapid, leaving little time for re-assembly of the nuclear membrane between cycles. If so, the appearance of nuclei at this stage—concurrent with the change to unequal cleavage—could reflect slowing of the cell cycle.

Boury-Esnault *et al.* (1999) were able to see polar bodies in some embryos and determined that the first cleavage is meridional, presumably in relation to the polar body and flagellated chambers. The second cleavage may be meridional or rotational (Leys *et al.*, 2006). The third cleavage is asynchronous, producing first six and then eight blastomeres. New blastomeres in the six-cell embryo are placed just above the cleavage plane of the first tier; similarly future blastomeres are lodged compactly over each previous cleavage plane giving the appearance of a type of spiral cleavage. Cells in the 16- and 32-cell blastula are thus geometrically arranged around a hollow centre.

The next cleavages are unequal producing small cells (micromeres) on the outside and larger cells (macromeres) on the inside. While some micromeres inherit yolk and lipid inclusions, most of the inclusions are retained by the macromeres. Some of the micromeres take up stains more intensely than others, possibly depending on their position within the new 'outer' layer of

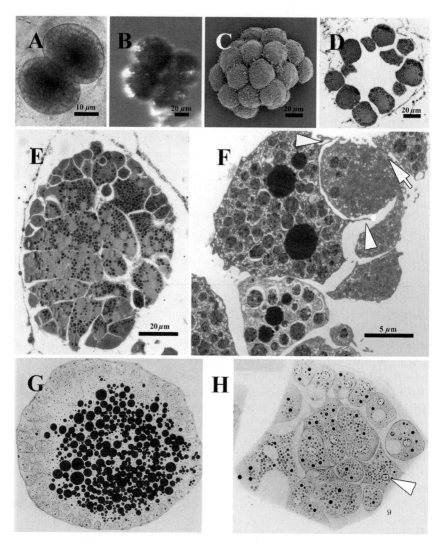

Figure 51 Embryogenesis. Stages in development of *Oopsacas minuta* as seen by
light microscopy (A, B, D, E), scanning (C), and transmission electron microscopy
(F) and as drawn by Okada (1928) (G, H) (A–F, Leys, unpublished data). (A, B)
Two- and four-cell embryo. (C, D) Thirty-two cell blastula. (E) Macromeres fuse to
form increasingly larger cells and eventually a single syncytial tissue, the trabecular
tissue. (F) Macromeres envelop micromeres (arrowheads), which are connected to
one-another by plugged cytoplasmic bridges (arrow). (G) Okada's drawing of an
embryo from *Farrea sollasii* at approximately the same stage as E (Plate 5, Figure 7).
(H) Okada's drawing of cells at the periphery of the embryo in G. Cells are enveloped
by a reticulate tissue (arrowhead by S.P.L.).

the embryo. Some micromeres remain wedged between macromeres, while others truly form an outer layer. All embryos are polarised at this stage: one pole lacks micromeres altogether while the opposite pole has a large number of micromeres; thus the embryo is not exactly bilayered at this point.

At this stage several remarkable developments occur that are quite unique to hexactinellids.

First, as soon as they are formed, the micromeres are connected to one-another by plugged cytoplasmic bridges (Figures 51F and 54G and H). Cytoplasmic bridges may occur parallel or perpendicular to the surface of the embryo. The first micromeres (i.e. at the sixth cycle) are connected to others by cytoplasmic bridges and have already differentiated a cilium. Second, the macromeres extend out filopodia and pseudopodia which interact and eventually fuse with filopodia from other macromeres. Third, macro-meres extend filopodia apically to surround and eventually completely envelop the micromeres above them, thus forming the surface membrane or epithelium of the new larva. Finally, the newly formed syncytial tissue forms cytoplasmic bridges with each micromere. Thus, from this point on the entire embryo is cytoplasmically connected; it consists of a reticular syncytial tissue—the 'inceptional trabecular system' (Ijima, 1904)—and a collection of uninucleate cells all of which are linked cytoplasmically by plugged junctions. Unequal cleavage of the 32-cell stage blastula has been termed 'gastrulation by delami-nation' (Boury-Esnault et al., 1999), but epithelialisation of the embryo and future larva occurs by the fusion of macromeres and their amoeboid envelop-ment of the micromeres; this is then the moment at which the larval tissue layers are formed.

8.1.4. Cellular differentiation

Interpretation of the continued differentiation of cellular components from thick or thin sections is difficult and varies depending on the perspective of the observer. In *F. sollasii*, Okada (1928) describes an outer layer of cells that surrounds a central mass of 'clear, transparent, jelly-like substance' containing amoeboid cells that have migrated in from the outer layer. We know from *V. fertilis* (Ijima, 1904) and *Oopsacas minuta* (Boury-Esnault et al., 1999) that the outer cells are multiciliated cells, which differentiated cilia in the early 'gastrula', as described above, and continue to produce cilia until there are some 50 per cell. The cilia are eventually completely surrounded by the newly formed trabecular epithelium so that they 'pierce' or project through the smooth syncytial epithelium. The multiciliated cells remain connected to one-another and to the trabecular tissue below them by plugged cytoplasmic bridges. According to Okada's description, the inner cells that lie among the trabecular tissue differentiate into the sclerocytes, skeletongenic cells and the choanocytes of the future flagellated chambers (Figures 52 and 54H). Sclerocytes are first evident at the periphery of the central mass. They have a large intracellular vacuole in which silica is deposited around a square axial filament (Leys, 2003a). Young sclerocytes have numerous filopodia, suggesting that they may, as Okada described, have wandered in to their present location. Alternatively they may use these filopodia to anchor themselves during spicule development and elongation. Boury-Esnault *et al.* (1999) suggest that sclerocytes in *Oopsacas minuta* originate from macromeres in the inner mass, but sections showing early sclerocytes at the periphery of the embryo (Leys *et al.*, 2006) indicate that they more likely originate from micromeres, as in *F. sollasii*. Without cell-tracing experiments we remain unsure about their precise origin. The sclerocytes are originally uninucleate, but become a multinucleate sclerosyncytium (Section 3.5) as they elongate the larval stauractin (four-rayed) spicule. At all times plugged cytoplasmic bridges connect sclerocytes to the trabecular tissue.

The last cell type to differentiate is the choanocyte. Although Boury-Esnault *et al.* (1999) suggest that choanoblasts arise from the yolk-rich tissue at the posterior pole, subsequent work (Leys *et al.*, 2006) concurs with Okada that they develop from the same population of amoeboid cells as do the sclerocytes. These amoeboid cells take up a central-posterior location when the embryo is still spherical. As the posterior pole extends with the elongation of spicules at the periphery of the embryo, the remaining amoeboid cells become rounded and arranged in small groups. In *Oopsacas minuta*, the first choanocytes possesses a collar, flagellum and single nucleus in a cell that is connected to the neighbouring cells and to the reticular tissue by plugged cytoplasmic bridges. Okada (1928) did not see either collars or flagella in the embryo of *F. sollasii* and concluded that these structures differentiate after the larva leaves the parent sponge, but given the minute size of collar and flagellum at this stage it is possible that they were not visible in his sections. He also described an invagination at the surface of the larva near the flagellated chambers, which he suggested could represent the start of canal formation; no equivalent invagination has been found in *Oopsacas minuta*. Ijima (1904) saw archaeocytes in the posterior region of *V. fertilis*, but did not believe they were the anlagen of flagellated chambers because they were still connected directly to other mesohyl spaces in the larva and because the centre of the presumptive chambers was traversed by the reticular tissue. TEMs show that the reticular tissue traverses the nonfunctional chambers of *Oopsacas minuta*, thus it is likely that chambers are already present in *V. fertilis* as they are in *F. sollasii* and *Oopsacas minuta*.

Stages in the development of *Oopsacas minuta* are summarised in Figure 54.

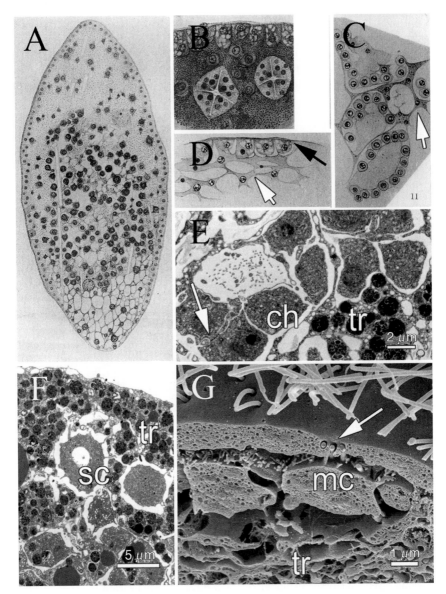

Figure 52   Larval differentiation. (A–D) Drawings of *Farrea sollasii* embryos from Okada (1928), (E–G) electron micrographs of *Oopsacas minuta* embryos (Leys, unpublished data). (A) Longitudinal section through an embryo approximately 70 μm long (Plate 6, Figure 1) with fully developed stauractin spicules. (B) A peripheral portion of embryo to show micro-scleroblast masses with young disco-hexasters (Plate 6, Figure 10). (C) Choanocytes in a reticular tissue (arrow by S.P.L.) (Plate 6, Figure 11). (D) Peripheral portion of the posterior region of the embryo,

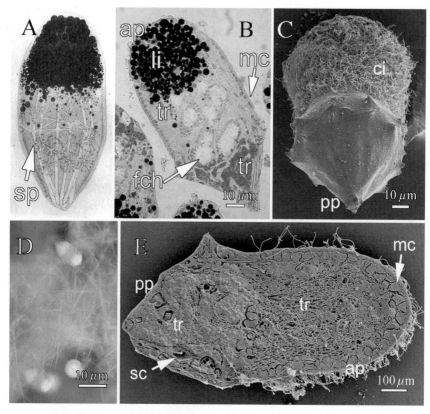

Figure 53 Hexactinellid larvae (A, Okada, 1928; B–D, Leys, unpublished data).
(A) Oldest embryo found by Okada (1928) in the parental tissue of *Farrea sollasii* (Plate 2,
Figure 6). Sp, spicules. (B) Thick plastic section of a larva from *Oopsacas minuta*.
Ap, anterior pole; li, lipid inclusions; mc, multiciliated cells; tr, trabecular reticulum;
fch, flagellated chamber. (C) View from the posterior pole (pp) of a larva from *Oopsacas
minuta* showing a smooth outer epithelium and a girdle of cilia (ci) at the equator of the
larva. (D) Light micrograph showing live larvae (*Oopsacas minuta*) suspended from
the trabecular reticulum of the parent sponge. (E) A longitudinal fracture of a larva
from *Oopsacas minuta* viewed by scanning electron microscopy. Ap, anterior pole; pp,
posterior pole; tr, trabecular reticulum; sc, sclerocyte; mc, multiciliated cell.

---

showing a reticular tissue (white arrow by S.P.L.) directly beneath the cells. A thin
tissue is shown directly over the cells (black arrow). (E) Transmission electron
micrograph (TEM) of a newly formed choanocyte chamber. Choanocytes (ch) are
connected to one-another by plugged cytoplasmic bridges (arrow) and are surrounded by strands of the trabecular syncytial tissue (tr). (F) TEM of a newly formed
sclerocyte (sc) within the trabecular syncytium (tr). (G) Scanning electron micrograph of the peripheral portion of the central region of the embryo showing multiciliated cells (mc) directly over strands of the syncytial trabecular reticulum (tr). The
cilia project through the smooth outer layer of the trabecular reticulum (arrow).

8.1.5. Larval structure

The unusual construction of the larva has caused quite a lot of confusion. Early accounts of embryos and larvae from hexactinellids describe a 'cellular' equatorial outer layer, and a 'jelly-like' reticular tissue that forms the inner mass (Ijima, 1904; Okada, 1928). The multiciliated cells were quite distinct in thick sections, but the reticular tissue defied precise definition. Ijima saw that the cellular region was absent from anterior and posterior poles of the larva where the inner mass was exposed to the environment and concluded that the reticular tissue could be traced from anterior to posterior regions. Both Okada and Ijima described additional cells among the reticular tissues at the posterior of the larva. It is likely because of this that Boury-Esnault *et al.* (1999) were inclined to describe the inner region of *Oopsacas minuta* as consisting of two types of cells, one that remains uninucleate, at the anterior end, and one that becomes syncytial and occupies most of the posterior pole. Re-examination of the tissue from *Oopsacas minuta* has shown the reticular tissue to be a single multinucleate syncytium that, as Ijima found, pervades the entire inner mass from anterior to posterior poles *and* forms the surface epithelium of the larva (Leys *et al.*, 2006) (Figure 53).

The reticular tissue is thicker at the posterior pole of the larva where it has numerous yolk-rich inclusions. It thins out towards the centre of the larva as it surrounds the incipient flagellated chambers, and remains very thread-like in the anterior region where it is no more than a sheath around nuclei and massive lipid inclusions. In all parts of the larva, the strands of the reticular tissue are interlaced with a distinct collagenous mesohyl, the presence of which Ijima inferred from its fluid like consistency (Ijima, 1904). The cellular components of the larva include: (1) the multiciliated cells that form a band around the middle third of the larva, lying directly underneath and piercing with their cilia, the syncytial epithelium; (2) the branched choanocytes, which form two to three flagellated chambers in the posterior–central region and (3) a type of cell with clear vacuolar inclusions. In both *F. sollasii* and *Oopsacas minuta*, sclerocytes are already multinucleate in the larvae (Okada, 1928; Boury-Esnault *et al.*, 1999).

The larva possesses a skeleton that is distinct from the adult. In *F. sollasii*, there are 12 oxystauractin (four-rayed) megascleres that lie 'paratangentially' to the surface of the larva as well as an unknown number of microscleres—discohexasters—that appear just under the multiciliated cells at the posterior pole of the larva (Okada, 1928). The anterior portion of stauractin longitudinal ray is 90 μm and the posterior portion is 50 μm; the transverse rays are each 20 μm. The *Oopsacas minuta* larva has up to 14 stauractin spicules but lacks microscleres altogether (Leys, 2003a). Measurements of the rays were made after dissolution of the larval tissue in nitric acid, but

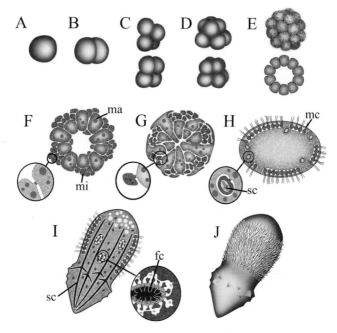

Figure 54 Stages in embryogenesis and larval development of *Oopsacas minuta* (after Leys *et al.*, 2006). (A) Oocyte, (B) two cells, (C) four cells, rotational or equatorial cleavage, (D) eight cells, (E) hollow blastula, (F) unequal cleavage to form micromeres (mi) and macromeres (ma), (G) gastrulation: fusion of macromeres (ma) to form the trabecular syncytium, and envelopment of micromeres by this tissue to form the outer epithelium. (H) Cellular differentiation: formation of multiciliated cells (mc) and sclerocytes (sc, inset). (I and J) Larva in longitudinal section (I), and external view (J) with sclerocytes (sc) and flagellated chambers (fc, inset).

lipid-rich material in the anterior pole obscured the ends of spicules in some preparations so that measurements were made from several different preparations. The longitudinal rays of the stauractins in *Oopsacas minuta* are estimated to be 40–80 μm, while the transverse rays are 27–50 μm.

8.1.6. Larval behaviour, settlement and metamorphosis

Until live larvae could be studied, it was thought, based on Okada's and Ijima's work, that the pointed pole was anterior; the contrary is true (Boury-Esnault *et al.*, 1999). The larva swims with the lipid-dense rounded anterior end upright, rotating in a right-handed direction (clockwise when viewed from the posterior pole). The live larva from *Oopsacas minuta* is white like the adult sponge, but under the optics of a stereomicroscope the lipid-dense region at the anterior pole appears nearly opaque.

Observations on larval behaviour, settlement and metamorphosis stem from work done by one of us on *Oopsacas minuta* (Leys, unpublished observations). In glass dishes in the laboratory, larvae swim slowly upward. If the surface is left open to the air, the larvae quickly get caught in the air–water interface, but if the surface is covered by a piece of Parafilm™ or by a cover slip, the larvae continue to swim at the surface and occasionally turn and swim back down to the bottom of the dish. Larvae will swim for up to 7 days if the dishes are continually disturbed, but most larvae settle and metamorphose into the juvenile sponge within 12–24 hours after release from the parent; they do this in dishes kept on the table top at room temperature and in dishes maintained at 14°C (the temperature in the cave from where the adult sponges live).

Oopsacas minuta larvae attach by the rounded anterior pole. One hour after attachment the anterior pole has flattened, spreading the still visible lipid-filled inclusions over a broad base while the former posterior pole remains conical. After 24 hours, the post-larva has become broader apically (former posterior pole) and the tissue is translucent. Stauractin spicules can still be seen like tentpoles around the circumference of the main body. Thick sections show that the lipid inclusions remain at the base of the post-larva, while the centre and former posterior pole undergo a massive change in density, from being nearly opaque after osmium fixation to nearly transparent. This change appears to be due to the radical reorganisation of the reticular tissue and disappearance of most of the yolk-filled inclusions. The multiciliated cells resorb their cilia, but the fate of these cells within the post-larva is not clear at this time. Flagellated chambers enlarge and become enveloped by the reticular tissue, as in the adult.

The fate of the larval sclerocytes in *Oopsacas minuta* is not yet known, but Okada (1928) reports that in juveniles of *F. sollasii* the stauractins are replaced by pentactins, which then form the dermalia (spicules supporting the dermal membrane). The smallest specimen observed by Okada was a 0.54 mm × 0.36 mm barrel-shaped juvenile found attached to the base of the adult. The choanocyte layer in this individual was a continuous sheet of choanosome, with 'evaginated protuberances' that lay suspended between the inner and outer trabecula reticula. Week-old juveniles from *Oopsacas minuta* are approximately 200 μm in diameter and barrel-shaped. Although the choanosome could not be seen in live specimens, the central region appears open and an osculum opens at the apical side (Leys, unpublished observations).

8.2. Asexual reproduction

The asexual production of buds is characteristic of several species of rossellid hexactinellid. Perhaps the most prolific 'budder' known is *Lophocalyx* (*Polylophus*) *philippinensis* (Schulze, 1887), although budding is also rampant in *Rossella racovitzae*, one of the most abundant of the glass sponges in Antarctica (Dayton, 1979, 1989; Barthel and Gutt, 1992). Buds are also common in the NE Pacific species *Rhabdocalyptus dawsoni*, where they arise from the base of these tube-shaped animals. Although the cavity of the bud is completely separate from the atrial cavity of the parent sponge, physiological experiments have demonstrated that stimuli to the parent tube cause the arrest of flow within that tube and in the bud. Thus, despite the separation of flow compartments, the cytoplasm of the trabecular syncytium connecting parent with bud is continuous (Mackie *et al.*, 1983).

The ability to reproduce asexually probably depends slightly on the morphology of different groups of hexactinellids because species with a fused skeleton are thought not to be able to modify the fused scaffold at their base once it is formed. Nevertheless, other means of 'budding' may be possible. Divers monitoring a population of *Aphrocallistes vastus* in Saanich Inlet, British Columbia, have observed that 'drips' of tissue arise from the palm-shaped processes of large animals (Austin, 2003). It is speculated that these 'drips' could attach and grow into a new sponge in much the way that portions of the tetractinomorph demosponge *Chondrosia reniformis* (Bavestrello *et al.*, 1998) is able to colonise new locations in the Mediterranean.

9. CLASSIFICATION AND PHYLOGENY

9.1. Classification of recent Hexactinellida

First awareness that hexactine sponge spicules might constitute a distinct spicule type is attributed to Bowerbank (1858). He made no proposition to group together the seven relevant sponges which were scattered across several taxa in arrangements of the time. By 1868, 14 species of future hexactinellids had been described when Thomson (1868) recognised their distinction in formal proposal of the taxon "Vitrea" to contain them, including in his definition that their spicules '... may all be referred to the hexradiate stellate type'. Thomson's concept thus served as the basis for Schmidt's proposal (1870) of the taxon 'Hexactinellidae' for sponges with triaxonal spicules. Claus (1872) proposed another formal taxon for this group, 'Group 6 Hyalospongiae, Glassschwämme', but he made no mention of their unique spicule symmetry in the group definition. Although Claus's Hyalospongiae and Schulze's name (1886), 'Triaxonia', introduced in a prelude to his famous 'Challenger' report (Schulze, 1887), were used by some workers through the 1960s, Schmidt's name ultimately gained favour and he is now universally acknowledged as authority for the class name, Hexactinellida.

Present classification of Recent Hexactinellida (Table 1) derives from the review of taxonomic literature in Systema Porifera (Hooper and Van Soest, 2002), recent addition of the order Fieldingida by Tabachnick and Janussen (2004),

Table 1 Classification of Recent Hexactinellida to family level, with number of genera (Gen), number of species (Spec), and main skeletal types (Skel) indicated

Rank	Taxon name	Gen	Spec	Skel[a]
Class	Hexactinellida Schmidt, 1870	119	531	L/D
Subclass 1/2	Amphidiscophora Schulze, 1886	12	158	Ln
Order 1/1	Amphidiscosida Schrammen, 1924	12	158	Ln
Family 1/3	Pheronematidae Gray, 1870	6	42	Ln
Family 2/3	Monorhaphididae Ijima, 1927	1	2	Ln
Family 3/3	Hyalonematidae Gray, 1857	5	114	Ln
Subclass 2/2	Hexasterophora Schulze, 1886	107	372	L/D
Order 1/5	Hexactinosida Schrammen, 1903	37	113	Ds
Family 1/7	Farreidae Gray, 1872	6	24	Ds
Family 2/7	Euretidae Zittel, 1877	16	45	Ds
Family 3/7	Dactylocalycidae Gray, 1867	3	8	Ds
Family 4/7	Tretodictyidae Schulze, 1886	8	23	Ds
Family 5/7	Aphrocallistidae Gray, 1867	2	7	Ds
Family 6/7	Craticulariidae Rauff, 1893	1	1	Ds
Family 7/7	Cribrospongiidae Roemer, 1864	1	1	Ds
Order 2/5	Aulocalycoida Tabachnick and Reiswig, 2000	8	10	Ds
Family 1/2	Aulocalycidae Ijima, 1927	6	7	Ds
Family 2/2	Uncinateridae Reiswig, 2002	2	3	Ds
Order 3/5	Fieldingida Tabachnick and Janussen, 2004	1	2	Ds
Family 1/1	Fieldingidae Tabachnick and Janussen, 2004	1	2	Ds
Order 4/5	Lychniscosida Schrammen, 1903	3	6	Dl
Family 1/2	Aulocystidae Sollas, 1887	2	4	Dl
Family 2/2	Diapleuridae Ijima, 1927	1	2	Dl
Order 5/5	Lyssacinosida Zittel, 1877	53	241	Ls
Family 1/3	Leucopsacidae Ijima, 1903	3	16	Ls
Family 2/3	Euplectellidae Gray, 1867	27	89	Ls
Family 3/3	Rossellidae Schulze, 1885	23	142	Ls
	Hexasterophora Incertae Sedis	5	5	L/D

[a]Skeleton types, degree of fusion indicated by shading, are designated by Ln, loose lyssacine network of separate megascleres never fused together; Ls, lyssacine network of usually separate megascleres which sometimes is rigidified by spicule fusion at contact points or by synapticulae but never involving fusion of hexactins; D, rigid dictyonine framework of fused simple hexactins (Ds) or lychnisc hexactins (Dl); L/D, includes both lyssacine and dictyonine members.

and review by Reiswig (2006). The first division as introduced by Schulze (1899) is based on types of microscleres: amphidiscs and no hexasters in sub-class Amphidiscophora; hexasters and no amphidiscs in subclass Hexaster-ophora. Before 1899 (e.g. Schulze, 1887), first division of the class was based on pattern of the primary skeleton: spicules either always isolated or subsequently united irregularly in Lyssacina; hexactine spicules fused together early to form a compact and more or less regular, rigid, dictyonal framework in Dicty-onina. In spite of additions and discoveries of new types of hexactinellid organ-isation since 1899, the subclass division seems very secure. Amphidisc-like microscleres occur in some hexasterophorans, but they are accompanied by hexasters and are considered a convergent development (Tabachnick and Lévi, 1997).

Subclass Amphidiscophora contains only one Recent order and three families (Figure 55) distinguished by their major choanosomal spicules: pentactins in Pheronematidae, tauactins in Monorhaphididae and diactins in Hyalonematidae. This arrangement, initiated by Ijima in 1927, replaced the two-family scheme used by Schulze (1904). Spicule fusion never occurs, by any method, in this subclass.

Subclass Hexasterophora contains five orders and several still unplaced species. Classification at both order and family level remain unsatisfactory, with several changes having occurred since the overall revision by Ijima (1927) and more expected in the future. Four of the five orders are dictyo-nines (Figure 56), which differ in gross structure of their fused rigid frame-works: the Hexactinosa framework is constructed of fused simple hexactins, all rays of which are approximately one mesh in length in regular arrange-ment with or without longitudinal strands (the pattern of this group requires re-examination and revision); the Aulocalycoida is constructed of simple hexactins with some rays forming longitudinal strands by continuous exten-sion; the Fieldingida have an irregular, large-mesh framework of simple hexactins with Weltner bodies and a surface crust of fused stauractins (regular tetractins); the Lychniscosida have a regular framework of lychnisc hexactins as dictyonalia and lack longitudinal strands. Within Hexactino-sida, families are distinguished primarily by distinctive types of gaps (chan-nels) in the rigid skeleton (see chapters in Hooper and Van Soest, 2002 for details), but Farreidae and Euretidae are differentiated by combination of their sceptrule spicules and thickness of framework at the growth margin. The two Aulocalycoida families differ in structure of their longitudinal strands, being composed of single dictyonal rays in Aulocalycidae and of overlapping rays in Uncinateridae. The two Lychniscosida families are dis-tinguished by frontal ranking of lychniscs in Aulocystidae and lack of such ranking in Diapleuridae.

The fifth order of Hexasterophora, the non-dictyonine Lyssacinosida, is both the most abundant and diverse of the orders, containing nearly

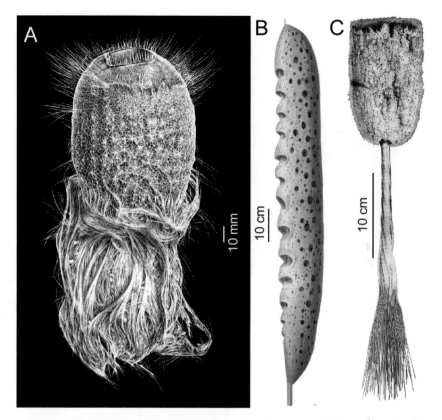

Figure 55 Representative species of the three Amphidiscophora families. (A) *Pheronema carpenteri* (Pheronematidae) from Thomson (1869). (B) *Monorhaphis chuni* (Monorhaphididae) from Schulze (1899). (C) *Hyalonema sieboldi* (Hyalonematidae) from Schultze (1860).

one-half of all known Recent species. Its classification, like that of Hexacti-nosa, has undergone recent major changes and cannot yet be considered stable. Presently, the group contains three families (Figure 57) differentiated by pentactine hypodermalia and diactine principalia in Rossellidae; no hypodermalia and hexactine principalia in Leucopsacidae; no hypodermalia and stauractine, tauactine and diactine principalia in Euplectellidae.

9.2. Classification of fossil Hexactinellida

Palaeontologists are faced with making order out of material usually far less favourable than that available to neontologists. Specimens of fossils are rare,

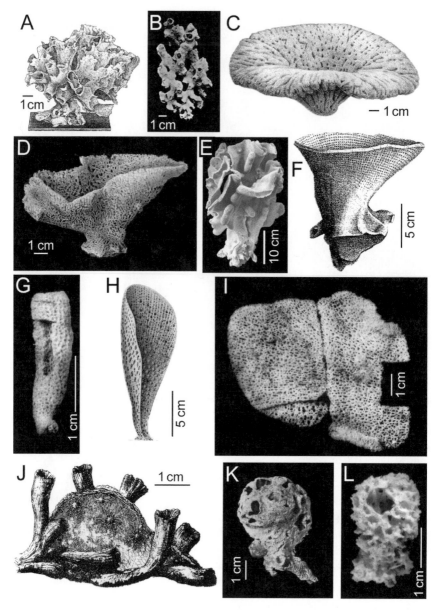

Figure 56 Representative species of the 12 families of dictyonine Hexastero-phora. (A–G) order Hexactinosida; (H–I) order Aulocalycoida; (J) order Fieldingida; (K–L) order Lychniscosida. (A) *Farrea occa* (Farreidae) from Carter (1885). (B) *Eurete simplicissima* (Euretidae) from Marshall (1875). (C) *Dactylocalyx pumiceous* (Dactylocalycidae) from Sollas (1879). (D) *Hexactinella ventilabrum* (Tretodictyidae) from Schulze (1887). (E) *Aphrocallistes vastus* (Aphrocallistidae) from Schulze (1899).

fragmentary and often remineralised. Whole body preservations virtually never contain the loose spicules on which Recent classifications are based, and in some strata, the only material that can be obtained are loose spicules extracted from sedimentary rocks. Faced with such limitations, palaeontologists have made astonishing progress in documenting the history of hexactinellids, but still have difficulty in attaining consensus within their own ranks and integrating their results to the classification schemes of Recent hexactinellids. Two classification schemes of fossil hexactinellids have recently been reported. Krautter (2002) gave a non-exhaustive overview that includes recent opinions of mainland European workers (Table 2). Rigby (2004), with input from Finks and Reid, put together an exhaustive classification in revision of the influential Treatise on Invertebrate Palaeontology (Table 3).

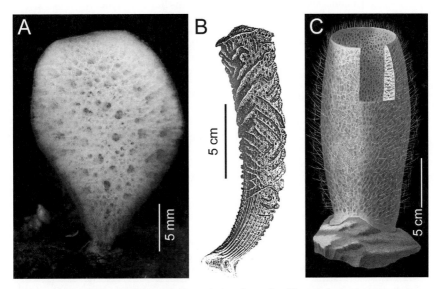

Figure 57 Representative species of the three families of lyssacine Hexasterophora. (A) *Leucopsacas scoliodocus* (Leucopsacidae). (B) *Euplectella aspergillum* (Euplectellidae) from Owen (1843). (C) *Staurocalyptus solidus* (Rossellidae) from Schulze (1899).

(F) *Laocoetis crassipes* (Craticulariidae; fossil species illustrated since the only Recent form is known as fragments) from Pomel (1872). (G) *Stereochlamis incerta* (Cribrospongiidae) from Ijima (1927). (H) *Euryplegma auriculare* (Aulocalycidae) from Schulze (1887). (I) *Tretopleura candelabrum* (Uncinateridae) from Ijima (1927). (J) *Fieldingia lagettoides* (Fieldingidae) from Kent (1870). (K) *Neoaulocystis zitteli* (Aulocystidae) from Ijima (1927). (L) *Scleroplegma lanterna* (Diapleuridae) from Reiswig (2002b).

Table 2 Classification of fossil Hexactinellida to family, from Krautter (2002) (taxa lacking authorities and stratigraphic ranges were inserted from text statements)

"Order Reticulosida" Reid, 1958 ?Neoproterozoic–Permian–?Late Jurassic
   Superfamily Protospongiodea Finks, 1960 ?Neoproterozoic–Middle
      Cambrian–Permian
      Family Protospongiidae Hinde, 1887 Middle Cambrian–Early Devonian
      Family Dictyospongiidae Hall, 1884 Neoproterozoic ?, Ordovician–Early
Carboniferous
      Family Stereodictyidae Finks, 1960 Permian
      Family Hintzespongiidae Finks, 1983 Ordovician
      Family Teganiidae
      Family Hyphanteniidae
   Superfamily Brachiospongioidea Finks, 1960 Ordovician–Permian–?Late Jurassic
      Family Brachiospongiidae Beecher, 1889 Ordovician–Silurian
      Family Stiodermatidae Finks, 1960 Permian
      Family Docodermatidae Finks, 1960 Permian–?Late Jurassic
      Family Stromatidiidae Finks, 1960 Permian
      Family Pileoutidae
Palaeozoic "Rossellimorpha" ?Late Vendian–Permian
Order Lyssacinosida Zittel, 1877 Neoproterozoic–Recent
Order Hexactinosida Schrammen, 1903 Late Devonian–Recent
      Family Euretidae Zittel, 1877 Late Jurassic–Recent
      Family Aphrocallistidae Gray, 1867 Late Cretaceous–Recent
      Family Craticulariidae Rauff, 1893 Middle Jurassic Recent
      Family Cribrospongiidae Roemer, 1864 Late Jurassic–Recent
      Family Staurodermatidae Zittel, 1878 Late Jurassic
      Hexactinosida *incertae sedis* Early Jurassic–Cretaceous
Order Lychniscosida Schrammen, 1902 Middle Jurassic–Recent
      Family Cypelliidae Schrammen, 1937 Late Jurassic
      Family Sporadoscinidae Schrammen, 1912 Late Jurassic–Tertiary
      Family Ventriculitidae Smith, 1848 Late Jurassic–Cretaceous
      Family Diapleuridae Ijima, 1927 Late Jurassic–Recent
      Family Neoaulocystidae Zhuravleva, 1962 Late Jurassic–Recent
      Family Becksiidae Schrammen, 1912 Cretaceous
      Family Coeloptychidae Zittel, 1877 Cretaceous
      Lychniscosida *incertae sedis*—Late Jurassic–Cretaceous

Both of these schemes were intended to bring classification of fossil forms into register with that of Recent forms, but they differ in some important ways that reflect the differences in philosophy of the two groups of workers.

Krautter's scheme, intentionally restricted, was arranged on the cautious premise that those groups of fossil forms that cannot convincingly be related to Recent groups should be maintained separate from those taxa and only those fossil groups that can convincingly be shown to be monophyletic should be retained as taxa. Rigby attempted to assign almost all taxa of

Table 3 Classification of fossil Hexactinellida to family, with number of fossil genera in parenthesis, from Rigby (2004)

Class Hexactinellida Schmidt, 1870 (432) Lower Cambrian–Holocene
  Subclass Amphidiscophora Schulze, 1887 (160) Lower Cambrian–Holocene
    Order Amphidiscosa Schrammen, 1924 (41) Lower Cambrian–Holocene
      Family Hyalonematidae Gray, 1857 (I) Cretaceous–Holocene
      Family Pattersoniidae Miller, 1889 (3) Middle Ordovician–Upper Ordovician
      Family Pelicaspongiidae Rigby, 1970 (24) Lower Ordovician–Triassic
      Family Stiodermatidae Finks, 1960 (13) Lower Cambrian–Permian
    Order Reticulosa Reid, 1958 (118) Ediacaran–Holocene
      Superfamily Protospongioidea Hinde, 1887 (20) Lower Cambrian–Jurassic
        Family Protospongiidae Hinde, 1887 (20) Lower Cambrian–Jurassic
      Superfamily Dierespongioidea Rigby and Gutschick, 1976 (24) M. Cambrian–Holocene
        Family Dierespongiidae Rigby and Gutschick, 1976 (6) Middle Ordovician–Permian
        Family Hydnodictyidae Rigby, 1971 (2) Middle Cambrian–Upper Ordovician
        Family Amphispongiidae Rauff, 1894 (1) Upper Silurian
        Family Multivasculatidae de Laubenfels, 1955 (1) Upper Cambrian
        Family Titusvillidae Caster, 1939 (6) Upper Devonian–Holocene
        Family Aglithodictyidae Hall and Clarke, 1899 (8) Upper Devonian–Carboniferous
      Superfamily Dictyospongioidea Hall and Clarke, 1899 (62) Ediacaran–Upper Triassic
        Family Dictyospongiidae Hall and Clarke, 1899 (55) Ediacaran–Permian
        Family Docodermatidae Finks. 1960 (5) Silurian–Permian
        Family Stereodictyidae Finks, 1960 (2) Carboniferous–Upper Triassic
      Superfamily Hintzespongioidea Finks. 1983 (12) Lower Cambrian–Carboniferous
        Family Hintzespongiidae Finks, 1983 (5) Lower Cambrian–Devonian
        Family Teganiidae de Laubenfels, 1955 (7) Cambrian–Carboniferous
    Order Hemidiscosa Schrammen, 1924 (1) Carboniferous
      Family Microhemidisciidae Finks and Rigby, 2004 (1) Carboniferous
  Subclass Hexasterophora Schulze, 1887 (272) Ordovician–Holocene
    Order Lyssacinosa Zittel, 1877 (36) Ordovician–Holocene
      Family Pheronematidae Gray (2) ?Upper Jurassic, Cretaceous–Holocene
      Family Euplectellidae Gray, 1867 (11) Lower Triassic–Holocene
      Family Asemematidae Schulze, 1887 (1) Palaeogene–Holocene
      Family Rossellidae Schulze, 1887 (1) ?Palaeogene–Holocene
      Family Stauractinellidae de Laubenfels, 1955 (1) Jurassic–Neogene
      Family Leucopsacidae Ijima, 1903 (1) Palaeogene
      Family Uncertain (6)
    Superfamily Crepospongioidea Finks and Rigby, 2004 (1) Triassic
      Family Crepospongiidae Finks and Rigby, 2004 (1) Triassic
    Superfamily Brachiospongioidea Beecher, 1889 (11) Upper Ordovician–Permian
      Family Brachiospongiidae Beecher, 1889 (4) Upper Ordovician–Silurian

---

(Continued)

Table 3   (Continued)

      Family Pyruspongiidae Rigby, 1971 (1) Upper Ordovician
      Family Malumispongiidae Rigby, 1967 (5) Upper Ordovician–
         Carboniferous
      Family Toomeyospongiidae Finks, herein (1) Permian
   Superfamily Lumectospongioidea Rigby and Chatterton, 1989 (1) Silurian
      Family Lumectospongiidae Rigby and Chatterton, 1989 (1) Silurian
   Superfamily Lumectospongioidea Rigby and Chatterton, 1989 (1) Silurian
      Family Lumectospongiidae Rigby and Chatterton, 1989 (1) Silurian
Order Hexactinosa Schrammen, 1903 (134) Upper Ordovician–Holocene
      Family Euryplegmatidae de Laubenfels, 1955 (1)?Cretaceous, Holocene
      Family Farreidae Schulze, 1885 (4) Cretaceous–Holocene
      Family Euretidae Zittel, 1877 (38) Triassic–Holocene
      Family Craticulariidae Rauff, 1893 (30) Triassic–Holocene
      Family Cribrospongiidae Roemer, 1864 (15) Middle Triassic–Holocene
      Family Staurodermatidae Zittel, 1877 (6) Jurassic–Neogene
      Family Aphrocallistidae Gray, 1867 (1) Lower Cretaceous–Holocene
      Family Tretodictyidae Schulze, 1887 (9) Upper Jurassic–Holocene
      Family Cystispongiidae Reid, herein (1) Upper Cretaceous–Neogene
      Family Aulocalycidae Ijima, 1927 (1) Upper Jurassic
      Family Emplocidae de Laubenfels, 1955 (1) Middle Jurassic
      Family Uncertain (16)
   Superfamily Pillaraspongioidea Rigby, 1986 (1) Devonian
      Family Pillaraspongiidae Rigby, 1986 (1) Devonian
   Superfamily Pileolitoidea Finks, 1960 (9) Upper Ordovician–Holocene
      Family Pileolitidae Finks, 1960 (2) Permian–Middle Triassic
      Family Wareembaiidae Finks and Rigby, 2004 (2) Upper Ordovician
      Family Euretidae Zittel, 1877 (2) Upper Devonian
      Family Craticulariidae Rauff, 1893 (5) Upper Devonian
      Family Pileospongiidae Rigby, Keyes and Horowirz, 1979 (1)
         Carboniferous
Order Lychniscosa Schrammen, 1903 (81) Jurassic–Holocene
      Family Calyptrellidae Schrammen, 1912 (1) Cretaceous
      Family Callodictyonidae Zittel, 1877 (23) Upper Jurassic–Holocene
      Family Callodictyonidae Zittel, 1877 (23) Upper Jurassic–Holocene
      Family Coeloptychidae F. A. Roemer, 1864 (4) Lower Cretaceous–Upper
         Cretaceous
      Family Ventriculitidae Smith, 1848 (21) Jurassic–Upper Cretaceous.
      Family Camerospongiidae Schrammen, 1912 (4) Lower Cretaceous–Upper
         Cretaceous
      Family Polyblastidiidae Schrammen, 1912 (2) Upper Jurassic–Cretaceous
      Family Dactylocalycidae Gray, 1867 (10) Jurassic–Holocene
      Family Sporadopylidae Schrammen, 1936 (3) Upper Jurassic–Cretaceous
      Family Pachyteichismatidae Schrammen, 1936 (3) Upper Jurassic–Lower
         Cretaceous
      Family Cypelliidae Schrammen, 1936 (5) Jurassic
      Family Uncertain (5)
Order Uncertain (20)

fossil forms to positions within the classification scheme of Recent hexactinellids, even where diagnoses of the fossil groups could not be encompassed
in diagnoses of the Recent taxa. Some major disagreements are obvious
between the two schemes. In Krautter's arrangement, the order Reticulosa
and two included superfamilies, Protospongioidea and Brachiospongioidea,
although listed in Table 2, were considered unrelated to Recent forms and,
indeed, not even defensible as taxa because of paraphyly and/or polyphyly
within those groups. Rigby, on the other hand, considered those taxa valid
and assigned the order and one of its superfamilies, Protospongioidea, to
subclass Amphidiscophora and the other subfamily, Brachiospongioidea, to
the subclass Hexasterophora. Krautter included a variety of fossil lyssacine
sponges to the very large informal grouping 'Rossellimorpha' (a name
introduced by Mehl, 1996) and indicated that convincing monophyletic
groupings could not yet be formed within this mixture, and were not yet
relatable to Recent taxa. Rigby, on the other hand, assigned almost all
genera of lyssacine fossil genera to positions within the Recent classification,
leaving no genera unplaced to subclass and only 20 of 432 unplaced to order.
Both of the schemes interface well with the Recent hexactinellid classification but perhaps Rigby has gone somewhat beyond credibility in this regard.
It is hoped that better communication between the two groups in the near
future will produce a consensus arrangement that will identify particular
fossil groups or strata that require more intensive examination and facilitate
more collaboration between taxonomists working on fossil and Recent
hexactinellids.

9.3. Phylogeny of Hexactinellida within Porifera

There is at present no doubt about the monophyly of Hexactinellida; indeed it is
the best characterised group of Porifera (Mehl, 1992). Monophyly of Porifera,
however, remains hotly contested, with those using morphological evidence,
including palaeontologists, accepting monophyly and those using molecular
techniques tending to support paraphyly of the phylum. Hexactinellids
are the first recognisable members of Porifera recorded in the fossil record.
Their hexactine spicules have been reported from rocks stratigraphically
corresponding to the Ediacaran Period (Neoproterozoic Era) of Mongolia by
Brasier et al. (1997) and from the similar age Shiobatan Member of South
China by Steiner et al. (1993). Gehling and Rigby (1996) also reported numerous body impressions of a small, globular sponge, *Palaeophragmodictya
reticulata*, from Ediacaran age deposits of South Africa. This species has a
surface reticulation reminiscent of two quadrules of spicule impressions which
prompted their assignment to Hexactinellida. Mehl et al. (1998) questioned

their hexactinellid assignment and suggested that the impressions may not have been of siliceous spicules, but rather of organic skeletal fibres, since the early ground plan of sponges developed by Reitner and Mehl (1996) excluded mineralised spicules. The first certain indications of Demospongiae and Calcarea by calthrops and advanced tetractinellid spicules, respectively, occur in the Lower Cambrian when hexactinellids have already reached a moderate diversity (Mehl *et al.*, 1998). According to Mehl *et al.* (1998), mineralised spicules were developed independently in the three classes of Porifera from aspiculate ancestral sponges (Figure 58), a hypothesis that contradicts the general convention that siliceous spicules of Hexactinellida and Demospongiae represent a shared homologous character.

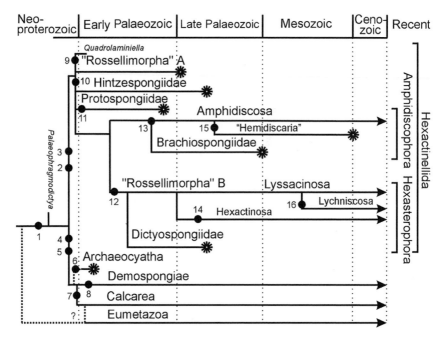

Figure 58 Diagram of the main features of Hexactinellida evolution summarised from Mehl-Janussen (1999), with modifications. Character states and innovations are indicated by numbers: 1, ancestral Ur-poriferan with sessile adult, planktonic larva, choanocytes in leuconoid flagellated chambers; 2, syncytial tissue organisation; 3, triaxial siliceous spicules; 4, pinacocytes; 5, secretion of basal calcareous skeleton; 6, regular archaeocyathid basal skeleton; 7, triactine magnesium–calcite spicules; 8, tetraxial siliceous spicules; 9, spicules reduced from triaxial to monaxial diactins; 10, differentiation of dermal and choanosomal spicules; 11, reduction of body spiculation to one-layer of dermal stauractins; 12, oxyhexaster; 13, amphidisc; 14, rigid skeleton of hexactins fused in regular cubic array; 15, hemidiscs derived by reduction from amphidiscs; 16, rigid skeleton of fused lychnisc hexactins. Group extinctions are indicated by asterisks.

Molecular sequence studies have so far been unable to unambiguously determine the early branching pattern of sponge classes, and thus whether the phylum Porifera is monophyletic or paraphyletic (Boury-Esnault and Sole-Cava, 2004; Nichols, 2005). Of molecular analyses published between 1996 and 2005 (duplicate reports grouped as one analysis), six supported or did not refute monophyly while seven supported or did not refute para-phyly of Porifera (Mehl *et al.*, 1998; Boury-Esnault and Sole-Cava, 2004; Nichols, 2005; Peterson and Butterfield, 2005). For more recent publica-tions, 2003–2005, these numbers are, respectively, two and three. Clearly, a consensus has not yet been reached regarding monophyly of Porifera from molecular work. A slightly different pattern is beginning to emerge regarding early branching within the phylum, only two of many possible alternatives of which have attracted support: (1) Hexactinellida and Demospongiae as sister groups of a clade excluding Calcarea (H+D/C), reported as support for a clade of siliceous sponges, or (2) Hexactinellida as a distinct lineage without sister grouping with either Demospongiae or Calcarea (H/D+C), interpreted as support for division of the Porifera into subphyla Symplasma (syncytial Hexactinellida) and Cellularia (Demospongiae and Calcarea), the proposal of Reiswig and Mackie (1983) based on patterns of tissue organisation. Of those same molecular analyses surveyed for monophyly of Porifera, for 1996–2005, five supported or did not refute Symplasma/Cellularia while seven supported or did not refute a siliceous grouping (Silicea of Bowerbank, 1864, being a preferable group name). For more recent publications, 2003–2005, these numbers are respectively zero and three. While recent results favour a clade of siliceous sponges, the level of statistical support for that position is gener-ally inadequate to rule out the independent status of Hexactinellida, as favoured by palaeontological and histological data.

Estimated dates of major branch points in early metazoan evolution from amino acid sequences do not yet specifically address divergence time of hexactinellids from the other sponge classes, but they bracket it. Divergence of Eumetazoa and demosponges (*Geodia*) occurred 800 Ma according to analysis of S-type lectins (Hirabayashi and Kasai, 1993), and 650–665 Ma according to analysis of class II tyrosine kinases (Müller *et al.*, 1995). Using amino acid sequences from seven nuclear-encoded housekeeping genes, Peterson and Butterfield (2005) estimated the divergence of Metazoa (including Porifera) from Fungi at between 664 and 867 Ma by minimum evolution and maximum likelihood methods, respectively. They also calcu-lated divergence of Eumetazoa from demosponge Porifera at between 634 and 826 Ma by the same methods. As their analysis favoured paraphyly of Porifera and sister grouping of Calcarea with Eumetazoa, both Demo-spongiae and Hexactinellida (no members included in the analysis) are presumed to have branched from the metazoan stem between these dates, either individually or together as a single clade. Since the earliest fossil sponge spicules have been dated to about 600 Ma (Mehl *et al.*, 1998), the 'missing' fossil record of early metazoan evolution may be relatively short, 34–64 Ma, or long, 226–267 Ma.

9.4. Phylogeny within Hexactinellida

Molecular studies of hexactinellid tissues, unfortunately, are still too few to enable formation of phylogenetic hypotheses within the group. Systematists working with morphological evidence of Recent hexactinellids have made very few speculations on their phylogeny. Schmidt (1880) warned that dictyonal skeletons have likely been developed independently several times, hence the grouping of all dictyonal sponges in one taxon, Dictyonina, was unrealistic. This view was supported by Ijima's division (1927) of the dictyonine hexactinellids into orders Hexactinosa and Lychniscosa on the basis of his contention that they had evolved independently from a protohexasterophoran ancestor. Mehl (1992) extended this separation, suggesting that dictyonine Lychniscosa and lyssacine Euplectellidae, two groups sharing graphiocome microscleres, were sister groups derived from a common ancestor. She grouped these together in the taxon Graphiocomida and pointed out that, if this grouping survived testing, the Recent order Lyssacina could no longer be considered monophyletic. Tabachnick and Menshinina (1999) developed a phylogeny within Amphidiscophora on the basis of spiculation and body form, concluding that the common ancestor, a cup-shaped form rooted in soft sediments, gave rise to two radiating lineages, a hyalonematid line and a pheronematid–monorhaphidid line. The long separation of the two main hexactinellid groupings, Amphidiscophora and Hexasterophora, continues to be supported by all workers and forms the cornerstone for speculation on phylogeny within the class.

More extensive hypotheses of hexactinellid phylogeny have been developed by palaeontologists attempting to integrate morphological information from both fossils and Recent specimens. Here we will not review these in detail, but concentrate on main aspects of interest to neontologists: characteristics of the basal group of hexactinellids and origin of the major recent groups.

In the classical scheme developed by Finks (1970, 1983, 2003a,b), the Middle Cambrian Protospongiidae, the first whole-body fossils of hexactinellids, are considered the basal stock from which all later groups derived. These were thin-walled, vasiform sponges with a large distal osculum and basal root tuft of simple diactins (Figure 59). Their body wall consisted of a single layer of stauractins or pentactins (never hexactins) of several discrete sizes arranged in parallel to form a characteristic pattern of quadrules of

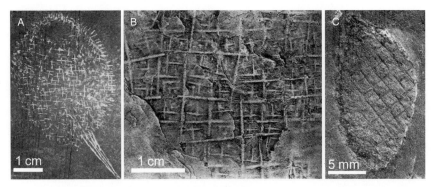

Figure 59 Representatives of the controversial Early Palaeozoic fossil family Protospongiidae. (A) *Protospongia tetranema* in which the spicules have been visually enhanced by white paint applied by the original describer, Sir W. Dawson (from Mehl, 1996). (B) Surface pattern of the single layer of stauractin spicules forming the characteristic quadrule pattern in *Protospongia hicksi*, from an enhanced photograph of a latex peel (from Rigby, 1986). (C) *Diagoniella robisoni*, a member of the second genus of Protospongiidae, in which the primary set of stauractins is oriented 45° from the body axis (from Mehl, 1996).

several size orders (Figure 59B). This structure is retained as the dermal skeleton to which several types of parenchymal skeletons were added internally in derived groups, Hintzespongioidea, Dictyospongioidea and Dierespongioidea. In the contrasting scheme developed by Mehl-Janussen and coworkers (Mehl, 1991, 1992; Mehl *et al.*, 1998; Mehl-Janussen, 1999), based on the phylogenetic systematic conclusions reached by Reitner and Mehl (1996), and the occurrence of isolated hexactin spicules in Late Proterozoic rocks (noted above), the basal stock of hexactinellids cannot presently be characterised, but contained hexactine spicules, not stauractins, as original skeletal elements (Figure 58). Several Lower Cambrian fossils originally considered demosponges, for example *Quadrolaminiella*, are reinterpreted to be likely members of an early radiation of hexactinellids with dictyospongiid skeletons—thin body walls composed of diactin and sometimes hexactin dermalia in a regular rectangular pattern of longitudinal and transverse spicules. In this scheme, the Protospongiidae, rather than being basal to the entire class, is considered an offshoot that went extinct in mid-Palaeozoic and led to no existing members.

Origin of Amphidiscophora is partially obviated in Finks' scheme by assignment of most early Palaeozoic groups to this subclass, but a path leading to the diagnostic amphidisc microscleres was suggested. The paraclavules (spicules with one hemispheric umbel on a tapering rod) which occur in the Mississippian dictyosponge *Griphiodictya* are suggested to have evolved to hemidiscs (tapering rod with unequal umbels) in the Late

Carboniferous *Microhemidiscia* and to true amphidiscs (rod with two equal-size umbels) in the similar-aged *Uralonema*. The taxon Hemidiscosa (with hemidiscs) is considered a valid, if extinct, order of Amphidiscophora. Mehl (1991) pointed out that the then known stratigraphic sequence of occurrence of these spicules, hemidiscs (isolated) from Upper Cambrian, amphidiscs (isolated) from Upper Silurian and paraclavules (in *Uralonema*) from Mississippian, did not support Finks' derivation. Reitner and Mehl (1995) concluded that these three spicule types probably developed independently and the early Upper Cambrian hemidiscs (considered to be tylodiscs by Mehl, 1992) were probably not homologous with later Carboniferous forms. Mehl-Janussen (1999) considered the only acceptable evidence of Amphidiscophora to be true amphidiscs which occur first as isolated spicules in Upper Silurian. In her scheme, Amphidiscophora have a common ancestry with Brachiospongiidae, a group characterised by thick walls, radial lobate, cup- or vase-like shape, dermal layer as a fine net of stauractins in a regular quadrangular orientation, with hypodermalia as large hexactins with long proximal rays protruding far into the body wall and parenchymal spicules as hexactins of different size in irregular orientation. In contrast, Finks placed the Brachiospongiidae as ancestral to modern Hexasterophora. Contrary to Finks' suggestion, Mehl-Janussen regarded the Late Carboniferous hemidiscs as most likely derived from amphidiscs by reduction of one umbel, a process that probably occurred many times in the past and can even be documented in Recent forms. She considered Hemidiscaria (or Hemidiscosa) to be unsupported as a monophyletic taxon.

Origin of Hexasterophora in Finks' scheme (2003b) is intertwined with origin of Amphidiscophora. He considered it likely that the two main hexactinellid lineages were not differentiated in the Early Palaeozoic lineages, and interpreted the Mississippian dictyosponge, *Griphodictya epiphanes*, with oxyhexasters and paraclavules, as possessing the distinguishing characters of both major subclasses. By the Permian, the Brachiospongioidea, had taken on new body organisation, including thicker body wall and large hypodermal pentactins, that can be related to those of lyssacine hexasterophorans. He interpreted the Permian genus *Pileolites* to be a true hexactinosan and indicated that the aulocalycoid framework was primitive and developed into more regular dictyonal forms later. Mehl-Janussen (1999) accepted the earlier occurrence of oxyhexasters as isolated spicules from Lower Ordovician of Sweden (Mostler, 1986) as evidence of the existence of Hexasterophora by that time. She accepted oxyhexasters in the dictyosponge *Griphodictya epiphanes* as indication that the species, and the Dictyospongiidae, were members of Hexasterophora, but, as indicated above, she rejected the interpretation that paraclavules in that species are homologous with amphidiscs. She considered the stem Hexasterophora to encompass a presently uncharacterisable group of genera referred to informally as "Rossellimorpha"

('Rossellimorpha' B in Figure 57), which gave rise to the Dictyospongiidae in the Ordovician, the dictyonine Hexactinosa in the Devonian, and the dictyonine Lychniscosa in the Middle Jurassic. The present Lyssacinosa arose as a gradual transition from the main stem group (Figure 58).

The overall pattern of diversification and extinction of hexactinellids is summarised by Krautter (2002). The class flourished and radiated rapidly during the Cambrian, giving rise to many new taxa and new skeletal plans, most of which diversified through the early Palaeozoic. In the later Palaeozoic, many families went extinct but the record between the Carboniferous and Triassic is very scarce. Most apparently died out by the Permian– Triassic boundary but at least four major lineages must have survived that barrier (Figure 58). The Hexactinosa underwent a major radiation and diversification during the Mesozoic, extending their distribution worldwide during the Late Triassic. They, together with the newly arisen and rapidly diversifying Lychniscosa, and lithistid demosponges, formed a discontinuous siliceous sponge reef belt spreading over more than 7000 km on the northern shelf of the Tethys Sea, the largest biological structure ever formed in the history of the Earth. Diversity of dictyonal hexactinellids continued to increase in the Cretaceous, but has gradually declined since the Late Cretaceous, leading to the present situation where hexactinellid faunas are generally dominated by lyssacine forms (Brückner and Janussen, 2005).

10. CONCLUSIONS

Hexactinellids are a very distinctive group with a long fossil record and have attracted continuous attention from taxonomists and palaeontologists since their first discovery, but it is only in the last 30 years that they have come into their own as living animals. This came about with the belated realisation that certain species could be collected alive and in good condition by SCUBA from shallow waters around Vancouver Island, and later from a subterranean cave in the Mediterranean. These findings immediately led to a variety of studies which have given us an increasingly clear picture of how glass sponges live, grow, feed, reproduce, and respond to environmental stimuli.

At the same time, advances in techniques for observing and monitoring the activities of benthic organisms *in situ* have contributed greatly to the overall picture, showing the importance of hexactinellids in deeper benthic ecosystems.

While all this has been going on, molecular biology has begun to make a growing impact in several areas. Molecular evidence, taken in conjunction with palaeontological discoveries, makes it very likely that hexactinellids, with the possible exception of Placozoa (Dellaporta *et al.*, 2006), are the basal extant animal group. As such they are of unique interest from almost every evolutionary point of view, as much for those interested in the early evolution of conduction systems as for those tracing the history of molecules that play key roles in human physiology.

Molecular approaches will undoubtedly be applied increasingly to aspects of glass sponge physiology, biochemistry and development in the future, but we hope and expect that study of the living animals will not be neglected and that methods for maintaining them in captivity will be improved to the point that one or more species will become available for long-term studies in the laboratory. The difficulty of maintaining captive hexactinellids in good condition is perhaps the greatest barrier currently facing workers in this field and merits much closer attention than it has yet received.

ACKNOWLEDGEMENTS

This work was supported by grants from the Natural Sciences and Engineering Research Council of Canada. We thank those who contributed original illustrations. G.O.M. thanks Dorte Janussen, Gitai Yahel and Kurt Pueschel for critically reading certain sections. The photographs and drawing by Jan Köster were posted by him to the *Porifera* internet discussion list (http://www.jiscmail.ac.uk/lists/porifera.html) where they are still archived (April 1999). Heidi Schomann of the Institut für Meereskunde, Universität Kiel, kindly sent a photocopy of the unpublished thesis from which they were taken. S.P.L. thanks Kim Conway for comments on the ecology section. Original material from *Oopsacas* derives from work done by S.P.L. in collaboration with Nicole Boury-Esnault.

REFERENCES

Aiello, E. (1974). Control of ciliary activity in Metazoa. *In* "Cilia and Flagella" (M. A. Sleigh, ed.), pp. 353–376. Academic Press, London.
Aizenberg, J., Sundar, V. C., Yablon, A. D., Weaver, J. C. and Chen, G. (2004). Biological glass fibers: Correlation between optical and structural properties. *Proceedings of the National Academy of Sciences of the United States of America* **101**, 3358–3363.
Aizenberg, J., Weaver, J. C., Thanwala, M. S., Sundar, V. C., Morse, D. E. and Fratzl, P. (2005). Skeleton of *Euplectella* sp.: Structural hierarchy from the nanoscale to the macroscale. *Science* **309**, 275–278.

Austin, W. C. (1983). Underwater birdwatching. *Canadian Technical Report of Hydrography and Ocean Sciences* **38**, 83–89.

Austin, W. C. (1999). The relationship of silicate levels to the shallow water distribution of hexactinellids in British Columbia. *Memoirs of the Queensland Museum* **44**, 44.

Austin, W. C. (2003). Sponge Gardens: A hidden treasure in British Columbia. http://mareco.org/khoyatan/spongegardens

Barthel, D. (1992). Do hexactinellids structure Antarctic sponge associations? *Ophelia* **36**, 111–118.

Barthel, D. (1995). Tissue composition of sponges from the Weddell Sea Antarctica: Not much meat on the bones. *Marine Ecology Progress Series* **123**, 149–153.

Barthel, D. and Gutt, J. (1992). Sponge associations in the eastern Weddell Sea. *Antarctic Science* **4**, 137–150.

Barthel, D. and Tendal, O. S. (1994). Antartic Hexactinellida. *In* "Synopses of the Antarctic Benthos", Vol. 6, p. 154. Koeltz Scientific Books, Koenigstein.

Bavestrello, G., Burlando, B. and Sarà, M. (1988). The architecture of the canal systems of *Petrosia ficiformis* and *Chondrosia reniformis* studied by corrosion casts (Porifera, Demospongiae). *Zoomorphology* **108**, 161–166.

Bavestrello, G., Benatti, U., Calcinai, B., Cattaneo-Vietti, R., Cerrano, C., Favre, A., Giovine, M., Lanza, S., Pronzato, R. and Sarà, M. (1998). Body polarity and mineral selectivity in the demosponge *Chondrosia reniformis*. *Biological Bulletin Marine Biological Laboratory, Woods Hole* **195**, 120–125.

Bavestrello, G., Arillo, A. and Calcinai, B. (2003). The aquiferous system of *Scolymastra joubini* (Porifera, Hexactinellida) studied by corrosion casts. *Zoomorphology* **122**, 119–123.

Bearer, E. L., DeGiorgis, J. A., Bodner, R. A., Kao, A. W. and Reese, T. S. (1993). Evidence for myosin motors on organelles in squid axoplasm. *Proceedings of the National Academy of Sciences of the United States of America* **90**, 11252–11256.

Beaulieu, S. E. (2001a). Colonization of habitat islands in the deep sea: Recruitment to glass sponge stalks. *Deep-sea Research I* **48**, 1121–1137.

Beaulieu, S. E. (2001b). Life on glass houses: Sponge stalk communities in the deep sea. *Marine Biology* **138**, 803–817.

Bett, B. J. and Rice, A. L. (1992). The influence of hexactinellid sponge *Pheronema carpenteri* spicules on the patchy distribution of macrobenthos in the porcupine seabight (Bathyl NE Atlantic). *Ophelia* **36**, 217–226.

Bidder, G. P. (1923). The relation of the form of a sponge to its currents. *Quarterly Journal of Microscopical Science* **67**, 293–323.

Boury-Esnault, N. and de Vos, L. (1988). *Caulophacus cyanae*, n. sp., une éponge hexactinellide des sources hydrothermales. Biogéographie du genre *Caulophacus* Schulze, 1887. *Oceanologica Acta* **8**, 51–60.

Boury-Esnault, N. and Rutzler, K. (1997). Thesaurus of sponge morphology. *Smithsonian Contributions to Zoology* **596**, 1–55.

Boury-Esnault, N. and Sole-Cava, A. M. (2004). Recent contributions of genetics to the study of sponge systematics and biology. *Bollettino dei Musei e degli Istituti Biologici dell'Università di Genova* **68**, 3–18.

Boury-Esnault, N. and Vacelet, J. (1994). Preliminary studies on the organization and development of a hexactinellid sponge from a Mediterranean cave, *Oopsacas minuta*. *In* "Sponges in Time and Space. Proceedings of the Fourth International Porifera Congress" (R. W. M. van Soest, T. M. G. van Kempen and J. Braekman, eds), pp. 407–416. AA Balkema, Rotterdam.

Boury-Esnault, N., Efremova, S., Bézac, C. and Vacelet, J. (1999). Reproduction of a hexactinellid sponge: First description of gastrulation by cellular delamination in the Porifera. *Invertebrate Reproduction and Development* **35**, 187–201.

Bowerbank, J. S. (1858). On the anatomy and physiology of the Spongiadae. Part I: On the spicules. *Philosophical Transactions of the Royal Society of London* **148**, 279–332.

Bowerbank, J. S. (1864). "A Monograph of the British Spongiadae", Vol. 1. Ray Society, London.

Boyd, I. (1981). The spicule jungle of *Rhabdocalyptus dawsoni*: A unique microhabitat. B.Sc. Thesis, University of Victoria, Victoria, BC, Canada.

Brasier, M. D., Green, O. and Shields, G. (1997). Ediacaran sponge spicule clusters from southwestern Mongolia and the origins of the Cambrian fauna. *Geology* **25**, 303–306.

Brückner, A. and Janussen, D. (2005). *Rossella bromleyi* n. sp.: The first entirely preserved fossil sponge species of the genus *Rossella* (Hexactinellida) from the Upper Cretaceous of Bornholm, Denmark. *Journal of Paleontology* **79**, 21–28.

Bütschli, O. (1901). Einige Beobachtungen über Kiesel- und kalknadeln von Spongien. *Zeitschrift für wissentschafliche Zoologie* **69**, 235–286.

Carter, H. J. (1885). Report on a collection of marine sponges from Japan made by Dr. J. Anderson. *Annals and Magazine of Natural History* **15**, 387–406.

Cattaneo-Vietti, R., Bavestrello, G., Cerrano, C., Sarà, M., Benatti, U., Giovine, M. and Gaino, E. (1996). Optical fibres in an Antarctic sponge. *Nature* **383**, 397–398.

Cerrano, C., Arillo, A., Bavestrello, G., Calcinai, B., Cattaneo-Vietti, R., Penna, A., Sarà, M. and Totti, C. (2000). Diatom invasion in the antarctic hexactinellid sponge *Scolymastra joubini*. *Polar Biology* **23**, 441–444.

Cha, J. N., Shimizu, K., Zhou, Y., Christiansen, S. C., Chmelka, B. F., Stucky, G. D. and Morse, D. E. (1999). Silicatein filaments and subunits from a marine sponge direct the polymerization of silica and silicones *in vitro*. *Proceedings of the National Academy of Sciences of the United States of America* **96**, 361–365.

Claus, C. F. W. (1868). Ueber *Euplectella aspergillum* (R. Owen). *In* "Ein Beitrag zur Naturgeschichte der Kieselschwämme", N.G. Elwert'sche Universitäts-Buchhandlung, Marburg.

Claus, C. F. W. (1872). "Grundzüge der Zoologie", 2nd edn. N.G. Elwert, Marburg und Leipzig.

Conway, K., Krautter, M., Vaughn Barrie, J., Whitney, F., Thomson, R. E., Reiswig, H. M., Lehnert, H., Mungov, G. and Bertram, M. (2005a). Sponge reefs in the Queen Charlotte Basin, Canada: Controls on distribution, growth and development. *In* "Cold-water Corals and Ecosystems" (A. Freiwald and J. M. Roberts, eds), pp. 605–621. Springer-Verlag, Berlin.

Conway, K., Vaughn Barrie, J. and Krautter, M. (2005b). Geomorphology of unique reefs on the western Canadian shelf: Sponge reefs mapped by multibeam bathymetry. *Geo-Marine Letters* **25**, 205–213.

Conway, K. W., Barrie, J. V., Austin, W. C. and Luternauer, J. L. (1991). Holocene sponge bioherms on the western Canadian continental shelf. *Continental Shelf Research* **11**, 771–790.

Conway, K. W., Krautter, M., Barrie, J. V. and Neuweiller, M. (2001). Hexactinellid sponge reefs on the Canadian continental shelf: A unique "living fossil". *Geoscience Canada* **28**, 71–78.

Conway, K. W., Vaughn Barrie, J. and Krautter, M. (2004). Modern siliceous sponge reefs in a turbid, siliciclastic setting: Fraser River delta, British Columbia, Canada. *Neues Jahrbuch für Geologie und Paläontologie Monatshefte* **6**, 335–350.

Cook, S. E. (2005). Ecology of the Hexactinellid sponge reefs on the Western Canadian continental shelf. M.Sc. Thesis, University of Victoria, Victoria.

Croce, G., Frache, A., Milanesio, M., Viterbo, D., Bavestrello, G., Benatti, U., Giovine, M. and Amenitsch, H. (2003). Fiber diffraction study of spicules from marine sponges. *Microscopy Research and Technique* **62**, 378–381.

Dahlgren, U. and Kepner, W. A. (1930). "Principles of Animal Histology", Macmillan, New York.

Dayton, P. K. (1979). Observations of growth, dispersal and population dynamics of some sponges in McMurdo Sound, Antarctica. *Colloques internationaux du C.N.R.S.* **291**, 271–282.

Dayton, P. K. (1989). Interdecadal variation in an Antarctic sponge and its predators from oceanographic climate shifts. *Science* **245**, 1484–1486.

Dayton, P. K., Robilliard, G. A., Paine, R. T. and Dayton, L. B. (1974). Biological accommodation in the benthic community at McMurdo Sound, Antarctica. *Ecological Monographs* **44**, 105–128.

Dellaporta, S. L., Xu, A., Sagasser, S., Jakob, W., Moreno, M. A., Buss, L. W. and Schierwater, B. (2006). Mitochondrial genome of *Trichoplax adhaerens* supports Placozoa as the basal lower metazoan phylum. *Proceedings of the National Academy of Sciences of the United States of America* **103**, 8751–8756.

Dendy, A. (1926). On the origin, growth and arrangement of sponge spicules: A study in symbiosis. *Quarterly Journal of Microscopical Science* **70**, 1–74.

Drum, R. W. (1968). Electron microscopy of siliceous spicules from the freshwater sponge *Heteromyenia. Journal of Ultrastructure Research* **22**, 12–21.

Ehrlich, H., Hanke, T., Simon, P., Goebel, C., Heinemann, S., Born, R. and Worch, H. (2005). Demineralization of natural silica based biomaterials: New strategy for the isolation of organic frameworks. *Biomaterialien* **6**, 297–302.

Fernandez-Busquets, X. and Burger, M. M. (1999). Cell adhesion and histocompatibility in sponges. *Microscopy Research and Technique* **44**, 204–218.

Finks, R. M. (1970). The evolution and ecologic history of sponges during Palaeozoic times. *In* "The Biology of the Porifera" (W. G. Fry, ed.), pp. 3–22. Academic Press, London.

Finks, R. M. (1983). Fossil Hexactinellida. *In* "Sponges and Spongiomorphs, Notes for a Short Course. Studies in Geology (7)" (J. K. Rigby and C. W. Stearn, eds), pp. 101–105. Department of Geological Sciences, University of Tennessee, Knoxville, Tennessee.

Finks, R. M. (2003a). Evolution and ecological history of sponges during Paleozoic times. *In* "Treatise on Invertebrate Paleontology, Part E, Porifera, Revised, Vol. 2: Introduction to the Porifera" (R. L. Kaesler, ed.), pp. 261–274. The Geological Society of America, Boulder, Colorado.

Finks, R. M. (2003b). Paleozoic Hexactinellida: Morphology and phylogeny. *In* "Treatise on Invertebrate Paleontology, Part E, Porifera, Revised, Vol. 2: Introduction to the Porifera" (R. L. Kaesler, ed.), pp. 135–154. The Geological Society of America, Boulder, Colorado.

Fraser, C. M. (1932). A comparison of the marine fauna of the Nanaimo region with that of the San Juan Archipelago. *Transactions of the Royal Society of Canada* **26**, 49–70.

Garrone, R. (1969). Collagène, spongine et squelette minéral chez l'éponge *Haliclona rosea* (O.S.) Démosponge, Haploscléride. *Journal de Microscopie* **8**, 581–598.

Garrone, R., Simpson, T. L. and Pottu-Boumendil, J. (1981). Ultrastructure and deposition of silica in sponges. *In* "Silicon and Siliceous Structures in Biological

Systems" (T. L. Simpson and B. E. Volcani, eds), pp. 495–525. Springer-Verlag, New York.

Gehling, J. G. and Rigby, J. K. (1996). Long expected sponges from the neoproterozoic Ediacara fauna of South Australia. *Journal of Paleontology* **70**, 185–195.

Ghiold, J. (1991). The sponges that spanned Europe. *New Scientist* **2**, 58–62.

Goodall, H. (1985). Membrane channels: Bridging the junctional gap. *Nature* **317**, 286–287.

Green, C. R. and Bergquist, P. R. (1982). Phylogenetic relationships within the invertebrata in relation to the structure of septate junctions and the development of occluding junctional types. *Journal of Cell Science* **53**, 279–305.

Guilbault, J.-P., Krautter, M., Conway, K. W. and Barrie, J. V. (2006). Modern foraminifera attached to Hexactinellid sponge meshwork on the West Canadian Shelf: Comparison with Jurassic Counterparts from Europe. *Palaeontologia Electronica* **9**(1)3A:48p, 6.3MB http://palaeo-electronica.org/paleo/2006_1/sponge/issue1_06.htm

Gundacker, D., Leys, S. P., Schröder, H. C., Müller, I. M. and Müller, W. E. G. (2001). Isolation and cloning of a C-type lectin from the hexactinellid sponge *Aphrocallistes vastus*: A putative aggregation factor. *Glycobiology* **11**, 21–29.

Harrison, F. W. and de Vos, L. (1991). Porifera. *In* "Microscopic Anatomy of Invertebrates" (F. W. Harrison and J. A. Westfall, eds), pp. 29–89. Wiley-Liss, New York.

Hartman, W. D. (1983). Modern Hexactinellida. *In* "Sponges and Spongiomorphs, Notes for a Short Course" (T. W. Broadhead, ed.). Department of Geological Sciences, University of Tennessee, Knoxville, Tennessee.

Hirabayashi, J. and Kasai, K. (1993). The family of metazoan metal-independent beta-galactoside-binding lectins: Structure, function and molecular evolution. *Glycobiology* **3**, 297–304.

Hooper, J. A. and Van Soest, R. W. M. (2002). Systema Porifera. *In* "A Guide to the Classification of Sponges". Kluwer Academic/Plenum Publishers, New York.

Hooper, J. N. A. and Wiedenmayer, F. (1994). Porifera. *In* "Zoological Catalogue of Australia" (A. Wells, ed.), pp. 1–632. CSIRO, Australia, Melbourne.

Ijima, I. (1901). Studies on the Hexactinellida. Contribution I (Euplectellidae). *Journal of the College of Science of the Imperial University of Tokyo* **15**, 1–299.

Ijima, I. (1904). Studies on the Hexactinellida. Contribution IV. (Rossellidae). *Journal of the College of Science of the Imperial University of Tokyo* **28**, 13–307.

Ijima, I. (1927). The Hexactinellida of the Siboga Expedition. *Siboga Expedition Reports* **6**, 1–383, 26 plates.

Imsiecke, G. (1993). Ingestion, digestion, and egestion in *Spongilla lacustris* (Porifera, Spongillidae) after pulse feeding with *Chlamydomonas reinhardtii*. *Zoomorphology* **113**, 233–244.

Imsiecke, G., Steffen, R., Custodio, M., Borojevic, R. and Müller, W. E. G. (1995). Formation of spicules by sclerocytes from the freshwater sponge *Ephydatia muelleri* in short-term cultures *in vitro*. *In vitro Cellular and Developmental Biology-Animal* **31**, 528–535.

Jamieson, G. S. and Chew, L. (2002). Hexactinellid sponge reefs: Areas of interest as marine protected areas in the north and central coast areas. *Canadian Science Advisory Secretariat. Department of Fisheries and Oceans,* 1–78.

Kachar, B. and Reese, T. S. (1988). The mechanism of cytoplasmic streaming in characean algal cells: Sliding of endoplasmic reticulum along actin filaments. *Journal of Cell Biology* **106**, 1545–1552.

Kent, W. S. (1870). On two new siliceous sponges taken in the late dredging expedition of the yacht 'Norna' off the coasts of Spain and Portugal. *Annals and Magazine of Natural History* **6**, 217–224.

Kirkpatrick, R. (1910). On Hexactinellida sponge spicules and their names. *Annals and Magazine of Natural History* **5**, 208–213, 347–350.

Koltun, V. M. (1967). Vitreous sponges of the northern and far-eastern seas of the USSR. *Opredeliteli po Faune S.S.S.R.* **94**, 1–124.

Koonce, M. P., Euteneuer, U., McDonald, K. L., Menzel, D. and Schliwa, M. (1986). Cytoskeletal architecture and motility in a giant freshwater amoeba, *Reticulomyxa*. *Cell Motility and the Cytoskeleton* **6**, 521–533.

Köster, J. (1997). Untersuchungen zur Ultrastuktur antarktischer Hexactinelliden (Porifera) Diploma Thesis, Institut für Meereskunde, Christian-Albrechts-Universität, Kiel, Germany.

Krasko, A., Lorenz, B., Batel, R., Schröder, H. C., Müller, I. M. and Müller, W. E. G. (2000). Expression of silicatein and collagen genes in the marine sponge *Suberites domuncula* is controlled by silicate and myotrophin. *European Journal of Biochemistry* **267**, 4878–4887.

Krautter, M. (2002). Fossil Hexactinellida: An Overview. *In* "Systema Porifera: A Guide to the Classification of Sponges" (J. N. A. Hooper and R. W. M. Van Soest, eds), pp. 1211–1223. Kluwer Academic/Plenum Publishers, New York.

Krautter, M., Conway, K. W., Barrie, J. V. and Neuweiller, M. (2001). Discovery of a "Living Dinosaur": Globally unique modern hexactinellid sponge reefs off British Columbia, Canada. *Facies* **44**, 265–282.

Kube-Granderath, E. and Schliwa, M. (1995). Gamma tubulin of *Reticulomyxa filosa*: Amino acid sequence and expression in bacteria. *European Journal of Cell Biology* **Suppl.**, 199a.

Lawn, I. D., Mackie, G. O. and Silver, G. (1981). Conduction system in a sponge. *Science* **211**, 1169–1171.

Leinfelder, R. R., Krautter, M., Laternser, R., Nose, M., Schmid, D. U., Schweigert, G., Werner, W., Keupp, H., Brugger, H., Herrmann, R., Rehfeld-Kiefer, U. Schröder, J. H. *et al.* (1994). The origin of Jurassic reefs: Current research developments and results. *Facies* **31**, 1–56.

Leys, S. P. (1995). Cytoskeletal architecture and organelle transport in giant syncytia formed by fusion of hexactinellid sponge tissues. *Biological Bulletin Marine Biological Laboratory, Woods Hole* **188**, 241–254.

Leys, S. P. (1996). Cytoskeletal architecture, organelle transport, and impulse conduction in hexactinellid sponge syncytia Doctoral Dissertation, University of Victoria, Victoria.

Leys, S. P. (1997). Sponge cell culture: A comparative evaluation of adhesion to a native tissue extract and other culture substrates. *Tissue and Cell* **29**, 77–87.

Leys, S. P. (1998). Fusion and cytoplasmic streaming are characteristics of at least two hexactinellids: Examination of cultured tissue from *Aphrocallistes vastus*. *In* "Sponge Sciences—Multidisciplinary Perspectives" (Y. Watanabe and N. Fusetani, eds), pp. 215–226. Springer-Verlag, Tokyo.

Leys, S. P. (1999). The choanosome of hexactinellid sponges. *Invertebrate Biology* **118**, 221–235.

Leys, S. P. (2003a). Comparative study of spiculogenesis in demosponge and hexactinellid Larvae. *Microscopy Research and Technique* **62**, 300–311.

Leys, S. P. (2003b). The significance of syncytial tissues for the position of the Hexactinellida in the Metazoa. *Integrative and Comparative Biology* **43**, 19–27.

Leys, S. P. and Lauzon, N. R. J. (1998). Hexactinellid sponge ecology: Growth rates and seasonality in deep water sponges. *Journal of Experimental Marine Biology and Ecology* **230**, 111–129.

Leys, S. P. and Mackie, G. O. (1994). Cytoplasmic streaming in the hexactinellid sponge *Rhabdocalyptus dawsoni* (Lambe 1873). *In* "Sponges in Time and Space" (R. W. M. van Soest, T. M. G. van Kempen and J. Braekman, eds), pp. 417–423. AA Balkema, Rotterdam.

Leys, S. P. and Mackie, G. O. (1997). Electrical recording from a glass sponge. *Nature* **387**, 29–30.

Leys, S. P. and Meech, R. W. (2006). Physiology of coordination in sponges. *Canadian Journal of Zoology* **84**, 288–306.

Leys, S. P. and Reiswig, H. M. (1998). Nutrient transport pathways in the neotropical sponge *Aplysina. Biological Bulletin Marine Biological Laboratory, Woods Hole* **195**, 30–42.

Leys, S. P. and Tompkins, G. J. (2004). Glass sponges arrest pumping in response to increased sediment loads. *Integrative and Comparative Biology* **44**, 719.

Leys, S. P., Mackie, G. O. and Meech, R. W. (1999). Impulse conduction in a sponge. *Journal of Experimental Biology* **202**, 1139–1150.

Leys, S. P., Wilson, K., Holeton, C., Reiswig, H. M., Austin, W. C. and Tunnicliffe, V. (2004). Patterns of glass sponge (Porifera, Hexatinellida) distribution in coastal waters of British Columbia, Canada. *Marine Ecology Progress Series* **283**, 133–149.

Leys, S. P., Cheung, E. and Boury-Esnault, N. (2006). Embryogenesis in the glass sponge *Oopsacas minuta*: Formation of syncytia by fusion of blastomeres. *Integrative and Comparative Biology* **46**(2), 104–117.

MacGinitie, G. E. (1939). The method of feeding of tunicates. *Biological Bulletin Marine Biological Laboratory, Woods Hole* **77**, 443–447.

Mackie, G. O. (1979). Is there a conduction system in sponges? *Colloques Internationaux du Centre National de la Recherche Scientifique* **291**, 145–151.

Mackie, G. O. (1981). Plugged syncytial interconnections in hexactinellid sponges. *Journal of Cell Biology* **91**, 103a.

Mackie, G. O. and Singla, C. L. (1983). Studies on hexactinellid sponges. I. Histology of *Rhabdocalyptus dawsoni* (Lambe, 1873). *Philosophical Transactions of the Royal Society of London Series B* **301**, 365–400.

Mackie, G. O., Lawn, I. D. and Pavans de Ceccatty, M. (1983). Studies on hexactinellid sponges. II. Excitability, conduction and coordination of responses in *Rhabdocalyptus dawsoni* (Lambe 1873). *Philosophical Transactions of the Royal Society of London Series B* **301**, 401–418.

Maldonado, M., Carmona, M. C., Uriz, M. J. and Cruzado, A. (1999). Decline in Mesozoic reef-building sponges explained by silicon limitation. *Nature* **401**, 785–788.

Marshall, W. (1875). Untersuchungen über Hexactinelliden. *Zeitschrift für wissentschafliche Zoologie* **25**, 142–243.

McClay, D. R. (1972). Cell aggregation: Properties of cell surface factors from five species of sponge. *Journal of Experimental Zoology* **186**, 89–102.

Mehl, D. (1991). Are protospongiidae the stem group of modern hexactinellids? *In* "Fossil and Recent Sponges" (J. Reitner and H. Keupp, eds), pp. 43–53. Springer-Verlag, Berlin.

Mehl, D. (1992). Die Entwicklung der Hexactinellida seit dem Mesozoikum. Paläobiologie, Phylogenie und Evolutionsökologie. *Berliner geowissenschaftliche Abhandlungen* **E2**, 1–164.

Mehl, D. (1996). Phylogenie und Evolutionsökologie der Hexactinellida (Porifera) im Paläozoikum. *Geologische-Paläontologische Mitteilungen Innsbruck* **4**, 1–55.

Mehl, D. and Fürsich, F. T. (1997). Middle Jurassic Porifera from Kachchh, western India. *Paläontologische Zeitschrift* **71**, 19–33.

Mehl, D. and Reiswig, H. M. (1991). The presence of flagellar vanes in choanomeres of Porifera and their possible phylogenetic implications. *Zeitschrift für zoologische und systematische Evolutionsforschung* **29**, 312–319.

Mehl, D., Reitner, J. and Reiswig, H. M. (1994). Soft tissue organization of the deep-water hexactinellid *Schaudinnia arctica* Schulze, 1900 from the arctic seamount Vesterisbanken (Central Greenland Sea). *Berliner geowissenschaftliche Abhandlungen* **E13**, 301–313.

Mehl, D., Müller, I. and Müller, W. E. G. (1998). Molecular biological and paleontological evidence that Eumetazoa, including Porifera (sponges), are of monophyletic origin. *In* "Sponge Sciences: Multidisciplinary Perspectives" (Y. Watanabe and N. Fusetani, eds), pp. 133–156. Springer-Verlag, Tokyo.

Mehl-Janussen, D. (1999). Die frühe Evolution der Porifera: Phylogenie und Evolutionsöekologie der Poriferen im Paläozoikum mit Schwerpunkt der desmentragenden Demospongiae ("Lithistide"). *Münchner Geowissenschaftliche Abhandlungen, Reihe A* **37**, 1–72.

Mehl-Janussen, D. (2000). Schwämme in der fossilen Uberlieferung. *Zentralblatt für Geologie und Paläontologie, Teil I* **1/2**, 15–26.

Mezitt, A. A. and Lucas, W. J. (1996). Plasmodesmal cell-to-cell transport of proteins and nucleic acids. *Plant Molecular Biology* **32**, 251–273.

Miller, K. E. and Joshi, H. C. (1996). Tubulin transport in neurons. *Journal of Cell Biology* **133**, 1355–1366.

Minchin, E. A. (1909). Sponge-Spicules. A summary of present knowledge. *Ergebnisse und Fortschritte der Zoologie* **2**, 171–274.

Morris, R. H., Abbott, D. P. and Haderlie, E. C. (1980). "Intertidal Invertebrates of California". Standford University Press, Stanford, CA.

Mostler, H. (1986). Beitrag zur stratigraphischen Verbreitung und phylogenetischen Stellung der Amphidiscophora und Hexasterophora (Hexactinellida, Porifera). *Mitteilungen der Österreichische Geologische Gesellschaft* **78**, 319–359.

Müller, W. E. G., Conrad, J., Zahn, R. K., Steffen, R., Uhlenbruck, G. and Müller, I. (1984). Cell adhesion molecule in the hexactinellid *Aphrocallistes vastus*: Species-unspecific aggregation factor. *Differentiation* **26**, 30–35.

Müller, W. E. G., Müller, I., Rinkevich, B. and Gamulin, V. (1995). Molecular evolution: Evidence for the monophyletic origin of multicellular animals. *Naturwissenschaften* **82**, 36–38.

Müller, W. E. G., Koziol, C., Müller, I. M. and Wiens, M. (1999). Towards an understanding of the molecular basis of immune responses in sponges: The marine demosponge *Geodia cydonium* as a Model. *Microscopy Research and Technique* **44**, 219–236.

Müller, W. E. G., Krasko, A., Le Pennec, G., Steffen, R., Wiens, M., Ammar, M. S. A., Müller, I. M. and Schröder, H. (2003). Molecular mechanism of spicule formation in the demosponge *Suberites domuncula*: Silicatein–collagen–myotrophin. *Progress in Molecular and Subcellular Biology* **33**, 195–221.

Müller, W. E.G, Wendt, K., Geppert, C., Wiens, M., Reiber, A. and Schröder, H. C. (2006). Novel photoreception system in sponges? Unique transmission properties of the stalk spicules from the hexactinellid *Hyalonema sieboldi*. *Biosensors and Bioelectronics* **21**, 1149–1155.

Neuweiler, M. (2000). Untersuchungen an Kieselnadeln rezenter hexactinellider Schwämme. Diplomarbeit. Universität Stuttgart, Stuttgart.

Nichols, S. A. (2005). An evaluation of support for order-level monophyly and interrelationships within the class Demospongiae using partial data from the large subunit rDNA and cytochrome oxidase subunit. I. *Molecular Phylogenetics and Evolution* **34**, 81–96.

Nicol, J. A. C. (1967). "The Biology of Marine Animals". Pitman, London.

Okada, Y. (1928). On the development of a hexactinellid sponge, *Farrea sollasii*. *Journal of the Faculty of Science of the Imperial University of Tokyo* **4**, 1–29.

Owen, R. (1843). Description of a new genus and species of sponge (*Euplectella aspergillum*, Ow.). *Transactions of the Royal Society of London* **3**, 203–206.

Pavans de Ceccatty, M. (1982). *In vitro* aggregation of syncytia and cells of a hexactinellida sponge. *Developmental and Comparative Immunology* **6**, 15–22.

Pavans de Ceccatty, M. (1989). Les éponges, à l'aube des communications cellulaires. *Pour la Science* **142**, 64–72.

Pavans de Ceccatty, M. and Mackie, G. (1982). Genèse et évolution des interconnexions syncytiales et cellulaires chez une éponge Hexactinellide en cours de réagrégation après dissociation *in vitro*. *Comptes Rendus de l'Academie des Sciences Paris* **294**, 939–944.

Pavans de Ceccatty, M., Gargouil, M. and Coraboeuf, E. (1960). Les réactions motrices de l'éponge siliceuse *Tethya lyncurium* à quelques stimulations expérimentales. *Vie et Milieu* **11**, 594–600.

Perez, T. (1996). La rétention de particules par une éponge hexactinellide, *Oopsacas minuta* (Leucopsacasidae): le rôle du réticulum. *Comptes Rendus de l'Academie de Sciences Paris, Sciences de la Vie* **319**, 385–391.

Peterson, K. J. and Butterfield, N. J. (2005). Origin of the Eumetazoa: Testing ecological predictions of molecular clocks agains the Proterozoic fossil record. *Proceedings of the National Academy of Sciences of the United States of America* **102**, 9547–9552.

Pisera, A. (1997). Upper Jurassic siliceous sponges from the Swabian Alb: Taxonomy and paleoecology. *Paleontologia Polonica* **57**, 3–216.

Pisera, A. and Lévi, C. (2002). 'Lithistid' Demospongiae. *In* "Systema Porifera: A Guide to the Classification of Sponges" (J. A. Hooper and R. W. M. Van Soest, eds), pp. 299–301. Kluwer Academic/Plenum Publishers, New York.

Pomel, A. (1872). Paléontologie ou Description des animaux fossiles de la province d'Oran, Zoophytes. 5é Fascicule, Spongiaires. A.D. Perrier, Oran.

Puce, S., Calcinai1, B.,Bavestrello,G., Cerrano, C., Gravili, C. and Boero, F. (2005). Hydrozoa (Cnidaria) symbiotic with Porifera: A review. *Marine Ecology* **26**, 73–81.

Pueschel, C. M. (1989). An expanded survey of the ultrastructure of red algal pit plugs. *Journal of Phycology* **25**, 625–636.

Reed, C., Greenberg, M. J. and Pierce, S. K. (1976). The effects of the cytochalasins on sponge cell reaggregation: New insights through the scanning electron microscope. *In* "Aspects of Sponge Biology" (F. W. Harrison and R. R. Cowden, eds), pp. 153–169. Academic Press, Inc., New York.

Reid, R. E. H. (1958). A monograph of the Upper Cretaceous Hexactinellida of Great Britain and Northern Ireland. Part I. *Palaeontographical Society Monograph* **111**, 1–46.

Reid, R. E. H. (2003). Hexactinellida: General morphology and classification. *In* "Treatise on Invertebrate Paleontology, Part E, Porifera, Revised, Vol. 2: Introduction to the Porifera" (R. L. Kaesler, ed.), pp. 127–134. The Geological Society of America, Boulder, Colorado.

Reiswig, H. M. (1971a). The axial symmetry of sponge spicules and its phylogenetic significance. *Cahiers de Biologie Marine* **12**, 505–514.

Reiswig, H. M. (1971b). *In situ* pumping activities of tropical demospongiae. *Marine Biology* **9**, 38–50.

Reiswig, H. M. (1979a). Histology of Hexactinellida (Porifera). *Colloques internationaux du Centre national du Recheches scientifiques* **291**, 173–180.

Reiswig, H. M. (1979b). A new sponge with rapid contraction systems. Annual Meeting of the Canadian Society of Zooology. May, 1979. Laval University, Quebec. p. 83a.

Reiswig, H. M. (1990). *In situ* feeding in two shallow water hexactinellid sponges. *In* "New Perspectives in Sponge Biology" (K. Rutzler, ed.), pp. 504–510. Smithsonian Institution Press, Washington, DC.

Reiswig, H. M. (1991). New perspectives on the hexactinellid genus *Dactylocalyx* Stuchbury. *In* "Fossil and Recent Sponges" (J. Reitner and H. Keupp, eds), pp. 7–20. Springer-Verlag, Berlin.

Reiswig, H. M. (1992). First Hexactinellida (Porifera) (glass sponges) from the Great Australian Bight. *Records of the South Australian Museum* **26**, 25–36.

Reiswig, H. M. (2002a). Family Diapleuridae Ijima, 1927. *In* "Systema Porifera: A Guide to the Classification of Sponges" (J. N. A. Hooper and R. W. M. Van Soest, eds), pp. 1383–1385. Kluwer Academic/Plenum Publishers, New York.

Reiswig, H. M. (2002b). Order Aulocalycoida Tabachnick and Reiswig, 2000. *In* "Systema Porifera: A Guide to the Classification of Sponges" (J. N. A. Hooper and R. W. M. Van Soest, eds), p. 1361. Kluwer Academic/Plenum Publishers, New York.

Reiswig, H. M. (2006). Classification and phylogency of Hexactinellida (Porifera). *Canadian Journal of Zoology* **84**(2), 195–204.

Reiswig, H. M. (2004). Hexactinellida after 132 years of study—what's new? *Bollettino dei Musei e degli Istituti Biologici dell'Università di Genova* **68**, 71–84.

Reiswig, H. M. and Mackie, G. O. (1983). Studies on hexactinellid sponges III. The taxonomic status of Hexactinellida within the Porifera. *Philosophical Transactions of the Royal Society of London Series B* **301**, 419–428.

Reiswig, H. M. and Mehl, D. (1991). Tissue organization of *Farrea occa* (Porifera, Hexactinellida). *Zoomorphology* **110**, 301–311.

Reiswig, H. M. and Mehl, D. (1994). Reevaluation of *Chonelasma* (Euretidae) and *Leptophragmella* (Craticulariidae) (Hexactinellida). *In* "Sponges in Time and Space" (R. W. M. van Soest, T. M. G. van Kempen and J.-C. Braekman, eds), pp. 151–165. AA Balkema, Rotterdam.

Reiswig, H. M. and Tsurumi, M. (1996). A new genus and species of Aulocalycidae, *Leioplegma polyphyllon* (Porifera: Hexactinellida) from the Blake Ridge off South Carolina. *Bulletin of Marine Science* **58**, 764–774.

Reitner, J. and Mehl, D. (1995). Early paleozoic diversification of sponges: New data and evidences. *Geologische-Paläontologische Mitteilungen Innsbruck* **20**, 335–347.

Reitner, J. and Mehl, D. (1996). Monophyly of the Porifera. *Verhandlungen des naturwissenschaftlichen Vereins in Hamburg* **36**, 5–32.

Rigby, J. K. (1986). Sponges of the Burgess Shale, Middle Cambrian, British Columbia, Canada. *Paleontographica Canadiana* **2**, 1–105.

Rigby, J. K. (2004). Classification. *In* "Treatise on Invertebrate Paleontology, Part E, Porifera, Revised. Vol. 3. Porifera (Demospongiae, Hexactinellida, Heteractinida, Calcarea)" (R. L. Kaesler, ed.), pp. 1–8. The Geological Society of America, Boulder, Colorado.

Robinow, C. and Kellenberger, E. (1994). The bacterial nucleoid revisited. *Microbiological Reviews* **58**, 211–232.

Salomon, D. and Barthel, D. (1990). External choanosome morphology of the hexactinellid sponge *Aulorosella vanhoeffeni* Schulze and Kirkpatrick 1910. *Senckenbergiana maritima* **21**, 87–99.

Sancetta, C. (1989). Spatial and temporal trends of diatom flux in B.C. fjords. *Journal of Plankton Research* **11**, 503–520.

Sandford, F. (2003). Physical and chemical analysis of the siliceous skeletons in six sponges of two groups (Demospongiae and Hexactinellida). *Microscopy Research and Technique* **62**, 336–355.

Sarà, M., Bavestrello, G., Cattaneo-Vietti, R. and Cerrano, C. (1998). Endosymbiosis in sponges: Relevance for epigenesis and evolution. *Symbiosis* **25**, 57–70.

Sarikaya, M., Fong, H., Sunderland, N., Flinn, B. D., Mayer, G., Mescher, A. and Gaino, E. (2001). Biomimetic model of a sponge-spicular optical fiber—mechanical properties and structure. *Journal of Materials Research* **16**, 1420–1428.

Schmid, V. and Bally, A. (1988). Species specificity in cell-substrate interactions in medusae. *Developmental Biology* **129**, 573–581.

Schmid, V., Bally, A., Beck, K., Haller, M., Schlage, W. K. and Weber, C. (1991). The extracellular matrix (mesoglea) of hydrozoan jellyfish and its ability to support cell adhesion and spreading. *Hydrobiologia* **216/217**, 3–10.

Schmidt, O. (1870). "Grundzüge einer Spongien-fauna des Atlantischen Gebietes". Engelmann, Leipzig.

Schmidt, O. (1880). "Die Spongien des Meerbusen von Mexico (und des Caraibischen Meeres)". Zweites (Schluss-) Heft. Gustav Fisher, Jena.

Schröder, H. C., Krasko, A., Gundacker, D., Leys, S. P., Müller, I. M. and Müller, W. E. (2003a). Molecular and functional analysis of the (6–4) photolyase from the hexactinellid *Aphrocallistes vastus*. *Biochimica et Biophysica Acta* **1651**, 41–49.

Schröder, H. C., Krasko, A., Le Pennec, G., Adell, T., Wiens, M., Hassanein, H., Müller, I. and Müller, W. E. G. (2003b). Silicase, an enzyme which degrades biogenous amorphous silica: Contribution to the metabolism of silica deposition in the demosponge *Suberites domuncula*. *Progress in Molecular and Subcellular Biology* **33**, 249–268.

Schuchert, P. and Reiswig, H. M. (2006). *Brinckmannia hexactinellidophila*, n. g., n. spec., a hydroid living in tissues of glass sponges of the reefs, fjords and seamounts of Pacific Canada and Alaska. *Canadian Journal of Zoology* **84**, 564–572.

Schultze, M. J. S. (1860). "Die Hyalonemen. Ein Beitrag zur Naturgeschichte der Spongien". Adolph Marcus, Bonn.

Schulze, F. E. (1880). On the structure and arrangement of the soft parts in *Euplectella aspergillum*. *Royal Society of Edinburgh Transactions* **29**, 661–673.

Schulze, F. E. (1886). Über den Bau und das System der Hexactinelliden. *Abhandlungen der Königlichen Akademie der Wissenschaften zu Berlin (Physikalisch-Mathematisch Classe)* **1886**, 1–97.

Schulze, F. E. (1887). Report on the Hexactinellida collected by H.M.S. "Challenger" during the years 1873–1876. *Zoology* **21**, 1–513,104 pls.

Schulze, F. E. (1899). Zur Histologie der Hexactinelliden. *Siztungsberichte der Deutches Akademie von Wissenschaft* **14**, 198–209.

Schulze, F. E. (1900). Berichte der Commission für oceanographische Forschungen. Zoologische Ergebnisse. XVI. Hexactinelliden des Rothen Meeres. *Denkschriften der Kaiserlichen Akademie der Wissenschaften. Mathematisch-Naturwissenschaftliche Classe* **69**, 311–324.

Schulze, F. E. (1904). Hexactinellida. *Wissenschaftliche ergebnisse der Deutschen Tiefsee-Expedition auf dem Dampfer "Valdivia" 1898–1899* **4**, 1–266, 52 pls.

Schwab, D. W. and Shore, R. E. (1971). Mechanism of internal stratification of siliceous sponge spicules. *Nature (London)* **232**, 501–502.

Sebens, K. P. (1987). The ecology of indeterminate growth in animals. *Annual Review of Ecology and Systematics* **18**, 371–407.

Shimizu, H., Cha, J., Stucky, G. D. and Morse, D. E. (1998). Silicatein alpha: Cathepsin L-like protein in sponge biosilica. *Proceedings of the National Academy of Sciences of the United States of America* **95**, 6234–6238.

Simpson, T. L. (1984). "The Cell Biology of Sponges". Springer-Verlag, New York.

Simpson, T. L., Garrone, R. and Mazzorana, M. (1983). Interaction of germanium with biosilicification in the freshwater sponge *Ephydatia mülleri*: Evidence of localized membrane domains in the silicalemma. *Journal of Ultrastructure Research* **85**, 159–174.

Simpson, T. L., Langenbruch, P. F. and Scalera-Liaci, L. (1985). Silica spicules and axial filaments of the marine sponge *Stelletta grubii* (Porifera, Demospongiae). *Zoomorphology* **105**, 375–382.

Sollas, W. J. (1879). Observations on *Dactylocalyx pumiceus* (Stutchbury), with a description of a new variety, *Dactylocalyx Stutchburyi*. *Journal of the Royal Microscopical Society* **2**, 122–133.

Steiner, M., Mehl, D., Reitner, J. and Erdtmann, B.-D. (1993). Oldest entirely preserved sponges and other fossils from the Lowermost Cambrian and new facies reconstruction of the Yangtze platform (China). *Berliner geowissenschaftliche Abhandlungen* **E9**, 293–329.

Sundar, V. C., Yablon, A. D., Grazul, J. L., Ilan, M. and Aizenberg, J. (2003). Fibre-optical features of a glass sponge. *Nature* **424**, 899–900.

Tabachnick, K. R. (1991). Adaptation of the hexactinellid sponges to deep-sea life. *In* "Fossil and Recent Sponges" (J. Reitner and H. Keupp, eds), pp. 378–386. Springer-Verlag, Berlin, Heidelberg.

Tabachnick, K. R. and Janussen, D. (2004). Description of a new species and subspecies of *Fieldingia*, errection of a new family fieldingidae and a new order Fieldingida (Porifera; Hexactinellida; Hexasterophora). *Bollettino dei Musei e degli Istituti Biologici dell'Università di Genova* **68**, 623–637.

Tabachnick, K. R. and Lévi, C. (1997). Amphidiscophoran Hexasterophora. *Berliner geowissenschaftliche Abhandlungen* **20**, 147–157.

Tabachnick, K. R. and Menshenina, L. L. (1999). An approach to the phylogenetic reconstruction of Amphidiscophora (Porifera: Hexactinellida). *Memoirs of the Queensland Museum* **44**, 607–615.

Tabachnick, K. R. and Reiswig, H. M. (2002). Dictionary of Hexactinellida. *In* "Systema Porifera: A Guide to the Classification of Sponges" (J. N. A. Hooper and R. W. M. Van Soest, eds), pp. 1224–1229. Kluwer Academic/Plenum Publishers, New York.

Takahashi, K., Baba, S. A. and Murakami, A. (1973). The 'excitable' cilia of the tunicate. *Ciona intestinalis. Journal of the Faculty of Sciences, University of Tokyo* **(Section IV)** **13**, 123–137.

Thomson, C. W. (1868). On the "vitreous" sponges. *Annals and Magazine of Natural History* **1**, 114–132.

Thomson, C. W. (1869). On *Holtenia*, a genus of vitreous sponges. *Philosophical Transactions of the Royal Society of London* **159**, 701–720.

Travis, K. D. F., Franciois, C. J., Bonar, L. C. and Glimcher, M. J. (1967). Comparative studies of the organic matrices of invertebrate mineralized tissues. *Journal of Ultrastructure Research* **18**, 519–550.

Uriz, M. J., Turon, X., Becerro, M. A. and Agell, G. (2003). Siliceous spicules and skeleton frameworks in sponges: Origin, diversity, ultrastructural patterns, and biological functions. *Microscopy Research and Technique* **62**, 279–299.

Vacelet, J. (1996). Deep-sea sponges in a Mediterranean cave. *In* "Deep-sea and Extreme Shallow-water Habitats: Affinities and Adaptations" (F. Uiblein, J. Ott and M. Stachowitsch, eds), Biosystematics and Ecology Series 11, pp. 299–312.

Vacelet, J., Boury-Esnault, N. and Harmelin, J. (1994). Hexactinellid Cave, a unique deep-sea habitat in the scuba zone. *Deep-sea Research* **41**, 965–973.

Van de Vyver, G. and Buscema, M. (1981). Capacités morphogènes des cellules d'éponges dissociées. *Annales de la Societé Royale Zoologique de Belgique* **111**, 9–19.

Van Tright, H. (1919). Contribution to the physiology of the fresh-water sponges (Spongillidae). *Tijdschrift voor Diergeneeskunde* **17**, 1–220.

Vosmaer, G. and Pekelharing, C. (1898). Über die Nahrungsaufnahme bei Schwämmen. *Archives für Anatomie und Physiologie*, 168–186.

Wachtmann, D. and Stockem, W. (1992a). Microtubule- and microfilament-based dynamic activities of the endoplasmic reticulum and the cell surface in epithelial cells of *Spongilla lacustris* (Porifera, Spongillidae). *Zoomorphology* **112**, 117–124.

Wachtmann, D. and Stockem, W. (1992b). Significance of the cytoskeleton for cytoplasmic organization and cell organelle dynamics in epithelial cells of fresh-water sponges. *Protoplasma* **169**, 107–119.

Wachtmann, D., Stockem, W. and Weissenfels, N. (1990). Cytoskeletal organization and cell organelle transport in basal epithelial cells of the freshwater sponge *Spongilla lacustris*. *Cell and Tissue Research* **261**, 145–154.

Wakelam, M. J. O. (1988). Myoblast fusion—A mechanistic analysis. *In* "Current Topics in Membranes and Transport" (F. Bronner and N. Duzgunes, eds), pp. 88–107. Academic Press, London.

Weaver, J. C. and Morse, D. E. (2003). Molecular biology of demosponge axial filaments and their roles in biosilicification. *Microscopy Research and Technique* **62**, 356–367.

Weissenfels, N. and Landschoff, H. W. (1977). Bau und Funktion des Süsswasserschwamms *Ephydatia fluviatilis* L. (Porifera). IV. Die Entwicklung der monaxialen SiO₂-Nadeln in Sandwich-Kulturen. *Zoologische Jahrbücher Abteilung für Anatomie* **98**, 355–371.

Weissenfels, N., Wachtmann, D. and Stockem, W. (1990). The role of microtubules for the movement of mitochondria in pinacocytes of fresh-water sponges (Spongillidae, Porifera). *European Journal of Cell Biology* **52**, 310–314.

Weltner, W. (1882). Beiträge zur Kenntnis des Spongien Inaugural-Dissertation, Universität Freiburg, Freiburg im Breisgau, Germany.

Werner, B. (1959). Das Prinzip des endlosen Schleimfilters beim Nahrungserwerb wirbelloser Meerestiere. *Internationale Revue des gesamten Hydrobiologie* **44**, 181–216.

Whitney, F., Conway, K., Thomson, R., Barrie, J. V., Krautter, M. and Mungov, G. (2005). Oceanographic habitat of sponge reefs on the Western Canadian Continental Shelf. *Continental Shelf Research* **25**, 211–226.

Wiens, M. and Müller, W. E. G. (2006). Cell death in Porifera: Molecular players in the game of apoptotic cell death in living fossils. *Canadian Journal of Zoology* **84**, 307–321.

Wildon, D. C., Thain, J. F., Minchin, P. E. H., Gubb, I. R., Reilly, A. J., Skipper, Y. D., Doherty, H. M., O'Donnell, P. J. and Bowles, D. J. (1992). Electrical signalling and systemic proteinase inhibitor induction in the wounded plant. *Nature* **360**, 62–65.

Willenz, P. and Van de Vyver, G. (1982). Endocytosis of latex beads by the exopinacoderm in the fresh water sponge *Ephydatia fluviatilis*: An *in vitro* and *in situ* study in SEM and TEM. *Journal of Ultrastructure Research* **79**, 294–306.

Willenz, P. and Van de Vyver, G. (1984). Ultrastructural localization of lysosomal digestion in the fresh water sponge *Ephydatia fluviatilis*. *Journal of Ultrastructure Research* **87**, 13–22.

Wilson, H. V. (1907). On some phenomena of coalescence and regeneration in sponges. *Journal of Experimental Zoology* **5**, 245–258.

Woodland, W. (1908). Some observations on the scleroblastic development of hexactinellid and other silicious sponge spicules. *Quarterly Journal of Microscopical Science* **52**, 139–157.

Wyeth, R. C. (1999). Video and electron microscopy of particle feeding in sandwich cultures of the hexactinellid sponge *Rhabdocalyptus dawsoni*. *Invertebrate Biology* **118**, 236–242.

Wyeth, R. C., Leys, S. P. and Mackie, G. O. (1996). Use of sandwich cultures for the study of feeding in the hexactinellid sponge *Rhabdocalyptus dawsoni* (Lambe, 1892). *Acta Zoologica* **77**, 227–232.

Yahel, G., Eerkes-Medrano, D. I. and Leys, S. P. (2007b). Size independent selective filtration of ultraplankton by hexactinellid glass sponges. *Aquatic Microbial Ecology* (in press).

Yahel, G., Whitney, F., Reiswig, H. M., Eerkes-Medrano, D. I. and Leys, S. P. (2007a). *In situ* feeding and metabolism of glass sponges (Hexactinellida, Porifera) studied in a deep temperate fjord with a remotely operated submersible. *Limnology and Oceanography* **52** (in press).

Yu, W., Centonze, V. E., Ahmad, F. J. and Baas, P. W. (1993). Microtubule nucleation and release from the neuronal centrosome. *Journal of Cell Biology* **122**, 349–359.

The Northern Shrimp (*Pandalus borealis*) Offshore Fishery in the Northeast Atlantic

Elena Guijarro Garcia

Marine Research Institute, 101 Reykjavík, Iceland

ADVANCES IN MARINE BIOLOGY VOL 52
© 2007 Elsevier Ltd. All rights reserved

0065-2881/07 $35.00
DOI: 10.1016/S0065-2881(06)52002-4

This chapter describes the development and current situation of the offshore shrimp fisheries in Iceland, Greenland, Svalbard, Jan Mayen and the Norwegian Barents Sea area, with information on the biology of Pandalus borealis *and its relation to the environment. Some additional information about the inshore shrimp fisheries of Iceland and Greenland of relevance to this study is also included. The Icelandic offshore shrimp fishery started in 1975 and has formed between 68% and 94% of the annual catch of shrimp since 1984. Landings peaked at 66,000 tons in 1997. The offshore fleet increased threefold from 1983 to 1987, and catch per unit of effort doubled. The first signs of overfishing were detected in 1987, when the first total allowable catch (TAC) was set, and catches decreased during the next few years despite the discovery of new fishing grounds. Good recruitment allowed catches to rise steadily from 1990 to 1996. However, catches and stock index have decreased markedly since then, with a minimum catch for the period 1998–2003 of 21,500 tons in 2000. It has been suggested that predation by cod is an important factor affecting shrimp stock size, but mortality from predation is slightly lower than fishing mortality, so that the impact of fishing cannot be disregarded. The Greenland offshore shrimp fishery is one of the largest in the North Atlantic and it generates 90% of the export value of the country. The fishery started in 1970 in West Greenland with landings of 1200 tons, but since 1974 it has formed between 59% and 89% of the annual shrimp catch. In 2004, landings reached 113,000 tons and the fishable stock was estimated at 300,000 tons. The significant spatial expansion of the fishery from the original fishing grounds off the Disko Island area to all of the West coast south of 75°N and the fleet improvement over the past three decades have made possible this spectacular growth. Other fishing grounds off the East coast have been fished since 1978, mostly by foreign vessels. Catches in this area oscillated between 5000 and 15,000 tons during the period 1980–2004. The main problem of the shrimp fishery in Greenland is its overlapping with nursery areas of redfish, Greenland halibut, cod and other groundfish species, some of which show declining trends of biomass and abundance. This led to the implementation in 2000 of sorting grids and laws that forbid fishing when the bycatch exceeds legal limits. However, it is likely that ecological processes only partially understood, such as the trophic web and hydrography of the area, greatly influence the stock abundance of the demersal community. The offshore Norwegian fishery started in 1973. The main fishing grounds are off Svalbard and in the Barents Sea. Catches at Jan Mayen have never exceeded 5% of the total annual catch of northern shrimp. Large fluctuations in catches and stock size are the main characteristic of this fishery. Stock size seems to be largely dependent on the annual hydrographic variability in the area and trends in abundance of predator species, especially cod. However, shrimp mortality due to predation has been estimated to be the same as fishing mortality, and therefore fishing probably accounts for part of the observed variability in stock size. Large populations of juvenile cod, haddock, redfish and Greenland halibut are often found on the shrimp fishing grounds. The implementation of sorting grids in 1991 and a bio-economical model in 1993 to estimate allowable maximum catches of the commercial bycatch species have not solved the bycatch problem. All the commercial fish species present on the shrimp grounds are currently below safe biological limits. This is the only fishery within the studied area that is not regulated by means of a TAC system.*

1. INTRODUCTION

Pandalid shrimp are exploited for food on a wide scale in the northern hemisphere. The main target is *Pandalus borealis*, although there are smaller coastal fisheries directed to *Pandalus montagui* in the United Kingdom and Canada (Mistakidis, 1957; Fisheries and Oceans Canada, 2003). International Council for the Exploration of the Sea (ICES) data on English and Scottish landings of *Pandalus borealis* in Subarea IV include small catches of other Pandalid shrimp (ICES, 2005). However, *Pandalus montagui* is caught as bycatch species and it constitutes a very small proportion of the inshore catch within the study area considered in this chapter. The expansion of the fishing grounds and technological advances have transformed the local shrimp fisheries of the North Atlantic into industrial fisheries, involving factory trawlers equipped with single, double and sometimes triple trawls that supply an important export market. On average, the annual landings of the offshore *Pandalus borealis* fisheries in the North Atlantic during the period 1996–2004 amounted to 245,400 tons, of which 197,700 (80.56%) were caught within the study area considered in this chapter, the Barents Sea and waters off Greenland and Iceland (ICES, 2006; NAFO, 2006).

Shrimp are mostly fished with otter trawls, which are among the most destructive fishing gears (see Jennings and Kaiser, 1998; Løkkeborg, 2005 for reviews; Watling and Norse, 1998). The main physical impacts of otter trawls are the alteration of bottom topography due to relocation of boulders and to the furrows and depressions created by the trawl doors and rockhoppers (Schwinghamer *et al.*, 1998; Humborstad *et al.*, 2004), in addition to the loss of sediment surface characteristics, such as traces of bioturbation, caused by the suspension and redeposition of sediment. The persistence of these alterations depends largely on the size of the gear, substrata type, depth, wave action, currents and biological activity. In general, the furrows will last longer in deep and sheltered areas with soft bottom substrata (Tuck *et al.*, 1998).

Among the biological impacts, the most important is the increased and differential mortality of both target and non-target species. Most demersal fish and invertebrate species are found in mixed assemblages of both target and non-target species. The bottom trawls are not very selective since they are designed to sweep and create a disturbance on the seabed that will lead the animals into the net. Consequently, animals in the path of the gear are at risk of being captured, displaced, killed or seriously injured (Bergman and Santbrink, 2000). Mortality caused by fishing operations is highly variable between different phyla and even between higher taxonomic groups (Lindeboom and De Groot, 1998; Bergman and Santbrink, 2000; Hill and Wassenberg, 2000; Pranovi et al., 2001), but species that reach a large size, have a slow growth rate, reach maturity at a large size and greater age are the most sensitive to fishing activities (Greenstreet and Rogers, 2000). Fishing mortality can be categorised in two groups: (a) direct mortality, caused to captured animals and (b) indirect mortality, caused to animals that are not caught but can be killed due to injuries and/or habitat alteration caused by the fishing gear and fishing activities.

Benthic macrofauna suffers substantial indirect mortality (Eleftheriou and Robertson, 1992; Jennings and Kaiser, 1998; Lindeboom and De Groot, 1998; Prena et al., 1999). Densities of many scavenging species increase greatly in recently trawled areas as they gather to predate on the wake of dead and injured animals left behind by fishing gear (Caddy, 1973; Kaiser and Spencer, 1994; Kaiser and Ramsay, 1997). In fact, it has been suggested that most of the damage inflicted to benthic macrofauna during fishing activities takes place on the seabed (Lindeboom and De Groot, 1998). The killing or injuring of animals at the surface and uppermost layers of the sediment is the most drastic impact in soft bottoms (Gislason, 1994). The removal of fragile and long-lived sessile habitat-forming epifaunal species is the longer lasting and most important impact in hard bottoms, as loss of three-dimensional biogenic structures can cause severe habitat degradation and renders it unsuitable for its associated fauna (Collie et al., 2000; Hall-Spencer and Moore, 2000; Kaiser et al., 2000).

Long-lasting alterations in benthic communities may take place shortly after fishing. It is suspected that in hard-bottom habitats, the major impact of towed fishing gear takes place the first time an area is fished (Jennings and Kaiser, 1998). This is especially true in the case of habitats formed by animals consisting of fragile structures such as corals, sponges and hydrozoans (Currie and Parry, 1999; Prena et al., 1999). In any case, the enhanced mortality of certain taxa has been shown to lead to decrease of species richness and diversity (Currie and Parry, 1996; Collie et al., 1997; Tuck et al., 1998), marked changes in faunal composition of benthic communities (Kaiser et al., 2000), alterations of predator–prey relationships (Pauly et al., 1998), species replacement (Jennings and Kaiser, 1998) and important changes in the ecosystem (Lindeboom and De Groot, 1998).

Kelleher (2005) estimated that at a global scale, shrimp trawls are the fishing gear with the largest discard rates (measured as proportion of the catch that is discarded) of 62.3% on average. In American waters, the shrimp fishery in the Gulf of Mexico and Pacific inshore shrimp fishery have the highest discard rate at approximately 44%. The shrimp beam trawl fishery in Canada has an estimated discard rate of 29.1%, but it is much lower in the shrimp otter trawl (7.8%). Kelleher (2005) identified Scandinavian countries as having low discard rates, thanks in part to low diversity catches and high manufacturing capacity to process pelagic species into fish meal. The discard rate of the Pandalid fisheries was estimated to be 5.4% in Canada, Norway and Iceland, equalling 13,000 tons of discarded bycatch. The use of sorting grids, stricter management, increased retention of bycatch species and reduced effort have recently led to a significant decrease of bycatch in Atlantic shrimp fisheries.

Bycatch is the most investigated impact of shrimp fishing within the study area covered by this chapter, with regular data collection and analysis carried out mostly over the past decade. Much of the available literature focuses on bycatch species of commercial interest, especially gadoids and redfish (Pálsson and Thorsteinsson, 1985; Hylen and Jacobsen, 1987; Carlsson and Kanneworff, 1998a; Rätz, 1998a; Reithe and Aschan, 2004; Storr-Paulsen and Jørgensen, 2005). Pedersen *et al.* (2000) described a study that aimed to investigate the benthic fauna in shrimp fishing areas in Nuuk (Greenland) and Kongsfjorden (Svalbard) to estimate the impact of shrimp trawling, but they did not present any results or conclusions in their report.

As described above, it is evident that shrimp fisheries can have dramatic and deleterious effects on benthic habitats both physically and biologically, with clear consequences for marine biodiversity. However, the shrimp fisheries operating off Iceland, Greenland and Svalbard have not been the subject of a single, comprehensive review covering activities from the early part of the twentieth century to the present day. This chapter addresses this need by describing in detail the shrimp fisheries in these key North Atlantic regions in relation to shrimp biology, ecology and environment to elucidate the nature, extent and impact of exploitation on wild shrimp populations.

2. THE ECOLOGY OF *PANDALUS BOREALIS* WITHIN THE STUDY AREA

2.1. Geographical distribution

Pandalus borealis Krøyer, 1838 (northern shrimp, deep-sea prawn, pink shrimp) was formerly regarded as a circumboreal species of discontinuous distribution, present in both the North Atlantic and North Pacific (Figure 1). The Pacific subspecies, *Pandalus borealis eous* Makarov 1935, was raised to

1 cm

Figure 1   Pandalus borealis.

species level as *Pandalus eous* by Squires (1992), but there is still some reluctance to accept this division (Bergström, 2000). In the North Atlantic, *Pandalus borealis* ranges along the American East coast from Cape Cod to Baffin Island. In West Greenland it is found from Cape Farewell (59°5′N) up to Melville Bay (75°N), whereas in East Greenland it is known to extend at least to Angmagssalik (just below 70°N). It is also present in the Davis Strait, Iceland, the Faeroes, the western fjords of Norway and Sweden, the Skagerrak, the northern North Sea, the Barents Sea, Svalbard, north of Novaya Zemla and the Kara Sea (Figure 2). In all, the species is present between 35°N and 82°N (Horsted and Smidt, 1956; Shumway *et al.*, 1985; Bergström, 2000). In the North Pacific, *Pandalus eous* is found from Point Barrow in Alaska to the Columbia River in Washington state, including the Aleutian Islands and the British Columbia coast. It is also present in the Bering and Chuchki seas. In the Northwest Pacific, it is found off the Kamchatka Peninsula, Southeast Siberia, Korea and south to Honshu, Japan (Figure 2) (Bergström, 2000). The geographical distribution of *Pandalus borealis* overlaps largely with that of *Pandalus montagui*, a smaller cold-water pandalid that ranges slightly farther south on the European coast to the Thames estuary and is mostly found in shallower waters than *Pandalus borealis* (A. J. Southward, personal communication; Folmer, 1996; Kanneworff, 2003).

At a small spatial scale, larvae, juveniles and adults of *Pandalus borealis* may show different distributions. In Greenland, larvae and juveniles are normally found in relatively shallow waters or near land. Preference for deeper waters shown by adult shrimp in the study area might be related to temperature, which is more stable as depth increases. Juveniles can withstand a wider range of salinity and temperature than adults (Horsted and Smidt, 1956; Aschan, 2000a). Larval drift is thought to be crucial for recruitment to certain areas, as reported in Disko Bay, Greenland (Carlsson and Smidt, 1978). However, a study in the Barents Sea showed that shrimp larvae are retained in the area where the adult stock is (Pedersen *et al.*, 2003). Shrimp are known to migrate vertically, probably for feeding purposes, ascending in the water column in the evening and returning to the sea bottom in the morning (Horsted and Smidt, 1956; Barr, 1970; Wienberg, 1981; Shumway *et al.*, 1985; Bergström, 2000).

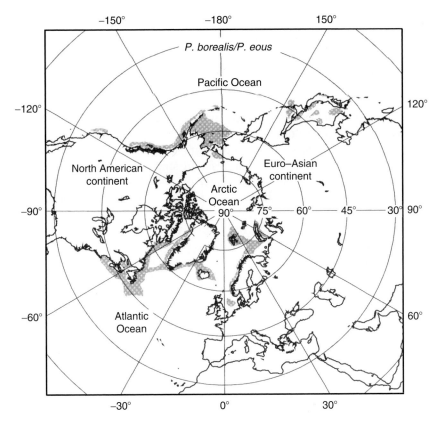

Figure 2 Geographical distribution of *Pandalus borealis* (Atlantic Ocean) and *Pandalus eous* (Pacific Ocean). Map from Bergström (2000).

The vertical distribution of *Pandalus borealis* ranges from 20 to 900 m. The depth of maximum abundance varies with latitude; the greatest concentrations being deeper at higher latitude (Horsted and Smidt, 1956; Shumway *et al.*, 1985). In West Greenland there are also differences in mean size at different depths, with larger shrimp normally being more abundant in deeper water (Smidt, 1969; Hassager, 1992). Additionally, there are seasonal migrations between different depths, probably a search for more favourable temperature conditions (Shumway *et al.*, 1985).

2.2. Environmental requirements

Temperature, substratum and salinity are important factors influencing the distribution of *Pandalus borealis* (Shumway *et al.*, 1985). The optimal temperature range seems to be between $-1.6°C$ and $8°C$, although they

are most common in waters above 0°C and they die at temperatures below −
1.6°C (Smidt, 1960; Shumway et al., 1985; Bergström, 2000). Temperatures
below 0°C do not impede spawning in the Barents Sea, but reproduction
appears to be erratic below −1°C. At these temperatures, there are very few
egg-bearing females and egg loss and embryo death are substantial (Lund,
1988a).

Temperature is a key factor for the survival of the Greenland shrimp stock
because here it is at the northern limit of its distribution. Therefore, even a
small but prolonged decrease in temperature can have drastic consequences.
Shrimp were indeed absent from Sisimiut and neighbouring fjords for
4–5 years after the unusually cold winter of 1948–1949 caused widespread
mortality (Horsted and Smidt, 1965; Smidt, 1965). The shrimp populations
in Greenland waters require stable water temperatures above 0°C and
therefore are most abundant in the deeper waters far from land (Horsted
and Smidt, 1956; Smidt, 1969). Development of larvae, juveniles and adults
is also positively correlated with water temperature, as higher temperatures
favour more frequent moulting. Length and age at female maturity are also
largely influenced by sea temperature. Shrimp in warm waters reach maturity
at a younger age and smaller size than shrimp in cold water (Shumway et al.,
1985; Nilssen and Hopkins, 1991; Skúladóttir and Pétursson, 1999).

Changes in environmental conditions may trigger a shift in local distribution of shrimp populations, and temperature has been linked to such changes
in Greenland and the Barents Sea, for example. In Greenland, it was observed that shrimp migrated into a fjord to avoid a mass of cold water
outside the fjord. There are also seasonal migrations of adults accompanying
water masses with optimal temperature, as is the case with berried females
(Horsted and Smidt, 1956; Carlsson and Smidt, 1978; Nilssen and Hopkins,
1991, Aschan, 2000a).

Salinity preferences appear to range from 33‰ to 35‰, but there are
records of Pandalus borealis found in areas with salinity as low as 23.4‰.
The preferred substratum is soft, muddy bottom or sandy silt (Horsted and
Smidt, 1956; Haynes and Wigley, 1969; Butler, 1971; Shumway et al., 1985;
Bergström, 2000). Shrimp seem to favour high particulate organic carbonate
(POC) in the sediment (Shumway et al., 1985, Bergström, 2000) and, in fact,
most fishing areas in Greenland are located within patches of high POC
(Ramseier et al., 2000).

Pedersen et al. (2002) did not find a good correlation between larval
shrimp abundance in West Greenland and temperature and salinity of
water masses, chlorophyll a concentrations or zooplankton abundance,
although Stickney and Perkins (1981) had found that diatoms and zooplankton constituted an important part of the stomach contents of Pandalus
borealis larvae. However, Pedersen et al. (2002) suggested that environmental
conditions affect year-class strength and patterns of larval transport in West

Greenland, as it has been shown for the *Pandalus borealis* larvae in the Barents Sea. In both areas there is large variability in year-class strength of age-1 shrimp, and no female stock-recruitment relationship. Pedersen *et al.* (2003) found that in the Barents Sea, shrimp settlement takes place every year in the Polar Front area, the location of which is influenced by the inflow of Atlantic water and varies annually; and the spatial distribution of modelled settled larvae showed a good correlation with the distribution of age-1 juvenile shrimp.

2.3. Biology

2.3.1. Nutrition and growth

Pandalus borealis are opportunist omnivores acting both as predators and scavengers (Shumway *et al.*, 1985). Stomach content analysis of shrimp in Greenland showed that they feed on small crustaceans, polychaetes, foraminifera, radiolarians, excretory pellets, sand and shrimp belonging to their own or other species (Horsted and Smidt, 1956). Samples from Norway contained, in addition to the former materials, remains of Holothuroidea, Porifera, Copepoda, tintinnids, diatoms and peridinians. This indicates that shrimp also feed in the water column during the nocturnal vertical migrations (Wollebæk, 1903; Horsted and Smidt, 1956).

Growth takes place all year round for males, although growth rates are lower in winter. All mature females grow during summer and sometimes also every second winter in areas where egg bearing occurs at 2-year intervals, such as in the offshore Icelandic populations (Skúladóttir *et al.*, 1991a). In inshore waters, growth may occur from spring to autumn, followed by a stagnation period during winter when no growth takes place (Nilssen and Hopkins, 1991). Growth rates depend on food availability, which shows wide seasonal variation in the Arctic and subarctic regions linked to temperature. Thus, growth rates are faster in warmer waters and slower in the northernmost locations such as Iceland, Svalbard and Greenland (Shumway *et al.*, 1985; Bergström, 2000). However, local and/or annual variations in water temperature, population density and recruitment may induce variability in growth rates and length at sex change between year-classes and/or close populations (Parsons *et al.*, 1989; Nilssen and Hopkins, 1991; Skúladóttir *et al.*, 1991a; Savard *et al.*, 1994; Skúladóttir, 1998a; Aschan, 2000a; Wieland, 2004a).

For example in the Barents Sea, where temperature is very variable between years and areas because it depends on the location of the Polar Front and the influx of warm Atlantic water, there are marked spatial and temporal variations in growth rates and age and/or length at sex change.

Local differences among water masses, such as those registered in several locations in Spitsbergen, seem to account for the 2-year difference in age at maturity observed on shrimp sampled in different locations (Rasmussen, 1953; Aschan, 2000a; Hansen and Aschan, 2000). Variable size and age at sex change has also been observed in Icelandic shrimp populations. Off the Northeast coast of Iceland, where north Icelandic winter water at 2–3°C and Arctic bottom water at <0°C are found under the warmer upper layer, shrimp spawn for the first time at 6 years of age and their L_{50} (length at which 50% of the females are mature) is 23.0–24.9 mm. On the other hand, Vestfirðir and Snæfellsnes, on the West coast, receive warm Atlantic water with the Irminger current at roughly 7°C and shrimp spawn at 3 years of age and L_{50} 17.6–19.8 mm and 17.2–22.1 mm, respectively. The largest shrimp are found in the Denmark Strait, where bottom temperature is 0–2°C and females attain maturity at L_{50} between 26.5 and 29.3 mm (Stefánsson, 1962; Skúladóttir et al., 1991a; Skúladóttir, 1993a, 1995a, 1998a; Vilhjálmsson, 1994; Skúladóttir and Pétursson, 1999). Similarly, the fishing grounds in Greenland show three different development patterns with sex change taking place at ages 4–6. The slowest type was found in the Disko Bay and Nuuk (Godthåb), with sex change at age 6, and the fastest type was found in Qaqortoq and Nanortalik, with sex change at age 4. In the intermediate area, Paamiut (Frederikshåb) and Aasiat (Egedesminde), sex change takes place at an intermediate age, as in Jan Mayen (Horsted and Smidt, 1956).

Age and growth estimates are based on length frequency data, as shrimp have no hard structures suitable for age estimation. The main problem of this method is the overlap between the older size classes due to lack of moulting during the ovigerous period, and the large variability of growth rates among regions, seasons, sexes and age classes (Shumway et al., 1985; Nilssen and Hopkins, 1991; Bergström, 2000). The deviation method from Sund (1930) as modified by Skúladóttir (1981) shows more year-classes than previous methods (Petersen, 1892; Sund, 1930). It consists of subtracting length frequency distribution (LFD) of a given sample from the mean LFD spanning several years, showing strong year-classes with positive deviations from the mean LFD.

2.3.2. Reproduction

A notable feature of most pandalid species is that they undergo a change of sex from male to female during their growth, as does the species under study in this chapter (Berkeley, 1930; Carlisle, 1959). The biology and taxonomy of *Pandalus borealis* and other species of the genus were reviewed by Bergström (2000).

Northern shrimp are protandric hermaphrodites, meaning they are born males and become females for the rest of their lives following a transitional phase, although it has been reported that some individuals may develop directly as females in the Gulf of Maine and also in a Swedish fjord (Haynes and Wigley, 1969; Bergström, 2000).

The males show two stages, either immature or mature, while the transitional phase has three stages and the female phase, three. The time at which sex change occurs seems to be related to individual size. As growth rates depend on food availability and temperature, age at sex change or maturity is very variable among the different shrimp populations, decreasing with increasing temperatures as described above (Allen, 1959; Horsted and Smidt, 1965; Shumway et al., 1985; Carlsson, 1997a; Carlsson and Kanneworff, 1999; Bergström, 2000). In the Denmark Strait, sex change starts at age 5 but takes place mostly at age 6 (Skúladóttir, 1997a), whereas shrimp off Svalbard change sex at age 5 compared to age 2 in southern Norway (Rasmussen, 1942).

Not much is known about mating, but it seems to take place offshore in summer or autumn within 36 hours after the female shrimp moult. Fertilisation is external and occurs right before the eggs are laid (Hoffman, 1973; Shumway et al., 1985; Bergström, 2000).

Spawning takes place once a year, between summer and early autumn, but in low temperature areas females may spawn every second year as a response to energetic demands derived from egg production (Teigsmark and Øynes, 1983a; Skúladóttir et al., 1991a). There are some variations in hatching between areas caused by differences in water temperature (Shumway et al., 1985). Thus, the Greenland stock between Disko Bay and Nanortalik spawns in July–August and hatches in April, but in areas where the water is colder, the period between spawning and egg hatching (ovigerous period) is longer (Horsted and Smidt, 1956; Bergström, 2000). Offshore shrimp populations in North Iceland have an ovigerous period of 10 months, compared to 5.5 months in the inshore populations off the West coast, and the female shrimp spawn every second year at bottom temperatures of 0°C (Skúladóttir et al., 1991a). Fecundity usually increases with body size as larger shrimp can carry more eggs, but varies with the size of the egg mass, that ranges between 600 and 5000 eggs, although figures around 2000 are most common. In some regions, size of the egg mass is inversely related to shrimp abundance and bottom temperature. Since shrimp attain larger sizes in lower temperatures, fecundity is higher in northernmost populations such as those in West Greenland and the Barents Sea (Horsted and Smidt, 1956; Shumway et al., 1985; Nilssen and Hopkins, 1991; Bergström, 1997). However, fecundity of the populations at their northernmost distribution limits may be negatively affected by low temperatures. At 73°N in West Greenland, only 50% of female shrimp carried eggs compared to 90% between 66°N and 68°5'N, although this difference may partly be due to the spawning being every second year. However, between 20% and 40% of the eggs north of 71 °N were dead. These differences are thought to be related to the lower water temperature in the locations where *Pandalus borealis* has its northernmost distribution limit off West Greenland (Lund, 1988a, 1990a).

2.3.3. Life cycle

The pelagic larvae go through six zoea stages before becoming post-larvae. These are found in shallow waters, although actual depth varies among locations. In the Barents Sea, the greatest density of larvae was found between 10 and 25 m (Horsted and Smidt, 1956; Lysy, 1978; Haynes, 1983; Shumway *et al.*, 1985).

Post-larvae need six moults before they metamorphose into juveniles. The onset of male sexual characteristics in juveniles can be observed from approximately 6-mm carapace length (CL), when the male copulative structure starts developing. Development of the appendix masculina on the second pleopod starts when juveniles reach 6.5-mm CL (Figure 3). The time at which external indications of sexual differentiation take place seems to be determined by size and not age (Stickney and Perkins, 1977).

The development of sexual characters and the period of sex change are usually classified into seven stages, juveniles being the first. The morphology of the endopod of the first pleopod (Figure 4) is used to distinguish among mature males, transitionals and females. Females that have never spawned may have head roe but they still have the ventral external spines intact just before spawning. These external spines disappear after carrying the eggs between the pleopods for some months prior to hatching (Figure 5). Females may then enter a resting period, characterised by lack of roe, or become ready to spawn again on the following year, in which case they develop head roe some months before spawning (McCrary, 1971). Average life expectancy is variable among regions but it ranges between 3 and 8.5 years (Shumway *et al.*, 1985).

Figure 3 Development of the appendix masculina on the second pleopod of juvenile *Pandalus borealis*. Modified from Stickney and Perkins (1977) in Shumway *et al.* (1985).

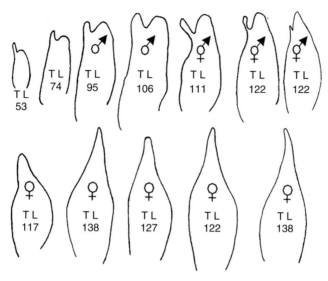

Figure 4 The first pair of pleopods change shape during the transition from male to female and can be used to determine sex of *Pandalus* shrimp. The illustration shows the shapes characteristic of males, transitional individuals and females. TL, total length measured in millimetre. Modified from Bergström (2000).

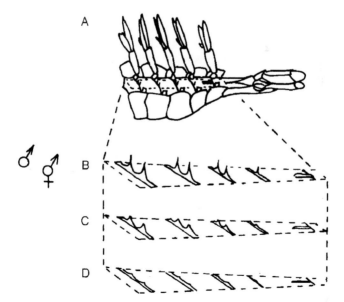

Figure 5 Diagrammatic view of *Pandalus borealis* showing the external spines. (A) Mature male with intact spines, (B) mature male and intersex with intact spines, (C) late-stage transitional, multiparous female with reduced spines and (D) multiparous female without spines. Modified from McCrary (1971) in Shumway *et al.* (1985).

2.3.4. Mortality

Mortality coefficient (Z) estimates range from 0.5 to 2.0 in the few available studies, with highest estimates in areas of high fishing intensity and predation. Mortality is also thought to be high among females following spawning (Shumway et al., 1985).

2.4. Population structure

The sex ratio and age composition of any population depends on natural and fishing mortality, recruitment, strength of each year-class and also season and location in those populations with horizontal migration patterns. Size composition within populations varies with depth, distance from shore, time of the day and season. Changes in abundance have been shown to be related to hydrographic conditions, predation and fishing pressure (Shumway et al., 1985).

2.5. Predators and parasites

Cod (*Gadus morhua*), Greenland halibut (*Reinhardtius hippoglossoides*), redfish (*Sebastes marinus*), eelpouts (*Lycodes* spp.), sea birds, that is Brünnich guillemot (*Uria lomvia*) and some marine mammals, such as harbour seals (*Phoca vitulina*), ringed seal (*Phoca hispida*), bearded seal (*Erignatus barbatus*), harp seal (*Phoca groenlandica*) and white or beluga whale (*Delphinapterus leucas*), are some of the main shrimp predators in the North Atlantic, with shrimp constituting an important part of their diets (Horsted and Smidt, 1956; Smidt, 1965; Carlsson and Smidt, 1978; Bowering et al., 1984; Shumway et al., 1985; Lilliendahl and Solmundsson, 1997; Bergström, 2000; Lilly et al., 2000).

   The parasite isopod *Phryxus abdominalis* has been identified from Greenland and Iceland samples. It is found under the tail in both sexes (Horsted and Smidt, 1956). Other known parasites are protozoans that infect muscle tissue and peridinians that cause egg mortality (Parsons and Khan, 1986; Stickney, 1978).

3. ICELAND

3.1. The fishery

Experimental shrimp fishing took place for the first time in Iceland in Ísafjarðardjúp, Vestfirðir (Northwest Iceland) in 1924 (Hallgrímsson, 1993a). However, the commercial fishery did not start until 1935, when facilities to process the catch in land became available. The fishery extended to Arnarfjörður, also in Vestfirðir, in 1938 and to Húnaflói (North Iceland) in 1965, after new fishing grounds had been found and processing facilities had been built (Hallgrímsson and Skúladóttir, 1979). The Norwegian trawl imported to fish shrimp was modified into the so-called 'Ísafjörður-trawl' by shortening the bridles significantly and keeping the headline 2.7–3.6 m from the bottom to minimise bycatch of juvenile and pelagic fish (Skúladóttir, 1969; Hallgrímsson, 1993b). Shrimp fishing boats of between 5 and 10 tons dominated in the early years of the fishery (Hermansson, 1980), increasing to around 13 tons on average during the period 1962–1970 (Skúladóttir, 1970).

Catches were limited by the number of people shelling the shrimp manually ashore, until the introduction of shelling machines in 1959 allowed the increase of fishing effort. Another significant change took place in 1967 when a bigger and more efficient trawl was introduced (22.9- to 27.4-m headline, compared to the 18.3-m headline of the old trawls). Fishing became profitable in areas where catchability had been very low with the old trawl, thus expanding the fishing grounds (Skúladóttir, 1970). The fishery gradually extended to other inshore areas around Iceland (Skúladóttir, 1971) and catches increased from <2000 tons during the period 1955–1968 to 2500–7800 tons between 1969 and 1984. From 1969 onwards the Marine Research Institute (MRI) of Iceland launched regular surveys with the goal of finding offshore shrimp grounds, as the species had been frequently recorded in deep-sea fishing grounds all around Iceland. The search was carried out between 1969 and 1993 with variable success (see Hallgrímsson, 1993b for review). Most new grounds were found off North and East Iceland (Eiríksson, 1971; Þorsteinsson and Eiríksson, 1971; Anonymous, 1975a, 1976a, 1978a, 1980a,b, 1984a, 1989; Einarsson, 1975; Hallgrímsson and Einarsson, 1978), although some very profitable fishing areas were also discovered in the Denmark Strait (Anonymous, 1977a). Fewer grounds were found south and west of Iceland where shrimp is absent or is too low in density to support a profitable fishery (Þorsteinsson, 1970; Anonymous, 1974a, 1977a, 1987a, 1994a; Hallgrímsson and Einarsson, 1978). A summary of the search surveys carried out during the period 1968–1993 is shown in Table 1.

The offshore shrimp fishery began in 1974. The shrimp catch from the inshore fishery provided 53–100% of the total annual shrimp landings until 1983, but since 1984 the offshore fishery has represented between 68% and 94% of annual catches every year. Shrimp catches increased by an order of magnitude, from the maximum of 7300 tons in 1973, prior to the onset of the offshore fishery, to 76,000 tons in 1995. The inshore landings peaked at nearly 12,000 tons in 1996, but they remained <8000 tons in most years, representing <15% of the total annual shrimp catch. Inshore shrimp catches have always been processed on land (data from Anonymous, 1979a, 1981a, 1982a, 1987b,

Table 1   The most relevant shrimp search surveys carried out in Icelandic waters

Year	Area	Main results	References
Inshore			
1968	East	150–773 kg hour^{-1} in Seyðisfjörður	Eiríksson, 1968; Skúladóttir, 1969
1969	East	Only one fjord with up to 60 kg hour^{-1}	Skúladóttir, 1970
	North	434 kg hour^{-1} in a small area	Eiríksson, 1968
	West	28–62 kg hour^{-1} in Jökuldjúp	Skúladóttir and Eiríksson, 1970; Þorsteinsson, 1970
1970	West	Shrimp found but widely distributed, possibilities of commercial fishing in Jökuldjúp	Þorsteinsson, 1970; Þorsteinsson and Eiríksson, 1971
1971	East	3300 kg in 20 min in Berufjörður	Anonymous, 1973
1973	North	100–200 kg hour^{-1} in Bakkaflóadjúp and Héraðsdjúp	Anonymous, 1974a
1974	West	>100 kg hour^{-1} in Kolluáll, Breiðafjörður 140 kg in 10 min in Fleteyjarsund	Anonymous, 1975a
	Northwest	640 kg hour^{-1} in Tálknafjörður	
1975	Northeast	Good fishing ground in Öxarfjörður	Anonymous, 1976a
1977	East	Good grounds in Reyðarfjörður and Seyðisfjörður	Anonymous, 1978a
1983	East	Good grounds in Reyðarfjörður	Anonymous, 1984a
1996	North	Up to 250 kg hour^{-1} in Eyjafjörður, but catch decreased quickly	Anonymous, 1997a
1997	North	157 kg hour^{-1} in a small area in Eyjafjörður, unsuitable for commercial fishing	Anonymous, 1998
Offshore			
1969	North	10–150 kg hour^{-1} in Eyjafjarðaráll	Skúladóttir, 1970
1970	North	250–300 kg hour^{-1} east off Grímsey	Þorsteinsson, 1970
1971	Northwest	320 kg hour^{-1} in Djúpáll	Anonymous, 1973
1973	Southwest	600 kg hour^{-1} west off Eldey	Anonymous, 1974a
1974	North	Large shrimp (140 individuals per kilogram) found in Reykjafjarðaráll	Anonymous, 1975a
1975	East	85–200 kg hour^{-1} in two areas off Vattarnes, very little or nothing found in other areas	Anonymous, 1976a
1976	North	20–300 kg hour^{-1}	Anonymous, 1977a
	East	14–160 kg hour^{-1}	
	West	20–130 kg hour^{-1}	

(Continued)

Table 1   (Continued)

Year	Area	Main results	References
	Denmark Strait	150 kg hour^{-1}, of very large shrimp, 50 individuals per kilogram	
1977	North and Northeast	Good banks on Norðurkantur and in Öxafjarðardjúp, 240 kg hour^{-1}	Anonymous, 1978a
1978	Denmark Strait	Good catches on Dohrn bank and Stredebanke	Anonymous, 1979a
1979	North	Good catches between Kolbeinsey and Grímsey	Anonymous, 1980b
1983	East	Shrimp grounds found on Tangaflak, Héraðsflóadjúp, Seyðisfjarðardjúp, Berufjarðarál, Bakkafjörður, Lónsdjúp and Hornafjarðardjúp	Anonymous, 1984a
1986	Southwest	Good catch in Grindavíkurdjúp	Anonymous, 1987a
1988	East	Up to 600 kg hour^{-1} on Rauðatorg	Anonymous, 1989
1993	Southwest	Large shrimp found in Grindavíkurdjúp and Skerjadjúp	Anonymous, 1994a

Note: The MRI of Iceland carried out several search surveys every year on inshore and offshore grounds, using both research and private vessels. An exhaustive list of all search surveys would be too extensive, but the literature listed in this table includes information on every search survey carried out.

1988a, 1996a, 2005). Most of the offshore catch is taken in waters off North Iceland (Figure 6). This chapter focuses on the offshore fishery because catches from the inshore fishery constitute a minor part of the annual landings.

The sudden increase of catches in 1984 was to a great extent stimulated by the catch restrictions that were applied to most fish stocks for the first time in 1984, encouraging many skippers to switch to shrimp trawling (Anonymous, 1985a). At the same time, new fishing grounds were discovered in East Iceland. These factors led to the growth of the fleet to 160 vessels, much more than in previous years. The number of vessels in the offshore shrimp fishery increased gradually until the MRI recommended decreasing effort because both biomass estimates and mean size of shrimp had declined (Anonymous, 1988a, 1990, 1991, 1993; Skúladóttir, 1989a).

The first trawl survey for stock assessment of offshore shrimp was carried out in 1987 and a total allowable catch (TAC) was established for the offshore grounds. By this time the first signs of overfishing had been detected, with

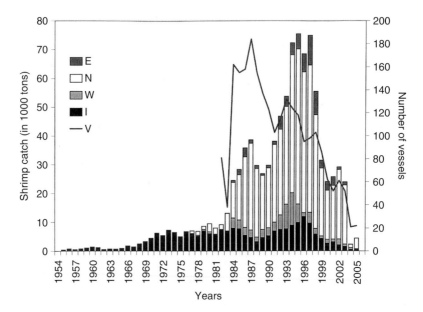

Figure 6 Annual shrimp catch (in thousands of tons) in Icelandic waters and number of vessels that took part in the offshore fishery since 1982. E, East; N, North; W, West; I, Inshore; V, Vessels in the fishery. Data from Anonymous (1979a, 1981a, 1982a, 1985a, 1986a, 1987b, 1988a, 1996a, 2005).

catches and standardised catch per unit of effort (CPUE) (estimated as catch per hour with a 1600-mesh trawl) declining despite the discovery of new shrimp grounds (Figure 7). Standardised CPUE decreased in North and Northwest Iceland from 126 kg hour^{-1} in 1985 to 94 kg hour^{-1} in 1989; and from 199 kg hour^{-1} in 1986 to 91 kg hour^{-1} in 1989 in Northeast Iceland (Anonymous, 2005).

Good recruitment from the year-classes 1987–1990 allowed the allocation of higher quotas from 1991 to 1994–1995, and catches, CPUE and fleet size rose accordingly (Figure 8) (Anonymous, 1994b). The discovery of shrimp grounds in the Flemish Cap in 1996 attracted part of the fleet and fishing pressure on the Icelandic offshore stock lightened somewhat during 1996 (Anonymous, 1997b). Nevertheless, much smaller quotas had to be implemented from 1997 to 1998 onwards. From 1990 to 1994, the stock production model seems to have underestimated the stock size, what probably benefited the shrimp stock. It must be borne in mind that the quota for any given year was calculated according to the stock size estimated the previous year. Thus, the 60,000 tons quota for the fishing year 1994–1995 was based on the stock and CPUE indices from 1993 to 1994 (roughly 70,000 tons). Actually, the MRI initially advised a TAC of 45,000 tons in spring

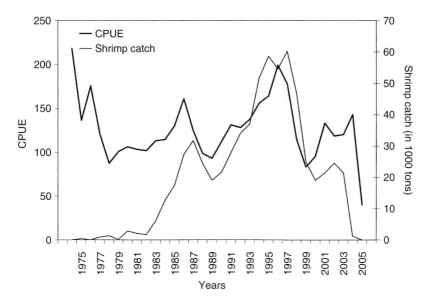

Figure 7 Trends of catch per unit of effort (CPUE), standardised to trawl size, and offshore shrimp catch from North and East Iceland during 1979–2003. Data from Anonymous (2005).

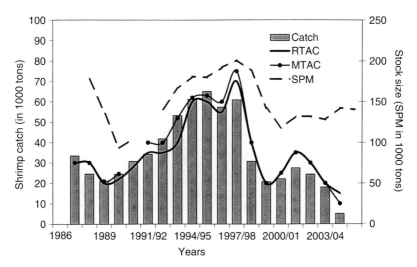

Figure 8 Annual offshore shrimp catch in Icelandic waters, total allowable catch (TAC) recommended by the Marine Research Institute (RTAC), TAC set by the Ministry of Fisheries (MTAC) and shrimp stock index as estimated with the stock production model (SPM). From 1991 onwards, TACs were set for fishing years, that is, from September to August. Data from Anonymous (2003a, 2004a).

1994, but since the stock showed signs of increase in 1993 and 1994 due to good recruitment and the decline of the immature cod stock, the TAC was raised to 60,000 tons in November 1994, assuming that the shrimp stock would remain large (Anonymous, 1994c, 1995). The original quota of 40,000 tons advised by the MRI for the fishing season 1995–1996 was increased by the Ministry of Fisheries to 63,000 without consulting the MRI (Anonymous, 1996b).

The time lag between data collection, estimation of quota and actual fishing explains partly why catches remained stable and quotas increased while the stock declined after 1998 according to the stock production model (Anonymous, 2003a). Another added problem that contributed to the very high quota in 1997–1998 was the underestimation of the cod stock. Cod was incorporated into the model in 1991, but only the immature part of the cod stock was considered. However, according to the annual report of the MRI, the increased migrations of cod to northern Icelandic waters led to an important decrease of the shrimp stock. In fact, the biomass of the cod and shrimp stocks follows asynchronous trends (Figure 9) and from 2000 onwards the total cod stock index estimated for North Iceland has been used in the stock production model (U. Skúladóttir, personal communication).

According to estimates obtained with a new statistical model for marine ecosystems, the Globally applicable Area-Disaggregated General Ecosystem Toolbox (GADGET), the average mortality rate of the shrimp stock due to

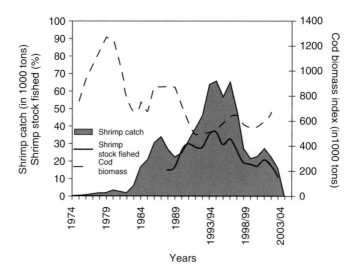

Figure 9   Trends of offshore shrimp catch in Iceland and percentage of the shrimp stock fished annually according to the stock production model (left *y*-axis) compared to cod biomass index for Iceland waters, including fish of age 3 and older (right *y*-axis). Data from Anonymous (2005).

predation by cod during the period 1983–2004 amounts to 0.41, whereas the average mortality due to fishing is 0.19. However, the estimates obtained in 2003 with the same model for the average mortality rates due to predation by cod and fishing during the period 1987–2003 were 0.26 and 0.46, respectively. With the figures used this year in the stock assessment, the mortality rates for the same period (1987–2003) are 0.52 (predation) and 0.31 (fishing mortality) (Anonymous, 2003a and 2006). Mortality due to predation is based on cod abundance as estimated from the annual groundfish survey. This large variation is due to the different values fed into the model for the back-calculations, and it is not possible to discern what estimate is more accurate (Figure 10). The trends of cod biomass, shrimp stock index and shrimp catch in the offshore grounds for the period 1983–2005 show that reductions in the cod stock are often synchronous with increases in the shrimp stock index and shrimp catches. It is, therefore, also possible that untimely increases in quota might have contributed partly to the observed stock reductions (Figure 11).

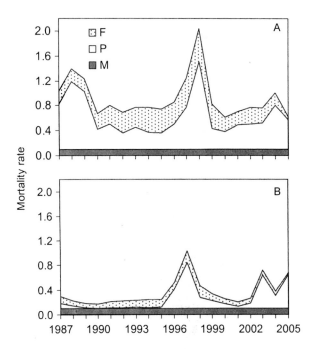

Figure 10 Shrimp mortality during the period 1987–2005 for individuals fully recruited to the fishery (>20-mm carapace length). (A) Mortality rates estimated with cod abundance data from the groundfish survey, (B) mortality rates estimated with cod abundance data from the shrimp survey. M, natural mortality; P, mortality due to predation by cod; F, fishing mortality. Modified from Anonymous (2006).

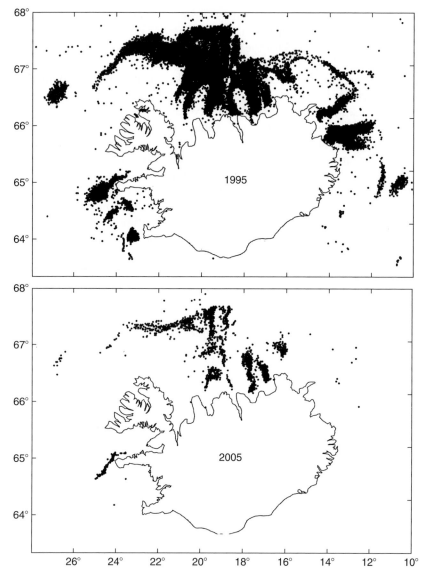

Figure 11   Distribution of the Icelandic shrimp fishery in 1995 and 2005.

3.2. Management

The Icelandic inshore and offshore shrimp fisheries have been managed since their inception by a license system, mandatory logbooks, minimum mesh size and minimum landing size (CL > 15 mm and CL > 13 mm in the offshore and inshore fisheries, respectively). The abundance of undersized specimens in landings must be below 30% (Anonymous, 1981b; U. Skúladóttir, personal communication).

A TAC system was implemented for the inshore fishery in 1964 (van den Hoonaard, 1992) and for the offshore fishery in 1987 (Table 2). In the inshore fishery, the quotas are based on the stock index estimated from survey data and they are distributed among the fleet. The vessels receive the same percentage of the TAC every year. In the offshore fishery, the TAC

Table 2   Overview of management measures for the inshore and offshore shrimp stocks in Iceland

Year	Management measure	References
Inshore		
1960	Logbooks mandatory	Hallgrímsson, 1977; Hallgrímsson and Skúladóttir, 1979
1962	Minimum mesh size 32 mm	Skúladóttir, 1969
	License mandatory	Hallgrímsson, 1977
1962–1967	TAC in Ísafjarðardjúp, Arnarfjörður and Húnaflói	Jónsson, 1990; Skúladóttir, personal communication
1974	TAC in all inshore shrimp fishing areas	Jónsson, 1990; van den Hoonaard, 1992
	Minimum mesh 36 mm	Hallgrímsson, 1977
1975	Closure of areas if bycatch of juvenile fish exceeds reference level	Anonymous, 1978a
1990	ITQ for all vessels	Skúladóttir, personal communication
1995	22-mm Nordmøre sorting grid used in Eldey, Skjálfanði and Snæfellsness	Anonymous, 1996a; Skúladóttir, personal communication
Offshore		
1970	Logbooks and licenses mandatory	Skúladóttir, personal communication
	32 mm minimum mesh size	Anonymous, 1981b
1974	36 mm minimum mesh size	Skúladóttir, personal communication
1987	TAC	Anonymous, 1993
1990	ITQ system	Skúladóttir, personal communication
1995	22-mm Nordmøre sorting grid mandatory	Skúladóttir, personal communication
1998–	7- to 9-mm sorting grid or 8-m long 40-mm square mesh cod-end mandatory in areas where >30% of shrimp is <15 mm	Skúladóttir, personal communication

recommended by the MRI is built on the stock production model and/or GADGET, which take into account cod predation, recruitment, CPUE and survey indices. The internal transferable quota (ITQ), which means that vessels can buy and sell quota among themselves, was implemented in Iceland in 1990. However, the ITQ is rather limited because the transference must be done among vessels registered in the same area (U. Skúladóttir, personal communication).

The use of 7- to 9-mm sorting grids for juvenile shrimp, or alternatively an 8-m long, 40-mm square mesh cod-end to minimise catches of juvenile shrimp is implemented by the Ministry of Fisheries. The regulations specifying the areas in which sorting grids or a square mesh cod-end must be used are released each year and have been published since the late 1990s (U. Skúladóttir, personal communication).

The bycatch of juvenile fish in the inshore fishery is limited by closing areas where the abundance of juvenile fish exceeds a quoted reference level that is estimated from the bycatch figures obtained during the inshore shrimp stock assessment survey. The acceptable maximum bycatch of juvenile fish is implemented by regulations from the Ministry of Fisheries and depends on species and age group. Bycatch of adult fish in the offhsore fishery seems to have ceased after the implementation of sorting grids in 1995. The Nordmøre grid is the most widely used sorting device in Iceland (Figure 12).

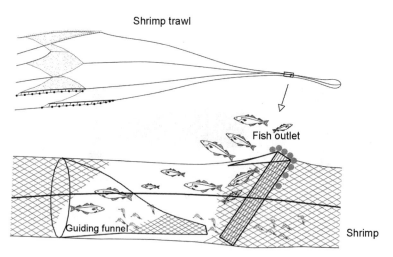

Shrimp trawl

Fish outlet

Guiding funnel

Shrimp

Figure 12 The Nordmøre grid. The funnel directs the fish and shrimp to the grid. The shrimp go through but fish are directed to the escape exit. Figure supplied by B. Isaksen.

3.3. Research

Research has focused mainly on stock assessment, analysis of biological data, compilation of fishery statistics dating to the early years of the inshore fishery, but stock identification based on genetic studies and size at sex change and the trophic relationship cod-shrimp have also been investigated (Table 3). Furthermore, in the early years of the offshore fishery, some research was also devoted to the use of shorting machines to discard undersized shrimp (Þorsteinsson, 1969; Eiríksson et al., 1974). The MRI carries out annual standardised inshore and offshore shrimp stock assessment surveys since 1973 and 1988, respectively. The main goal of the surveys is to collect biological data to investigate the population structure and to estimate the stock biomass index. Sampling of juvenile shrimp to estimate the abundance of shrimp 1- and 2-year old is carried out with a bag attached to the side of the trawl. In addition, surveys gather information on fish bycatch, especially gadoids, Greenland halibut and redfish. Fishing vessels also supply shrimp samples quite often, although this is not mandatory.

As the inshore fishery grew during the early years, the development of more selective and efficient gear to prevent the catch of juvenile fish and undersized shrimp became a priority because the use of bigger trawls and the expansion of the fishery into new areas increased greatly the bycatch of juvenile fish, especially gadoids (Pálsson and Thorsteinsson, 1985). Pálsson (1976) estimated that the shrimp fishery caused the loss of 32% and 25% (or 2.5 and 1.3 million individuals), respectively, of the year-class 1974 of cod and haddock in Ísafjarðardjúp. New management measures were introduced to minimise bycatch. Closure of certain inshore areas where bycatch is large was regulated following the introduction to Iceland in 1967 of larger trawls to fish in deeper waters that required a higher towing speed (Thorsteinsson, 1992). Closures did not affect total shrimp catch and reduced markedly the mortality of juvenile fish (Pálsson and Thorsteinsson, 1985).

The first deep-sea shrimp fishing experiments carried out in Iceland took place in 1969 and 1970, and used a trawl with two cod-ends, one above the other, separated by a fine mesh. The shrimp would be channelled to the upper cod-end, whereas fish would move to the lower one. The larger mesh size of the lower cod-end would allow smaller benthic invertebrates and juvenile fish to escape. This gear proved to be very successful after some minor modifications, although in muddy bottoms it was necessary to trawl with the lower cod-end open. However, results were not as good when this same gear was tried in deeper waters for a combined Nephrops-shrimp fishery with harder substrata, and no further research was done on the double cod-end trawl (Þorsteinsson, 1969, 1970; Anonymous, 1971; Thorsteinsson, 1973).

More shrimp trawls of both Icelandic and foreign design were tested in the following years. Some of them had bobbins that decreased the bycatch of

Table 3 Research carried out on the Icelandic shrimp stock

Year	Research	References
1965–1969	Marking-recapture to investigate migrations	Skúladóttir, 1969
1967	Sorting machine to discard smaller shrimp	Þorsteinsson, 1969
1968–	Improvement of fishing gear to decrease catches of juvenile shrimp and fish	
	Double cod-end trawl[1]	Þorsteinsson, 1970[1], 1976[2,4],
	Trial of different trawl designs[2]	1980[4,6], 1981[6], 1991[8], 1993[8], 1994a[8], 1994b[8], 1995[6];
	Shrimp trawl without wings[3]	Anonymous, 1971[1], 1973[2],
	Mesh size experiments[4]	1974b[4,5], 1975b[4], 1976b[4,6],
	Shortening of ground ropes and bridles[5]	1977b[3,6], 1978b[3]; Thorsteinsson, 1973[1,2,5], 1974[4],
	Increased net slack in side panels[6]	1980[4,6], 1981[6], 1989[7], 1992[7]
	Use of square mesh in cod-end[7]	
	Use of sorting grids[8]	
1968–	Inshore and offshore surveys to collect biological and catch data	Anonymous (MRI annual reports Hafrannsóknir and Fjölrit, since 1968).
1968–1993	Search for fishing grounds	See Table 1
1974	Trawl with double cod-end to fish shrimp and fish simultaneously	Eiríksson *et al.*, 1974
1974–1983	Shrimp larvae	Anonymous, 1976a, 1979a, 1981c
1976–1995	Estimation, management and survival of bycatch	Pálsson, 1976; Pálsson and Thorsteinsson, 1985; Thorsteinsson, 1995
1980–	Shrimp sampling from offshore grounds	Anonymous, 1981b
1981–	Fisheries data from Denmark Strait (Icelandic EEZ)	Jónsson and Hallgrímsson, 1981; Hallgrímsson and Skúladóttir, 1984, 1985, 1988; Skúladóttir, 1989b, 1991a, 1992a,b, 1993b, 1995b, 1996a, 1997a, 1998b, 1999, 2004
1984	Tagging experiments in West Iceland	Skúladóttir, 1985a
1987–	Stock assessment for offshore grounds	Skúladóttir, 1989a; Hallgrímsson, 1993b; Skúladóttir *et al.*, 1995; Anonymous (Annual reports Hafrannsóknir) since 1987
1989–1991	Trophic relationship cod-shrimp	Magnússon and Pálsson, 1989, 1991

(Continued)

Table 3   (Continued)

Year	Research	References
1993–	Shrimp growth and female sexual maturity	Skúladóttir and Stefánsson, 1985; Skúladóttir *et al.*, 1991a; Skúladóttir, 1993a, 1995a, 1998a; Skúladóttir and Pétursson, 1999
1994–	Multi-species models for stock assessment	Stefánsson *et al.*, 1994, 1998; Stefánsson and Pálsson, 1997, 1998; Begley and Howell, 2004; Taylor and Stefánsson, 2004
1998	Management of the cod-capelin-shrimp complex	Jakobsson and Stefánsson, 1998
	Genetic population studies	Jónsdóttir *et al.*, 1998

demersal species by keeping the towed net above the bottom. Results were not very satisfactory, however, because the gear that reduced most the bycatch of juvenile fish was rather heavy, had an excessive drag when used from smaller boats and it captured between 30% and 40% less shrimp than other gears (Anonymous, 1973; Thorsteinsson, 1973, Þorsteinsson, 1976). A shrimp trawl without wings, designed to minimise catch of juvenile gadoids, was tried in 1975 and 1976 in several fishing grounds with varying results (Anonymous, 1977b, 1978b).

Further research on gear selectivity included the following studies:

1. Effect of using different mesh sizes to avoid the capture of undersized shrimp without jeopardizing catches (Anonymous, 1974b, 1975b, 1976b; Thorsteinsson, 1974, 1980; Þorsteinsson, 1976, 1980).

2. Shortening of ground ropes and bridles, which resulted in decreased catch of both shrimp and bycatch species (Thorsteinsson, 1973; Anonymous, 1974b).

3. Increasing the net slack of the side panels. The idea was to let the net become loose, so the mesh would not close during towing and small shrimp and fish could escape through the side panels. With 10% net slack in the side panels, the escape rate of small shrimp increased without loss of bigger shrimp, and, quite unexpectedly, average catches were higher than with the conventional trawl. This gear was successfully used for 10–20 years in the inshore fishery (Anonymous, 1976b, 1977b; Þorsteinsson, 1980, 1981, 1995; Thorsteinsson, 1980, 1981).

4. Use of square mesh in the cod-end. Even during towing, square mesh remains open, facilitating the escape of small shrimp and 0-group gadoids. The results were highly satisfactory. Average size of shrimp in the catch

increased, and many fishermen adopted this gear (Thorsteinsson, 1989, 1992). Square mesh cod-end became mandatory in all inshore and certain offshore fishing grounds in 2003 (Table 2).

Further development to prevent fish bycatch consisted of equipping shrimp trawls with sorting grids with vertical bars spaced 22 mm apart. The Nordmøre grid is the most widely used and it is installed several metres ahead of the cod-end. It leads to an opening in the upper part of the trawl. Shrimp pass through the grid and into the cod-end, whereas fish are directed to the opening and are thus able to escape. The use of the grid was strongly recommended except in areas where shrimp are very large, as in the Dohrn Bank in the Denmark Strait (Þorsteinsson, 1991, 1993, 1994a,b,c).

As the importance of cod in shrimp management became clear (Pálsson, 1983, 1994; Magnússon and Pálsson, 1989, 1991), the modelling department of the MRI in Iceland was working on the development of BORMICON, a multi-species model that uses ecological information and variables such as cannibalism, predation and growth, among other variables, to investigate stock trends in Boreal ecosystems (Stefánsson and Pálsson, 1998). A newer model, GADGET (Begley and Howell, 2004), has also been used over the last years in shrimp stock assessment.

Genetic studies on shrimp samples from the inshore and offshore fisheries and from the Denmark Strait revealed significant differences between the three stocks, suggesting that they be managed as three biological units (Jónsdóttir et al., 1998). Research on trophic relationships of cod revealed that shrimp is an important prey, although its consumption varies with cod size and age, season and availability of other prey species (Pálsson, 1983; Magnússon and Pálsson, 1989, 1991).

3.4. Bycatch in the Icelandic shrimp fishery

The available bycatch data from the commercial fleet are restricted to the marketable fish species above the legal minimum capture size that were landed by vessels with quota for those species (Table 4). Shrimp fishing vessels without quota for species other than shrimp had to purchase quota before landing the marketable bycatch. There are no records on catches of under-sized fish or non-marketable species by the shrimp fleet. Bycatch of market-able species above the minimum capture size seems to have ceased after the implementation of Nordmøre sorting grids in 1995, although this does not mean that there is no bycatch. It is most likely that small-sized fish species and juvenile fish are caught in shrimp trawls as the mesh size is 40 mm and the sorting grids are designed to release fish >15–20 cm approximately. However, scientific or fishery observers sampled the bycatch of the offshore commercial

Table 4   Accidental catch, expressed in tons, declared by the offshore shrimp fleet during the period 1983–2005 in North, West and East Iceland

Species	Common name	North	East	West
Amblyraja radiata	Starry skate	0.55		
Anarhichas lupus	Atlantic wolffish	5.84	0.09	10.52
Anarhichas minor	Spotted wolffish	118.93	0.58	29.21
Clupea harengus	Herring			1.88
Cyclopterus lumpus	Lumpsucker	0.34		0.41
Dipturus batis	Skate	0.13		0.28
Gadus morhua	Cod	4057.03	49.46	461.18
Glyptocephalus cynoglossus	Witch		0.18	
Hippoglossoides platessoides	Long rough dab	5.66	0.40	4.70
Hippoglossus hippoglossus	Halibut	116.49	3.08	28.17
Limanda limanda	Dab	0.20		
Lophius piscatorius	Monkfish	1.96	0.18	1.89
Mallotus villosus	Capelin	38.90		
Melanogrammus aeglefinus	Haddock	191.08	10.01	467.78
Merlangius merlangus	Whiting	33.55		
Microstomus kitt	Lemon sole	0.06	0.50	
Molva dypterigya	Blue ling		0.18	
Molva molva	Ling	0.40	7.86	0.05
Pleuronectes platessa	Plaice	2.56	0.15	6.16
Pollachius virens	Saithe	19.28	0.77	0.92
Reinhardtius hippoglossoides	Greenland halibut	5277.26	57.59	328.46
Sebastes marinus	Redfish	265.39	21.07	51.84
Sebastes mentella oceanus	Deep-water redfish	0.10		
Unidentified		1236.49	10.32	236.46
Total		11,372.18	162.41	1629.89

Note: Only 0.5% and 0.003% of the landings in North and West Iceland respectively were captured after 1995, when the use of the sorting grid was made mandatory.
Soruce: Data from the MRI database.

fleet very rarely. This was done sporadically before the implementation of the sorting grid, and it focused on counting juvenile redfish.

Landings of bycatch by the commercial fleet and shrimp survey records of the most abundant and marketable demersal fish species during the period 1983–2005 are shown in Figure 13. Survey data are not directly comparable to data from the commercial fleet because most species are not weighed during surveys and only the number of individuals caught in each haul is recorded. Both biomass and abundance trends show great variability, and they are often asynchronous. For some species, such as spotted wolffish and long rough dab, the bycatch trends in the survey mirror those of the commercial bycatch but with 1-year delay despite the fact that most shrimp fishing and the surveys take place in summer and therefore the time lag between fishery and survey data is at most 3 months. Low bycatch of fish by the commercial fleet in years of high abundance of a given species can be explained by the fact that the

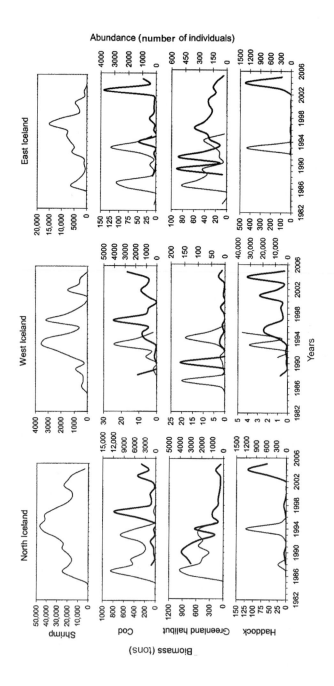

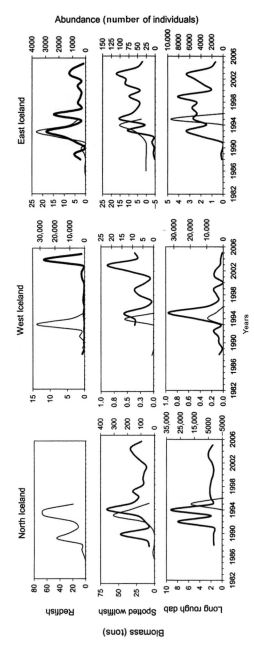

Figure 13 Trends of the most abundant marketable demersal fish species captured as bycatch in the offshore shrimp fishery and shrimp surveys off North, West and East Iceland during the period 1983–2005. Landings of the commercial shrimp fleet are expressed as biomass (tons, left *y*-axis) and they are represented by the thin line. Captures recorded during offshore shrimp surveys are expressed as abundance (number of individuals, right *y*-axis) and they are represented by the thick line. Data collection of bycatch species is available since 1988.

biomass data we have refer only to landings and not to the actual catch. Another possibility is that the fish were undersized. The low abundance of certain species registered during shrimp surveys in years of increased bycatch landings from the commercial fleet, as is the case of cod in North Iceland between 1994 and 1998 or haddock in East Iceland between 1992 and 1994, can be due to a mismatch of sampling locations and fishing marks. Bycatch trends in the Icelandic shrimp fishery are currently being investigated in detail and therefore further analysis of these data is out of the scope of this chapter (E. Guijarro Garcia *et al.*, unpublished data).

Greenland halibut, cod and redfish made up most landings of bycatch species by the commercial fleet in North and West Iceland, or 46.4%, 35.7% and 2.3% respectively in North Iceland, and 35.4%, 30.5% and 13.0% respectively in West Iceland. In East Iceland, haddock and cod were the most abundant bycatch species (28.7% and 28.3% of the bycatch landings, respectively), followed by Greenland halibut (20.1%) (Figure 14).

The shrimp trawl used in shrimp stock assessment surveys has the same mesh size as the trawl used in the commercial fleet, but it lacks the sorting grid. The bycatch recorded during surveys is therefore more diverse than the bycatch landed by the commercial fleet and it also includes adult fish (Table 5).

The bycatch recorded in the surveys includes several pelagic species that are listed in Table 5, and their proportion in the bycatch is shown in Figure 15. However, there is no further mention of these species because it is considered that the shrimp trawl does not sample them adequately. Considering only demersal fish species, the bycatch recorded in shrimp surveys off North Iceland during the period 1988–2005 amounted to 1,146,317 individuals

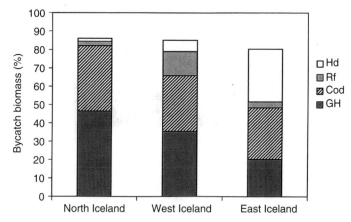

Figure 14 Relative biomass of the main bycatch species caught by the Icelandic shrimp fleet off North, West and East Iceland during the period 1983–2005. GH, Greenland halibut; Rf, redfish; Hd, haddock.

Table 5 Fish taxa recorded as bycatch in the Icelandic offshore shrimp surveys carried out in waters off North, West and East of the country during the period 1988–2005

Fish taxa	En	North			West			East		
		N	Mean (SD)	%S	N	Mean (SD)	%S	N	Mean (SD)	%S
Anarhichas denticulatus	BP	74	1.3 (1.5)	2.2	91	5.7 (16.1)	2.9	37	1.0 (0.3)	4.5
Beryx splendens	BP	22,424	32.6 (63.9)	27.7				1869	20.8 (27.2)	10.6
Cyclopterus lumpus	BP	162	5.4 (10.7)	1.1	88	2.0 (1.2)	7.7	16	1.3 (0.9)	1.4
Paraliparis bathybius	BP	49	8.2 (10.7)	0.2						
Agonus cataphractus	D	8	2.7 (2.1)	0.1				60	20.2 (26.0)	0.4
Amblyraja hyperborea	D	141	1.7 (2.2)	3.3				29	1.7 (1.2)	2.0
Amblyraja radiata	D	13,131	13.5 (67.2)	38.4	1700	7.3 (17.2)	41.7	1270	3.4 (4.1)	43.2
Ammodytes marinus	D	2	1.0 (0.0)	0.1						
Ammodytes tobianus	D	13	3.3 (2.1)	0.2	22	11.0 (11.3)	0.4			
Anarhichas lupus	D	695	6.5 (11.9)	4.3	57	1.2 (0.7)	8.4	298	2.4 (3.1)	14.6
Anarhichas minor	D	2239	3.2 (6.8)	28.1	127	2 (1.7)	11.6	880	2.6 (2.3)	39.3
Argentina silus	D	231	17.8 (39.1)	0.5	36,062	166.2 (350.1)	38.7			
Artediellus atlanticus	D	44,514	43.9 (91.2)	39.7	11,117	139.0 (180.0)	14.3	9896	35.5 (81.1)	32.7
Bathyraja spinicauda	D	1		0.0				122	30.5 (55.0)	0.5
Boreogadus saida	D	157,916	87.4 (199.4)	71.4	6611	81.6 (137.7)	14.4	5009	17.5 (57.1)	33.8
Brosme brosme	D	7	1.1 (0.0)	0.3	10	1.1 (0.6)	1.6	16	1.5 (0.9)	1.3
Careproctus micropus	D	3	1.5 (0.7)	0.1						
Careproctus reinhardti	D	11,482	12.2 (28.0)	37.4	391	8.5 (14.9)	8.2	718	5.1 (5.5)	16.6
Chimaera monstrosa	D	5		0.0	991	22.5 (31.6)	7.8			
Chirolophis ascanii	D	289		0.0				86		0.1
Ciliata mustela	D	15		0.0	2		0.2			
Coryphaenoides rupestris	D				24	3.4 (6.4)	1.2			
Cottunculus microps	D	208	3.7 (4.5)	2.3	1		0.2	10	2.5 (1.9)	0.5
Cottunculus thomsonii	D	8	2.7 (1.5)	0.1						

(Continued)

Table 5 (Continued)

Fish taxa	En	North			West			East		
		N	Mean (SD)	%S	N	Mean (SD)	%S	N	Mean (SD)	%S
Dipturus batis	D	18	1.2 (0.4)	0.6	3	0.8 (0.5)	0.7	2	1.0 (0.0)	0.2
Entelurus aequoreus	D	1		0.0						
Etmopterus princeps	D							3		0.1
Eutrigla gurnardus	D	25		0.0	11	3.7 (4.6)	0.5			
Gadus morhua	D	38,475	32.3 (152.9)	47.5	18,762	40.8 (106.7)	82	8615	21.2 (65.0)	47.6
Gaidropsarus vulgaris	D	1		0.0						
Glyptocephalus cynoglossus	D	102	2.5 (2.6)	1.7	5653	19.5 (37.3)	51.7	8	2.7 (2.9)	0.3
Gymnelus retrodorsalis	D	50	2.5 (2.7)	0.8	127	42.3 (66.5)	0.5	4	1.3 (0.6)	0.3
Helicolenus dactylopterus	D				1					
Hippoglossoides platessoides	D	98,110	75.8 (227.0)	50.6	104,370	208.7 (417.0)	89.1	65,219	105.2 (166.9)	72.3
Hippoglossus hippoglossus	D				831	11.7 (85.0)	12.7			
Hyperoplus lanceolatus	D	17		0.0	121	15.1 (23.5)	1.4	5	1.6 (1.2)	0.3
Icelus bicornis	D	70	6.4 (3.8)	0.4	1		0.2	5	2.5 (2.1)	0.2
Lamna nasus	D				2		0.2			
Lepidorhombus whiffiagonis	D				123	3.7 (4.5)	5.9			
Leptagonus decagonus	D	25,764	22.4 (43.8)	45.2	3593	50.6 (114.0)	12.7	2167	12.8 (24.2)	19.9
Leptoclinus maculatus	D	569	6.3 (9.7)	3.6	180	16.4 (25.0)	2	3726	120.2 (605.0)	3.6
Limanda limanda	D	2	1.0 (0.0)	0.1	572	9.1 (22.1)	11.2	14	7.0 (8.5)	0.2
Liparis fabricii	D	1323	6.1 (18.6)	8.8	68	6.8 (10.4)	1.8	19	1.5 (1.0)	1.5
Liparis liparis	D	1		0.0						
Liparis montagui	D	394	11.9 (14.2)	1.3	27	3.9 (2.7)	1.2	3	1.0 (0.0)	0.3
Lophius piscatorius	D	12	4.0 (5.3)	0.1	89	1.4 (0.8)	11.6	1		0.1
Lumpenidae	D	4581	50.3 (74)	3.7	3747	85.2 (140.9)	7.8	5238	45.2 (71.5)	13.6
Lumpenus lampretaeformis	D	4328	34.9 (53.6)	5.0	4431	44.3 (85.3)	17.8	23,238	83.6 (123.2)	32.7
Lycenchelys kolthoffi	D	197	2.7 (2.1)	2.9	38	4.2 (5.6)	1.6	16	3.2 (2.28)	0.6
Lycenchelys muraena	D	36	4.5 (4.3)	0.3				1		0.1

Lycodes esmarki	D	2350	6.2 (10.4)	15.2	1094	21.9 (37.5)	8.9	270	3.4 (3.4)	9.3
Lycodes eudipleurostictus	D	17,560	23.9 (57.7)	29.8	212	5.9 (8.4)	6.4	3008	27.1 (36.3)	13.0
Lycodes frigidus	D	77	8.6 (6.6)	0.4						
Lycodes reticulatus	D	8718	17.3 (28.4)	20.2	78	4.9 (5.7)	2.9	189	5.4 (11.5)	4.1
Lycodes rossi	D	409	14.1 (33.7)	1.2				3		0.1
Lycodes seminudus	D	7777	17.8 (34.6)	17.6	11	3.7 (2.5)	0.5	183	5.4 (6.6)	4.0
Lycodes spp.	D	140,080	175.5 (463.2)	31.1	7460	95.6 (237.0)	13.9	6780	46.4 (96.9)	17.2
Lycodes squamiventer	D	49	24.5 (10.6)	0.1				6	3.0 (2.8)	0.2
Lycodes vahli	D	10,899	19.2 (50.1)	23.0	5131	31.7 (69.0)	28.9	5032	18.1 (34.9)	32.7
Lycodonus flagellicauda	D	38	4.2 (5.5)	0.4	12	3.0 (4.0)	0.7	8	1.6 (1.3)	0.6
Macrourus berglax	D				13	6.5 (2.1)	0.4			
Melanogrammus aeglefinus	D	2534	11 (37.6)	9.3	141,521	306.3 (668.5)	82.4	3268	49.5 (146.9)	7.8
Merlangius merlangus	D	47	2.9 (4.7)	0.6	30,519	93 (263.9)	58.5	304	60.8 (128.1)	0.6
Microstomus kitt	D	13	2.6 (2.6)	0.2	1147	5.6 (8.5)	36.5			
Molva dypterygia	D	9	3 (3.5)	0.1	486	5.5 (6.4)	15.9			
Molva molva	D	7		0.0	119	3.3 (4.9)	6.4			
Myoxocephalus scorpius	D	13	3.3 (1.3)	0.2	8		0.2	5	1.2 (0.5)	0.5
Onogadus argentatus	D	690	3 (3.9)	9.2	140	3.3 (4.4)	7.7	120	2.1 (1.7)	6.6
Petromyzon marinus	D				18		0.2			
Phycis blennoides	D			0.1				1		0.1
Pleuronectes platessa	D	16	2.0 (2.9)	0.3	347	5.0 (13.3)	12.5	925	21.0 (41.7)	4.9
Pollachius virens	D	26	1.4 (0.9)	0.7	582	4.4 (15.4)	23.7	37	2.3 (3.7)	1.9
Raja clavata	D	2		0.0						
Raja lintea	D	2	1.0 (0.0)	0.1						
Reinhardtius hippoglossoides	D	25,063	13.8 (25.6)	71.5	290	5.7 (9.5)	9.1	3672	8.4 (15.2)	51.5
Rhinonemus cimbrius	D	1		0.0	3746	16.6 (25.1)	40.1	56	4.3 (5.4)	1.5
Sebastes marinus	D	523,563	398.4 (2840.4)	53.3	40,287	113.5 (1076.7)	63.3	15,330	31.2 (68.3)	57.6
Sebastes viviparus	D	98	2.2 (2.0)	1.8	3356	13.8 (31.2)	43.5	33	1.9 (3.9)	2.0
Sonniosus microcephalus	D	1		0.0						
Squalus acanthias	D	1		0.0	11	1.1 (0.3)	1.8			
Triglops murrayi	D	576	3.5 (5.9)	6.7	539	28.4 (46.8)	3.4	190	4.4 (7.0)	5.1

(Continued)

Table 5 (Continued)

Fish taxa	En	North			West			East		
		N	Mean (SD)	%S	N	Mean (SD)	%S	N	Mean (SD)	%S
Trisopterus esmarkii	D	707	15.4 (50.8)	1.9	130,937	509.5 (1489.2)	45.8	13	1.6 (0.9)	0.9
Liparidae	DP	1583	8.5 (17.7)	7.5	192	10.7 (19.6)	3.2	35	3.5 (3.7)	1.2
Macrouridae	DP	1		0.0						
Sebastes spp.	DP	2583	117.4 (224.9)	0.9						
Arctozenus risso	P	174	2.1 (1.7)	3.4	1		0.2	37	1.9 (1.7)	2.2
Cetorhinus maximus	P				1		0.2			
Chauliodus sloani	P				13		0.2			
Clupea harengus	P	39,546	227.3 (1323.4)	7.1	19,042	81.4 (174.3)	41.7	22,884	240.9 (770.8)	11.2
Hoplostethus atlanticus	P						0.2			
Lampanyctus crocodilus	P	83	2.2 (2.2)	1.5						
Lampanyctus intricarius	P	6		0.0						
Lampanyctus macdonaldi	P	52	17.3 (7.6)	0.1						
Mallotus villosus	P	1,844,480	852.0 (5133.5)	84.8	19,986	98.9 (187.2)	36	416,558	734.7 (4406.8)	66.6
Maurolicus muelleri	P	5	1.3 (0.5)	0.2						
Micromesistius poutassou	P	31,988	211.8 (830.2)	6.1	88,399	601.4 (1831.5)	26.2	53,457	396.0 (2025.5)	15.9
Myctophidae	P			38.1	487	14.8 (52.7)	5.9	10,608	50.3 (144.0)	24.8
Notoscopelus kroeyeri	P	4	2.0 (1.4)	0.1						
Paralepididae	P	91	5.1 (10.1)	0.7				12	2.4 (1.7)	0.6
Paralepis coregonoides	P	91	2.0 (1.9)	1.9				50	2.2 (1.2)	2.7
Platyberyx opalescens	P	1		0.0						
Scomber scombrus	P							2	1.0 (0.0)	0.2
Sebastes mentella	P	8844	13.7 (30.2)	26.1	96	3.7 (4.0)	4.6	4373	29.2 (66.4)	17.6
Sebastes mentella oceanus	P							7		0.1

En, environment; BP, benthopelagic, D, demersal; P, pelagic; DP, genera including demersal and pelagic species; N, total number of individuals recorded; mean (SD), mean number of individuals caught per sample and standard deviation (only for species found in more than one station); %S, percentage of samples in which each taxa was found.

Note: The number of samples for North Iceland is 2466; West Iceland, 561 and East Iceland, 850.

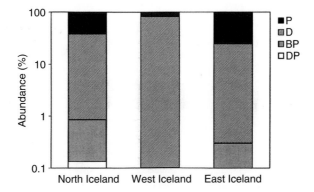

Figure 15 Abundance (in percentages) of the different fish, grouped according to their habitat, caught as bycatch during the shrimp surveys off North, West and East Iceland. Note that the *y*-axis has logarithmic scale. DP, demersal-pelagic; BP, benthopelagic; D, demersal; P, pelagic.

belonging to 70 species. The average capture of bycatch species per haul was 465 individuals. The most abundant species, in number of individuals, were *Sebastes marinus* (45.7% of the total), *Boreogadus saida* (13.8%), *Lycodes* spp. (12.2%) and *Hippoglossoides platessoides* (8.5%). These four species represented 80% of the bycatch (Figure 16). *G. morhua* and *R. hippoglossoides* were among the most abundant commercial species but they represented only 3.4% and 2.2% of the bycatch, respectively (Table 5). In West Iceland, 570,057 individuals belonging to 60 species were recorded, but the average abundance of bycatch per haul was 1016 individuals. Three species made up 66% of the bycatch: *Melanogrammus aeglefinus* (24.8%), *Trisopterus esmarkii* (23.0%) and *H. platessoides* (18.3%). *G. morhua* and *R. hippoglossoides* were scarce and constituted 3.3% and 0.05% of the bycatch, respectively (Table 5, Figure 16). In East Iceland, 157,712 individuals belonging to 52 species were recorded, and the average catch per haul was 187 individuals. Three species made up 64.6% of the bycatch: *H. platessoides* (40.7%), *Lumpenus lampretaeformis* (14.2%) and *Sebastes marinus* (9.7%). *G. morhua* and *R. hippoglossoides* represented 5.4% and 2.3%, respectively (Table 5, Figure 16).

4. GREENLAND

4.1. The fishery

After the collapse of the cod, halibut and redfish fisheries in the late 1960s, shrimp became the main marine resource in Greenland, making up 90% of the fishery products of the export value (Carlsson and Smidt, 1976;

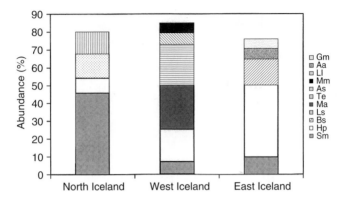

Figure 16 Abundance (in percentages) of the most abundant fish species caught as bycatch during the shrimp surveys off North, West and East Iceland. Sm, *Sebastes marinus*; Hp, *Hippoglossoides platessoides*; Bs, *Boreogadus saida*; Ls, *Lycodes* spp.; Ma, *Melanogrammus aeglefinus*; Te, *Trisopterus esmarkii*; As, *Argentina silus*; Mm, *Merlangius merlangus*; Ll, *Lumpenus lampretaeformis*; Aa, *Artediellus atlanticus*; Gm, *Gadus morhua*.

Pedersen, 1994; Rätz, 1997a). Most of the research done on the Greenland shrimp stock refers to the spatial division adopted by the North Atlantic Fisheries Organisation (NAFO) and studies from this region form the basis of this chapter (Figure 17).

4.1.1. West Greenland

The fishery started in 1935 at a very small scale in the Sisimiut (Holsteinsborg) area (NAFO Subarea 1B) as an alternative to the collapsing Atlantic halibut, cod and redfish fisheries (Horsted, 1978). Three or four vessels took part in the fishery between the years 1936 and 1939. Fishing ceased due to World War II and was resumed in 1947 with much lower catches because the shrimp were smaller. Extremely cold weather conditions that led to bottom water temperature to reach $-1.6\,^{\circ}\mathrm{C}$ in summer caused widespread mortality of the local stock in 1949. The fishery moved then to the nearby Qasigiannguit (Christianhåb) in Disko Bay (Subarea 1A), with the catch being landed and processed in Sisimiut. Fishing took place between June and October. Five vessels trawled within the 400-m depth contour. However, transport inflicted great losses on the catch and the processing plant in Sisimiut was finally closed in 1950 (Smidt, 1965, 1969; Carlsson, 1976). In the same year, two Danish cutters fished at 500 m in Disko Bay and obtained good catches but the trawl was damaged often (Horsted and Smidt, 1956).

Also in 1950, there was shrimp fishing in Qaqortoq (Julianehåb, Subarea 1F) with two trawlers. The establishment of freezing and canning plants

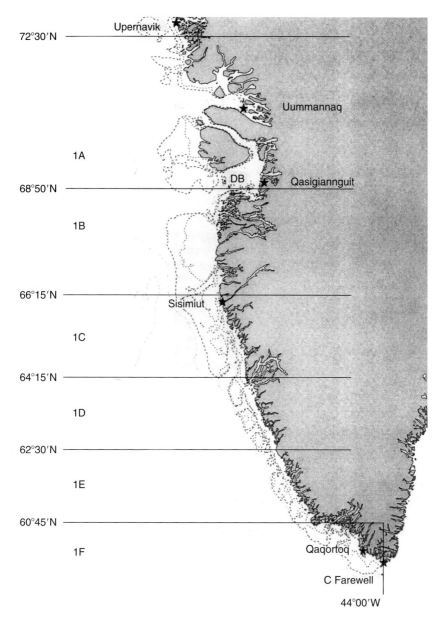

Figure 17 Spatial division of West Greenland used by the North Atlantic Fisheries Organisation (NAFO) showing the localities mentioned in this chapter. DB, Disko Bay. Modified from map by M. Storr-Paulsen.

in the town during the same year allowed an increase in catches and fleet size, with four to five extra vessels that operated all the year round if catches were large enough (Horsted and Smidt, 1956). Shrimp catches started to increase in the late 1960s as cod landings commenced their decline and by 1970, 170 trawlers between 20 and 150 gross registered tons (GRT) were involved in the fishery. Until then, the best fishing grounds were those in the shallow waters of Disko Bay. Catches there were stable, but fishing was not possible during winter. The southern fjords were open all the year round, but catches were lower and very variable (Smidt, 1965, 1969; Carlsson, 1976). The landings of the inshore fleet still represented over 88% of the total annual landings, but from 1970 onwards, catches in the offshore shrimp fishery in West Greenland increased greatly. Inshore catches represented 56% by 1973 and from 1974 onwards, the offshore fishery has contributed 56–89% of the total annual shrimp landings (Figure 18). The entire catch was processed on land until the introduction of factory trawlers (Horsted and Smidt, 1956; Smidt, 1965, 1969; Carlsson, 1976; Andersen, 1994; Hvingel, 2004a).

The offshore grounds are those more than 3 miles from the coast. The inshore fleet concentrates mostly off Disko Island in waters <400-m depth

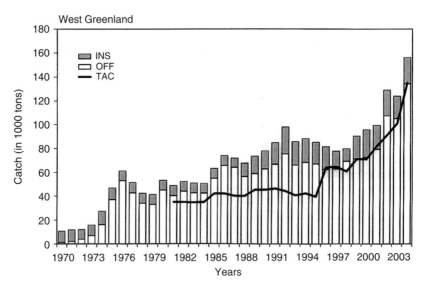

Figure 18 Annual shrimp catches (in thousands of tons) in West Greenland during the period 1970–2004. INS, inshore fishery; OFF, offshore fishery; TAC, total allowable catch for the offshore fishery. After introducing the ITQ system in 1991, catches decreased between 1993 and 1998 and the fleet was reorganised. Between 85% and 97% of the inshore catch is from Disko Bay (Andersen, 1994; Hvingel, 2002a). Data from Kingsley and Hvingel (2005).

and consists of vessels under 80 GRT, plus a few larger vessels with quotas to fish in both areas (Andersen, 1994; Hvingel, 2002a, 2004a). Both the inshore and offshore fisheries operate on the continental shelf, between 59°N and 74°N and 150–600 m (Figure 19).

Offshore catches south of 71°N within Subarea 1 increased in the early 1970s, from less than 1000 tons to roughly 60,000 tons in 1976. Catches were largest in Subarea 1B in all but 2 years during the period 1970–1990 (Carlsson, 1991). The Greenland fleet increased to 27 vessels after several trawlers from the cod fishery joined the vessels of the Royal Greenland Trade Department (RGTD), which at the time accumulated the largest fishing effort in the offshore fishery. There were also 56 vessels from Denmark, Faeroe Islands, Norway, France and Canada, although their number decreased to 30 in 1981. In 1975, the fishing effort of the RGTD vessels concentrated within Subarea 1B, mostly between 67°N and 68°N. During the years 1976–1978, fishing effort increased and expanded between 66°N

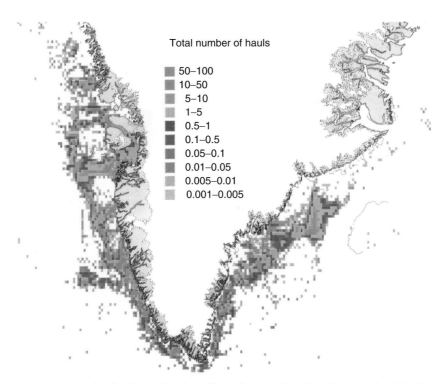

Total number of hauls

- 50–100
- 10–50
- 5–10
- 1–5
- 0.5–1
- 0.1–0.5
- 0.05–0.1
- 0.01–0.05
- 0.005–0.01
- 0.001–0.005

Figure 19 Distribution of fishing effort of the shrimp fleet in Greenland during the period 1982–2002 according to fishery logbooks. The colour code indicates total number of hauls in each square. Map supplied by M. Storr-Paulsen.

and 69°N, but was concentrated around areas at 69°N (Carlsson, 1980a, 1981a).

New shrimp grounds were found during 1984 in the Uummannaq and Upernavik municipalities (Subarea 1A). The fleet was restricted in 1986 to a maximum catch of 400 tons per trip north of 72°52′N, and in 1987 the Greenland Home Rule set a TAC for the area comprised between 71°N and 72°52′N. The distribution of fishing effort in these new grounds was very uneven and the CPUE decreased every year (600 kg hour^{-1} in 1987 vs 300 in 1988). It was feared that fishing effort had been too high considering that low water temperature in the area might hamper recruitment (Lund, 1988a,b, 1989a, 1990b). Catches north of 71°N decreased from 11,000 tons in 1986–1987 to 1000 tons in 1991 (Parsons et al., 1992).

South of Subarea 1A, the logbooks from the vessels >50 GRT in the offshore fleet show a southward shift of fishing effort (Figure 20). Shrimp grounds north of 65°N (Subarea 1B) were preferred in the 1970s and early

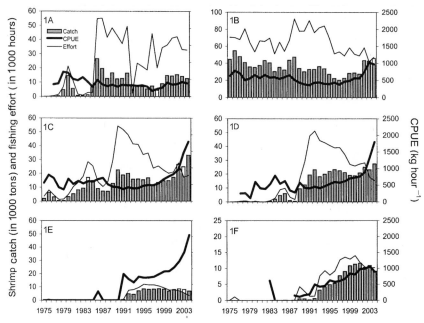

Figure 20 Distribution of the annual offshore shrimp catch in West Greenland showing the southward displacement of the fleet (see map in Figure 17). Shrimp catch (in thousand tons, represented by columns) and fishing effort (thin line, in thousands of trawling hours) refer to the left *y*-axis. Unstandardised CPUE refers to the right *y*-axis. It is represented by the thick line and measured in kilogram of shrimp per trawling hour. Note that the scales in the left *y*-axis are different for Subareas 1B and 1F. Data from Hvingel (2004a).

1980s. Landings in Subarea 1 reached 98,000 tons in 1992 (after corrections for over-packing and unreported catch, see Hvingel, 2003a, 2004a), 50% of which were from Subareas 1C–1F. The southward shift of the fishery could be explained by changes in the distribution of shrimp biomass and/or improvements in gear that permitted trawling in areas that were not accessible before. An important consequence of this movement of the fleet is that effort might be concentrating in smaller areas, since the shelf narrows towards the south of Greenland. Fishing takes place all the year round but maximum catches are taken in midsummer, although in some years there is another peak in autumn (Lund, 1989a; Carlsson, 1991, 1993; Parsons *et al.*, 1992; Hvingel, 1996, 2002a, 2004a; Hvingel *et al.*, 1997; Hvingel and Folmer, 1998). Nevertheless, fishing effort has decreased in the southern grounds in recent years and the fishery is moving northwards (Kingsley and Hvingel, 2005).

The shift southward reduced fishing effort of the shrimp fleet in Subareas 1A and 1B to 32% and 37%, respectively by 1996 compared to the 1990 values of 42% and 47%. However, overall fishing effort for Subarea 1 remained rather stable. Landings declined during 1993–1995 following the extinction of the strong year-class from 1985 that had kept landings increasing during the years 1989–1993. The good year-classes from 1990 and 1991 made possible the increase of catch rates in 1995 and 1996 despite the large variability of biomass observed during the period 1976–1988 (Folmer *et al.*, 1996a; Hvingel *et al.*, 1996; Siegstad, 1996). The increasing trend in shrimp biomass that started in 1998 reached a record biomass of 650,000 tons in 2004 (Siegstad, 1999; Kanneworff and Wieland, 2001, 2002, 2003; Wieland *et al.*, 2004), permitting the rise of landings, inshore and offshore catches combined, from 80,000 tons in 1998 to 141,000 tons in 2004 (Hvingel, 2004a). However, the recruitment index in 2003 and 2004 were below average, and fishable biomass may decline after 2005 (NAFO, 2004). Overall, catches in the West Greenland shelf, including Division 0A and Subarea 1, have always surpassed the recommended TAC. The standardised CPUE index (Section 4.3) has followed a sharp decreasing trend following the peak in catches reached in 1992, although it increased somewhat between 1998 and 2002 (Figure 21).

4.1.2. East Greenland and Denmark Strait

Shrimp are present along the East Coast from Cape Farewell to 70°N, down to 800-m depth, but the highest shrimp densities are found at 150- to 600-m depth (Hvingel, 1999). An Icelandic research vessel found the Dohrn Bank grounds in 1976, but they were not surveyed until 1978. Fishing started north of 65°N following the survey (Hallgrímsson, 1981; Jónsson and Hallgrímsson, 1981; Skúladóttir, 1989c). Norway started fishing in this area in 1979 and Greenland, Denmark, Faeroe Islands and France joined the fishery in 1980

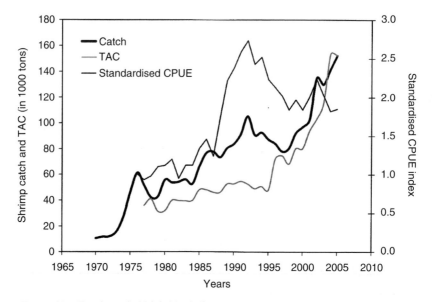

Figure 21 Total catch (thick black line), TAC (thick grey line), both in thousand tons (left *y*-axis), and standardised CPUE index (thin black line, rigth *y*-axis) for the West Greenland shelf, including Division 0A and Subarea 1. The standardised CPUE index for all West Greenland is estimated from four separate indices, one for each fleet. Each of the indices includes four variables: vessel fishing power, seasonal availability of shrimp, spatial availability of shrimp and annual mean CPUE. Data from Hvingel (2004a) and Kingsley and Hvingel (2005).

(Skúladóttir, 1995c). The main fishing grounds were in Strede and Dohrn Banks and on the slopes of the Storfjord Deep (Figure 22), and fishing took place mostly from January to June (Carlsson, 1990a). Catches had reached 15,000 tons by 1988 (Figure 23), exceeding sustainable levels and causing the biomass to decline below the minimum required to produce the maximum sustainable yield (B_{MSY}). In fact, there was an increase in the proportion of males in commercial catches during the period 1988–1990, suggesting that the fishery had reduced female abundance or that stronger year-classes were recruiting. Both fishing effort and catches decreased by 75% during the 1990s on the banks north of 65°N, mainly due to the discovery of new shrimp grounds south of 65°N, where CPUE was much higher, and the onset of shrimp fishing at Flemish Cap (off Newfoundland). This decline of fishing effort allowed for some recovery of the stock (Figure 24). The new fishery starting in 1993 in two areas south of 65°N, extending to Cape Farewell (Figure 25), kept landings above 8000 tons for more than 10 years (Figure 24). These new banks, discovered by Greenlandic fishing vessels, supported higher catch rates than the northern grounds from 1996 to 2003, and vessels from

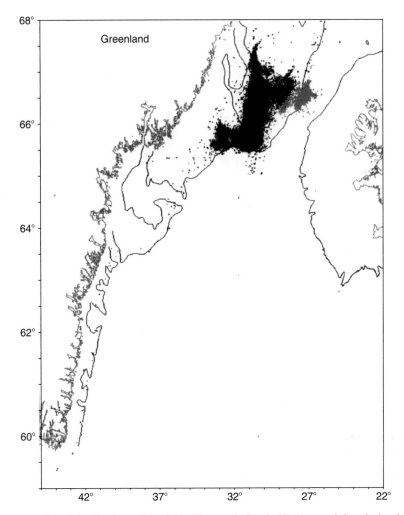

Figure 22   Distribution of hauls in Denmark Strait (Dohrn and Strede banks), East Greenland. Black dots correspond to Greenlandic, Faeroese and Danish trawlers fishing within the Greenland EEZ during the period 1987–1992, grey dots correspond to Icelandic trawlers fishing within the Icelandic EEZ during the period 1990–1992. The black line shows the 400-m isobath. Map from Hvingel (2004b).

Denmark, Norway and the Faeroe Islands started fishing there as well (Figure 26). CPUE south of 65°N has followed an increasing trend despite a large decrease in 2001 (Figure 24), which was nevertheless compensated for by the dominance of large shrimp in the catch according to the skippers. On the other hand, standardised fishing effort has decreased since 1993, with slight increases during 2003 and 2004 (Skúladóttir, 1990a, 1995c, 1998c; Smedstad

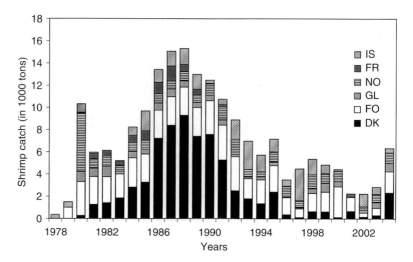

Figure 23 Annual shrimp catch (in thousands of tons) in East Greenland north of 65°N during the period 1978–2004. DK, Denmark; FO, Faeroe Islands; GL, Greenland; NO, Norway; FR, France; IS, Iceland. It should be noted that the Icelandic fleet operates within their EEZ, although the shrimp stock is the same. Data from Hvingel (2004b).

and Torheim, 1990; Skúladóttir *et al.*, 1991b; Cadrin and Skúladóttir, 1998; Hvingel, 1999, 2001, 2004b; NAFO, 2004). The TAC for East Greenland and Denmark Strait (Greenland EEZ) has often been set beyond the figures advised by scientists (Figure 27).

The trawlers that took part in the fishery during the first years were from 130 to 750 GRT, but by 2001 they were between 1000 and 3000 GRT and had increased to 1000–4000 GRT in 2004. These factory trawlers were from Norway, Greenland, Denmark and Faeroe Islands, but they operated in Greenland waters, where 70–90% of the total catches are obtained. The vessels operated throughout the year, but during the 1990s the fishing season gradually reduced to focus on the period from November to April when catches were highest (Carlsson, 1981b; Skúladóttir, 1998c; Hvingel, 1999, 2001, 2004b).

4.2. Management

4.2.1. West Greenland

For management purposes, the shrimp stock in West Greenland is assessed as one population within NAFO Subarea 1 and Division 0A (Hvingel *et al.*, 1997). Management advice has been based, as for other North Atlantic

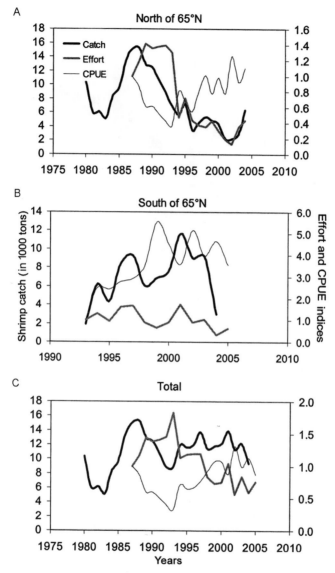

Figure 24 Shrimp catch (in thousands of tons) and standardised CPUE and effort indices for the shrimp fishery off East Greenland. (A) Catches from the Dohrn and Strede banks, (B) catches from the new banks found in 1993, (C) combined catch, effort and CPUE for all East Greenland. Data from Siegstad and Hvingel (2005).

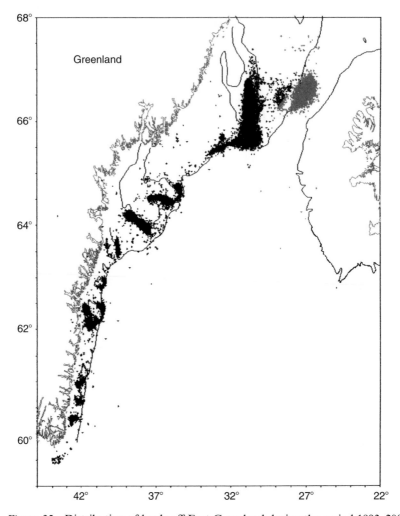

Figure 25 Distribution of hauls off East Greenland during the period 1993–2002. Black dots represent Greenlandic, Faeroese and Danish trawlers operating within the Greenland EEZ, grey dots represent the Icelandic fleet, that operates within the Icelandic EEZ. The black line shows the 400-m isobath. Map from Hvingel (2004b).

shrimp stocks, on qualitative assessment of biomass and length distribution trends in relation to catches, as described in Section 3.3.

The need for fishing regulations arose due to fluctuation in catches in the early years of the fishery. At first, it seemed unlikely that the main stock in Disko Bay could be overfished, since it was surrounded by large virgin shrimp grounds. Yet a decrease of mean size of shrimp was observed in

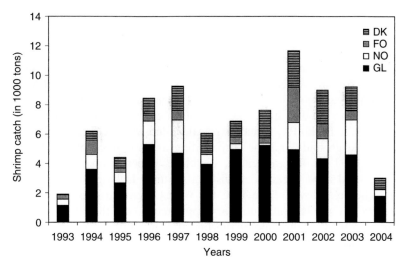

Figure 26   Annual shrimp catch (in thousands of tons) of the different fleets in the banks off East Greenland, south of 65°N. DK, Denmark; FO, Faeroe Islands; NO, Norway; GL, Greenland. Data from Hvingel (2004b).

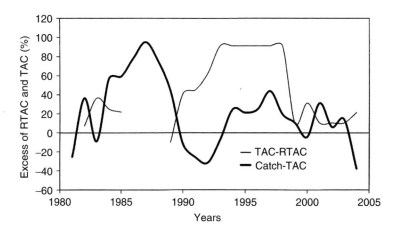

Figure 27   Difference between the TAC recommended by scientists for the Denmark Strait and East Greenland (RTAC) and the TAC adopted by managers (TAC-RTAC, thin line), and difference between shrimp catches and the official TAC (Catch-TAC, thick line), both expressed as percentages. Scientists lacked enough data to estimate the TAC during the years 1986–1988. In the Denmark Strait only the Greenland EEZ is subjected to TAC. Data from Carlsson (1988), Skúladóttir (1993c), Hvingel (2004b) and Siegstad and Hvingel (2005).

the catches. Experimental fishing with different mesh sizes to avoid capturing undersize shrimp was not satisfactory and it was suggested to fish at greater depth (Smidt, 1965, 1969).

Diverse management tools have been implemented since 1975 such as excluding foreign vessels from the fishery, limiting catches by TAC and licenses, increasing mesh size, regulating shrimp discards and recording all catches (Table 6). The first TAC, based on catch per unit area of shrimp ground in the Disko Bay fishery, was calculated with yield estimations from the fishable grounds and the surrounding areas that supplied adults and larvae. The TAC for each area, therefore, depends on its geographical position, and it was reduced in fishing grounds lacking shrimp-supplying areas. The breakdown of the TAC for West Greenland into smaller spatial units was advised as a precautionary approach in case there was larval drift or migration of adults among the inshore and offshore grounds. Hence, it was thought dividing the quota would counteract the concentration of effort within small areas and the possibility that these areas might be overfished (Carlsson and Smidt, 1976; Horsted, 1991).

Logbooks and TAC were applied to the inshore fishery in 1997, whereas they had been implemented in the offshore fishery in 1986. Until 1997, inshore catches of vessels <50 GRT, which lack processing facilities on board and were not subjected to offshore TAC regulations, had been estimated from the sales slips of landings. This method caused an uncertainty in estimations of catches and effort because some of these vessels fished offshore but their catch was not included in the offshore TAC (Horsted, 1991; Hvingel, 2002a).

Shrimp discards have been another cause for concern because they are difficult to estimate. Discarding of shrimp with weight over 2 g was prohibited by law in 1979 (Table 6), but larger shrimp have a higher market price and are therefore preferred by fishermen, especially by those with low quotas. Discards were explicitly included in the TAC during the first years, and until 1979, when there was no mention of them in the paragraph with the recommended TAC. Reports from subsequent years failed to include unreported shrimp discards in the TAC, thus assuming that they remained constant. The discard issue was raised again in 1990 and observers were allocated to trawlers from 1991 onwards. This measure, together with the increase of price for smaller shrimp and the obligation to take part of the catch to processing plants on land, has contributed to the decrease of discards (Anonymous, 1977c; Horsted, 1991).

Another measure set to achieve a more accurate record of catches was the implementation of a new law in 2004, mandating that landings have to be reported as live weight rather than product weight. This law allows correcting the underestimations of catches caused by the extra weight in packaging units that was unreported until 2003 (Hvingel, 2004a).

Table 6   Regulations for the management of the Greenland shrimp stock

Year	Regulation	References
1976	Mesh size fixed at 20–22 mm	Carlsson and Smidt, 1976; Horsted, 1991
	Fishing at greater depths	
	Joint TAC for Canada and Greenland in offshore grounds within Subarea 1 according to catch per area unit	
1977	Spatial breakdown of TAC	Anonymous, 1977c; Horsted, 1991
	Creation of shrimp box (68°–69°30′N east of 59°W)	
	Division A, north of 69°30′N	
	Division B, south of 68°N plus Div. 1C	
	Divisions 1D–1F	
	Minimum mesh size of 40 mm	
1979	Law forbidding discards of shrimp <2 g	Horsted, 1991
1980	Shrimp box decreased (east of 56°W)	Horsted, 1991
1981	Fishery closed to all but Greenlandic and Canadian fleets	Horsted, 1991; Hvingel, 2002a; Siegstad, 1998
	Separate TAC for Canada and Greenland due to EEC issues, Canadian fleet restricted to Division 0A	
1986	Logbooks mandatory for all ships >50 GRT (most of them in offshore fishery)	Hvingel *et al.*, 1997
1990	New TAC breakdown	Horsted, 1991
	Northwest Greenland (71°–72°52.5′N, east of 58°W plus 69°30′–72°52.5′N, west of 58°W	
	West Greenland (offshore south of 71°N, east of 58°W and south of 69°30′, west of 58°W)	
	Area between 68° and 70°45′N east of 56°W	
	Area 69°30′–70°45′N, east of 56°W (only ships 75–250 GRT allowed)	
	Quotas allocated to individual vessels through licenses	

(Continued)

Table 6   (Continued)

Year	Regulation	References
1991	Individual Transferable Quotas (ITQ) for offshore fleet	Arnason and Friis, 1995; Hvingel, 2000; Schmidt, personal communication
	Introduction of observers in offshore fleet	
	Mandatory to land part of catch in processing plants	
1992	Further restrictions to Canadian fleet	Hvingel, 2000
1997	Logbooks mandatory for all ships TAC extended to all ships <50 GRT (most of them in inshore fishery)	Hvingel *et al.*, 1997
2000	Sorting grids with bars spaced 22 mm in all shrimp grounds Ships must go to fishing grounds at least 5 nm away if bycatch exceeds legal limits (10% of catch)	Engelstoft, 1996, 2002
2004	Law to report catches in units of live weight instead of packaged product weight	Hvingel, 2004a

Bycatch of finfish is discussed in detail in the next section, so only regulations concerning this problem are addressed here. Sorting grids with 22 mm of grid space have been mandatory since 2000. Nevertheless, the maximum catch of fish bycatch per haul is further regulated by legislation, and vessels are required to move to different fishing grounds by at least 5 miles when the legal limit of bycatch is exceeded (Engelstoft, 2002). The Ministry of Fisheries also limits the fleet size, which had become a problem because the expansion of the fishing grounds encouraged the industry to invest in vessels beyond what could be supported by the fishery. Thus, new regulations were issued during the 1990s (Table 6) to adjust the fleet size and optimise short- and long-term economy while maintaining the stock size (Horsted, 1991).

4.2.2. East Greenland and Denmark Strait

During 1978–1980, the International Council for Northwest Atlantic Fisheries (ICNAF, predecessor of NAFO) advised a cautious approach to shrimp fishing as representatives of the different nations involved in the fishery could not agree on the stock size. The first estimates were unrealistic, based only on catches taken during a few months of the year. For this reason the council advised the countries interested in taking part in the fishery to carry out stock assessment surveys to enlarge the available database (Carlsson, 1980b, 1981b; Anonymous, 1982b).

The stock off East Greenland and in Denmark Strait has always been assessed as a single population, even after the discovery of the southernmost banks. The first TAC was issued in 1981, but it only applies to the shrimp fishing grounds within Greenland waters. Management of the shrimp fishing grounds within the Icelandic EEZ was considered unnecessary because ice conditions prevent fishing in most years, and nevertheless catches rarely exceed 2000 tons (Anonymous, 1983; Carlsson, 1983a; Hvingel, 1999, 2004b).

Reporting shrimp catches to Greenland for all vessels >80 GRT was mandatory between the beginning of the fishery in 1978 until 1987, when logbooks were implemented. The legal mesh size was 40 mm, as in West Greenland (Hvingel, 1999; Carlsson and Hvingel, 2000).

4.3. Research

Shrimp were found in West Greenland during two different expeditions carried out by Professor Adolf Jensen in 1908 and 1925 west of the Great Hellefiske Bank (south of Disko Island, Subarea 1A). In 1935, shrimp were found in Nuuk Fjord (Godthåb, Subarea 1E), Maniitsoq district (Sukkertoppen, Subarea 1C) and Disko Bay (Subarea 1A), but the survey was confined to waters shallower than 250 m. Research on the West Greenland shrimp stock started after World War II, with mapping of the trawling grounds and studies on shrimp biology (Horsted and Smidt, 1956; Table 7). This early work continued with the discovery of new grounds and mark-recapture experiments to determine whether shrimp would migrate with the masses of warm bottom water (Smidt, 1960). The surveys carried out between 1946 and 1954 provided descriptions of shrimp grounds, biology of *Pandalus borealis*, detailed lists of associated fauna in the different districts and hydrological information from the Qaqortoq and Nanortalik areas (Horsted and Smidt, 1956).

After 1963, research on the West Greenland stock concentrated on the offshore grounds and from the 1970s onwards focused on stock assessment and improvement in survey design. Additional research effort has been devoted to improvement of gear selectivity, sampling of juveniles and bycatch of demersal fish, which is discussed below (Table 8).

Stock assessment of shrimp is jeopardised by the biology of the species. First, the spawning stock biomass is made-up entirely of females and it has to be assumed that there are enough males. Furthermore, the stock-recruitment relationship is not known. Second, there are no hard parts to measure age,

Table 7 Surveys carried out between 1928 and 1954 and their main results (Horsted and Smidt, 1956)

Area	Year	Expedition	Main results
Upernavik and	1928	Godthaab	Shrimp widely distributed
Uummannak	1948–1949	Adolf Jensen	Bottom unsuitable for trawling
Sisimiut	1934–1938	Private	Experimental trawling, discovery of good shrimp grounds
	1949	Adolf Jensen	Failed to find new trawling areas
	1951–1954	Adolf Jensen	Recovery of stock after its collapse in 1949, by 1954 large shrimp dominated catches
	1955	Adolf Jensen	Survey in Itivdleq fjord, unsuitable for fishing
Maniitsoq	1935	Hansen	Experimental trawling
	1950	Adolf Jensen	Low catches, bottom unsuitable for trawling
Disko Bay	1935	Hansen	Found shrimp
	1947	Adolf Jensen	Confirmed existence of good trawling areas
	1948–1953		Experimental trawling, 100–200 kg hour^{-1}, best areas Ilulissat, Qasigiangguit and Qeqertasuaq
Nuuk	1946	Hansen	Suitable areas for trawling found
	1950–1953	Adolf Jensen	Irregular catches, 0.5–270 kg hour^{-1}, better in Narssaq area
Qaqortoq	1946–1953	Adolf Jensen	Found extensive areas suitable for commercial fishing
Vaigat	1948–1949	Adolf Jensen	Low catch, unsuitable bottom
Disko Fjord	1948–1949	Adolf Jensen	Few *P. borealis*; other shrimp species present
Aasiat	1949	Adolf Jensen	60 kg hour^{-1}, low catches, especially with temperature $<0\,^{\circ}$C
Paamiut	1951	Adolf Jensen	Experimental trawling, 6–52 kg hour^{-1}, shrimp grounds too small for commercial use
Nanortalik	1951–1954		Found areas suitable for commercial fishing
Qeqertasuaq (Disko Island)	1954	Dana	Experimental trawling, 130–200 kg hour^{-1}

and growth is stepwise because of moulting. Since there is no moulting during the 9 months of egg bearing, several year-classes can belong to the same size group. These aspects prevent the use of virtual population analysis (VPA), a method widely applied to other species (Carlsson and Smidt, 1976; Horsted, 1991).

Table 8 Research carried out on the Greenlandic shrimp stock since 1960

Year	Research	Main results	References
West Greenland			
1963–1964	Search for new fishing grounds	Offshore grounds found	Horsted, 1969a
1969	Mesh size	Mesh >24 mm reduced catches greatly, continue use of 20- to 22-mm mesh	Smidt, 1969
1969	Tagging experiments	Low recovery but tags did not affect shrimp	Horsted, 1969a,b
1977–1985	Photographic surveys	Disagreement with trawl survey	Kanneworff, 1979a,b, 1981, 1983, 1984, 1986; Jørgensen and Kanneworff, 1980; Carlsson and Kanneworff, 1987a
1979–1988	Foreign surveys	Biomass estimates, biological information, effort, yield, discards and bycatch	*Norway* Ulltang and Torheim, 1979; Jakobsen and Torheim, 1980, 1981a, 1983a; Smedstad and Torheim, 1984a *France* Dupouy and Fréchette, 1979; Fréchette and Dupouy, 1979; Minet, 1979; Derible *et al.*, 1980; Dupouy *et al.*, 1981a, 1983; Dupouy and Fontaine, 1983; Bertrand *et al.*, 1988a,b *Canada* Parsons, 1979, 1980
1980	Fisheries data	Shrimp growth Catches and CPUE estimated from landings and logbooks	Carlsson, 1980a, 1983b, 1984a, 1988, 1990b, 1993; Lund, 1988b, 1990b; Carlsson and Kanneworff, 1989a; Carlsson, 1990b; Lassen and Carlsson, 1990; Carlsson and Lassen, 1991; Carlsson and Kanneworff, 1991a, 1992a, 1993a; Carlsson *et al.*, 1993a; Siegstad, 1994a, 1996; Siegstad *et al.*, 1995

(Continued)

Table 8 (Continued)

Year	Research	Main results	References
1988	Offshore survey	Biomass estimations, length distribution, biological information, assessment	Carlsson and Kanneworff, 1989b, 1991b, 1992b, 1997a, 2000; Lund, 1989b; Carlsson et al., 1990, 1993b, 1995a, 1999; Andersen et al., 1993a; Folmer et al., 1996a; Hvingel, 2002b, 2003b
1989	Gear selectivity	43-mm mesh the most selective	Christensen, 1990
1990–1999	Growth rates	New analysis of survey samples	Carlsson, 1997a
		Increased longevity and slower growth in northernmost locations	Savard et al., 1994
		Growth may be correlated to bottom temperature	Carlsson and Kanneworf, 1999
		Decrease of length at sex change in relation to increased water temperature	Wieland, 2004a
		Decrease in mean size at age during years of increased stock density	Wieland, 2004b
1997	Relationship between catch and depth	Catches seemed to increase with depth	Carlsson, 1997b
		Shrimp most abundant at 400-m depth	Kingsley and Carlsson, 1998
1991	Inshore annual stock assessment (Disko Bay)	Biomass estimations, length distribution, biological information	Carlsson et al., 1992, 1995b; Andersen et al., 1993b, 1994a; Carlsson and Kanneworff, 1993b, 1997b, 1998a; Folmer et al., 1996b
1993	Grid devices	Problems with grid selectivity, flat fish blocked grid	Valdemarsen and Misund, 1993
1993–1994	Gear selectivity	Mesh size of 45 and 60 mm decreased catch of shrimp <22-mm CL by 40–50% and larger shrimp by 15–30%	Lehmann et al., 1993
		No difference between diamond and square mesh of 45 mm	
		High variation in selectivity, but mesh selection was detected	Boje and Lehmann, 1994

Year	Topic	Description	Reference
1997	Estimation of year-class strength and biomass index from commercial fishery data	First year-class is overestimated and the fourth year-class is underestimated	Carlsson and Kanneworff, 1997c
1998–2001	Survey improvement	The new biomass index indicates biomass of shrimp >17 mm	Hvingel et al., 1996, 1998
		Reduction of strata and towing time, sampling stations placed randomly, enlargement of study area and reduction of cod-end liner mesh to 20 mm allow more efficient sampling and better statistical analysis	Carlsson et al., 1998, 2000; Kingsley et al., 1999a; Folmer and Pennington, 2000; Kingsley, 2000, 2001
1998	Recruitment, larval distribution and condition	Recruitment in Store Hellefiske Bank probably dependent mostly on own larval source	Pedersen, 1998
		Larval transport seems to span 500 km Variability in growth and survival expectation of larvae seem to be related to food availability	Storm and Pedersen, 2003 Pedersen and Storm, 2002
		Low recruitment despite high female biomass may be related to decrease of length at sex transition	Wieland, 2004c
1991, 2000	Sampling of juveniles	There are no concentrations of age groups 1 and 2	Carlsson and Kanneworff, 1997d
		Side bags with mesh size 6 mm improve one-group sampling	Wieland, 2002a
2001	Precision of length measurements	Training of personnel and standardisation of data collection needed to avoid bias	Wieland, 2001
	Maturity stages	Classification method	Hansen and Aschan, 2001

(Continued)

Table 8 (Continued)

Year	Research	Main results	References
2002–2003	Juvenile abundance and growth	Mean length seems to be correlated to bottom temperature	Wieland, 2002b, 2003
2002	Model for stock dynamics	Includes all available biotic and abiotic information	Hvingel and Kingsley, 1998, 2002
2002	Recruitment	Abiotic and biotic factors affecting recruitment	Pedersen and Storm, 2002; Pedersen *et al.*, 2002; Storm and Pedersen, 2003
Denmark Strait and East Greenland			
1980–1992	Foreign surveys	Biomass estimates, biological information, effort, yield, shrimp discards and bycatch	*Norway* Jakobsen, 1980; Jakobsen and Torheim, 1981b, 1983b; Smedstad and Torheim, 1984b, 1985, 1986, 1987, 1989, 1990, 1991; Smedstad, 1985, 1988, 1989, 1992 *France* Minet *et al.*, 1980 Dupouy *et al.*, 1981b
		French research also included fisheries and biological data collected during fishing trips	Biseau *et al.*, 1984; Poulard and Fontaine, 1985; Poulard *et al.*, 1986; Bertrand *et al.*, 1988a,b *Faeroe Islands* Hoydal, 1980 *Iceland* Hallgrímsson, 1980, 1981
1980–1988	Fisheries data	Between 63°–68°30'N Estimation of TAC and sustainable yield	Carlsson, 1980b, 1981b, 1983a, 1984b, 1985, 1986, 1988, 2000; Skúladóttir, 1985b, 1990a,b, 1991b, 1992a, 1993c, 1994a,b, 1995c,d, 1996b,c, 1997b,c, 1998c,d; Skúladóttir and Hallgrímsson, 1986; Carlsson and Kanneworff, 1987b; Skúladóttir *et al.*, 1991b, 1993; Siegstad, 1994b; Hvingel, 1999, 2001, 2003c, 2004b; Carlsson and Hvingel, 2000; Siegstad and Hvingel, 2005

1989–1990	Search surveys for new fishing grounds off Southwest, Southeast and East Greenland	Unexploited areas investigated in Qaqortoq, from Cape Farewell to 65°30'N and 67°30'–77°36'N, catches too small for commercial fishing, no need to enlarge stock assessment survey area	Lehmann, 1989, 1990
1989–	Stock assessment survey	Biomass estimations, biological information, length distribution	Carlsson and Kanneworff, 1989c, 1990, 1991c, 1992c, 1993c,d, 1995; Andersen et al., 1994b; Carlsson, 1996
1995–	Sexual maturity of females	Relationship between sea temperature and size at maturity and sex transition	Skúladóttir, 1995a, 1998a
1998	Surplus production model	Based on the assumption that the net growth rate of a stock is related to its biomass. Allows better estimations of maximum sustainable yield (MSY), biomass needed to produce MSY (B_{MSY}) and fishing mortality to maintain MSY (F_{MSY}). Provides better management advice	Cadrin and Skúladóttir, 1998
1998	Genetic population studies	Shrimp from Dorhn Bank constitute a biological unit distinguishable from inshore and offshore shrimp in Icelandic waters	Jónsdóttir et al., 1998
2001	Length distribution	Shrimp samples from the commercial fishery	Kingsley and Siegstad, 2001

The first stock estimations were made by calculating the yield per unit area based on the stable catches from Disko Bay. The calculation would include fishable grounds and surrounding areas to supply larvae and adults, and it would consider only the spawning stock biomass. This method was applied to new grounds found in 1975, and the fishable stock was estimated at 17,000 tons. By the end of 1976, this figure had risen to 40,000 tons due to all the new fishing grounds found during that period (Carlsson and Smidt, 1976; Horsted, 1991).

CPUE data were used in combination with survey results, although it must be taken into account that they do not give an accurate idea of the stock size, as increasing experience of the skippers and technical improvements can raise CPUE indices independently of stock size trends (Horsted, 1991; Hvingel, 1996). Photographic surveys of the offshore fishing grounds were carried out during the period 1977–1985 to estimate total biomass and changes in distribution, but they were interrupted because their results did not agree with those from the trawl survey (Kanneworff, 1986; Carlsson and Kanneworff, 1987a).

The nations that took part in the fishery during the first years also surveyed the fishing grounds off West and East Greenland until 1988 and 1992, respectively, but very few foreign surveys were conducted in West Greenland after the fishery was closed to all nations in 1981 except Canada and Greenland (Table 8). The first annual trawl survey off East Greenland focused only on the fishing grounds, but extended to cover the main stock distribution area north of 65°N in the following years (Andersen *et al.*, 1994b).

Stock assessment and estimation of TAC for West Greenland are currently based on biological as well as catch data. There have been no surveys in East Greenland since 1996 and the TAC is estimated from quality indicators obtained from fishery data. Assessment for West Greenland is made with a surplus production model (ASPIC) that includes predation by cod, survey biomass indices of shrimp with carapace length >17 mm, unstandardised effort for the period 1978–1987, standardised catch rates (CPUE since 1987), cod biomass and cod predation on shrimp, and is based on the precautionary approach (Cadrin and Skúladóttir, 1998). Standardised CPUE indices are estimated with a model that takes into account seasonal and spatial availability of shrimp, annual mean CPUE and the number and size of the trawlers fishing in the different subareas each month and year, therefore accounting for fleet efficiency and preventing that increases in CPUE are wrongly attributed to a biomass build up (Hvingel *et al.*, 1998, 2000). The Bayesian method is used to estimate likelihood distributions of parameters relevant to management, that is, investigate the effect of five different TACs on the stock. The assessment assumes that increases in CPUE may be derived from an increase in efficiency rather than in shrimp biomass, and it consists on estimating the optimal fishing mortality to achieve maximum sustainable yield of the fishery ($F_{LIM} = F_{MSY}$). Fishing mortality is reduced if the stock biomass decreases below the biomass required to obtain the B_{MSY} (Cadrin and Skúladóttir, 1998; Anonymous, 1999; Hvingel, 2002b; Hvingel and Kingsley, 2002; NAFO, 2006).

Additional research has been carried out on larval condition and abundance in relation to recruitment processes. Pedersen (1998) observed that hatching is delayed in a south-north direction, in correlation with the later warming of water and the onset of primary and secondary production. The predominance of younger larvae in the Store Hellefiske Bank suggested that the shrimp stock there is mostly self-recruiting, despite the presence of older larvae that had drifted from more southern localities. The estimation of larval survival is also relevant to predict recruitment, and with this purpose the triacylglycerids (TAG):wet weight ratios are used as an index for larval feeding condition. Using these indices, Pedersen and Storm (2002) found that the variability in larval shrimp growth and survival correlated well to the annual variability in plankton production. Good feeding conditions enhance the probability of larval survival and therefore of good recruitment. Storm and Pedersen (2003) found that larval abundance at hatching correlated well with the size of the adult female spawning stock. Furthermore, egg production seems to be positively correlated to length at sex transition. When water temperature increased in West Greenland in the mid-1990s, stock biomass increased significantly but length at sex transition decreased, resulting in an unexpected reduced year-class (Wieland, 2004a,c).

4.4. Bycatch in the Greenland shrimp fishery

The presence of many benthic species was noted in the first surveys off West Greenland. Horsted and Smidt (1956) made a detailed list of the animal life found on the shrimp grounds they investigated. They recorded sponges, anthozoans, hydroids, polychaetes, nemerteans, jellyfish (2 species), amphipods (3), isopods (1), mysids (1), euphausids, shrimp (11), crabs (5), echinoderms (17), bivalves (14), gastropods (13), chaetognaths (1), ascidians and 40 fish species. In general, the most common fish bycatch species in shrimp grounds at 300- to 400-m depth with soft bottom and water temperature above 0°C were Greenland halibut (*R. hippoglossoides*), redfish (*Sebastes marinus*), longear eelpout (*Lycodes seminudus*) and Vahl's eelpout (*Lycodes vahlii*). In smaller numbers there were Arctic cod (*B. saida*), Atlantic cod (*G. morhua*), Atlantic poacher (*Leptagonus decagonus*), seasnail (*Liparis* sp.) and sea tadpole (*Careproctus reinhardti*), among others.

In threshold fjords with water below 0°C and 150-m depth, shrimp were smaller and were not the dominant species. Fish bycatch was dominated by long rough dab (*H. platessoides*) and longear eelpout (*Lycodes seminudus*). Invertebrates dominated areas of hard bottom (Horsted and Smidt, 1956). Smidt (1965, 1970, 1971) also found that the most abundant fish bycatch species were Greenland halibut, long rough dab, redfish and eelpouts, but he did not provide figures in his papers.

Available bycatch records for the period 1970–1990 were not comparable to recent data because they correspond to only part of the fleet and the data from several subareas were often pooled together. Eight Greenlandic trawlers fishing in Areas 0 and 1 reported that bycatch ranged from 0.8% to 23.1% of the shrimp catch during the years 1977–1983. Redfish made up most of the bycatch and it was the only bycatch species recorded separately (Carlsson, 1984a). Other bycatch data collected during surveys include redfish and Greenland halibut (Pedersen and Lehmann, 1989). Norwegian surveys reported bycatch of mostly redfish, but also Greenland halibut and cod in both West and East Greenland (Jakobsen and Torheim, 1980, 1983b; Smedstad and Torheim, 1984a, 1987). The first survey to assess the stocks of bycatch species took place in 1987 and identified 27 taxa. However, the coefficient of variation was over 25% for many of them and the biomass of some of the species was clearly underestimated. Cod biomass was the largest bycatch, close to 430,000 tons, followed by Greenland halibut (58,000 tons) and roundnose grenadier (*Coryphaenoides rupestris*, 43,000 tons). However, biomass of deep-sea redfish (*Sebastes mentella*) was estimated at 8000 tons despite being the most abundant bycatch species recorded in all other surveys (Yamada *et al.*, 1988).

More recently, nearly 130 fish bycatch species have been recorded in the shrimp survey off West Greenland. The bycatch was dominated in both abundance and biomass by juvenile Greenland halibut and redfish. Other common species, but not so abundant, were long rough dab, Atlantic wolffish (*Anarhichas lupus*), spotted wolfish (*Anarhichas minor*), cod and starry skate (*Amblyraja radiata*) (Carlsson and Kanneworff, 1998b; Kingsley *et al.*, 1999b; Storr-Paulsen, 2004). Trends in abundance and biomass of these species are given in Figure 28.

Greenland halibut has always been most abundant in Disko Bay and NAFO Subarea 1BN (between 67°–68°50′N and 50°–59°45′W), and it is also found further north (beyond 71°N) than redfish and in the eastern part of the area covered by the shrimp survey (Pedersen and Lehmann, 1989; Pedersen and Kanneworff, 1995). During the period 1988–2002, length ranged from 5 to 75 cm, but with frequency modes around 11–13 cm and 18–25 cm. The largest fish was found in the deeper strata, at 400- to 600-m depth. Variability in abundance between years was considerable, although biomass has tended to increase (Pedersen and Lehmann, 1989; Engelstoft, 1996, 1997, 2002; Jørgensen and Carlsson, 1998; Storr-Paulsen, 2004) (Figure 28A).

Redfish were found mostly in NAFO Subareas 1B and 1C (67°–68°50′N and 64°15′–66°15′N, respectively, both 50°–57°W). Redfish dominate in the

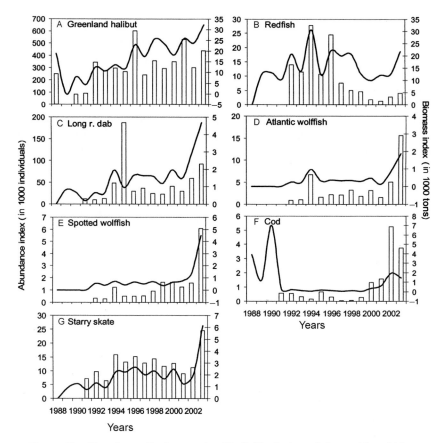

Figure 28 Abundance (in thousands of individuals, bars, left *y*-axis) and biomass (in thousands of tons, lines, right *y*-axis) indices of several demersal fish species off West Greenland, based on data from the annual shrimp assessment survey. Note that the scales of *y*-axis differ between graphs. Long r. dab, long rough dab. Data from Storr-Paulsen (2004).

western part of the area covered by the shrimp survey. Length ranked from 4 to 57 cm, but frequency modes in most years were 7 cm and around 13 cm. The largest fish were found at 400- to 600-m depth with both abundance and biomass varying greatly among years (Figure 28B) (Pedersen and Lehmann, 1989; Pedersen and Kanneworff, 1995; Engelstoft, 1996, 1997, 2002; Jørgensen and Carlsson, 1998; Storr-Paulsen, 2004). Differences in length distribution between sampling areas and different depth strata suggest that juvenile redfish and Greenland halibut migrate from shallow to deeper water, and westward and southward from an area between 67°N and 69°30′N (Kanneworff and Pedersen, 1992; Pedersen and Kanneworff, 1992).

Available information for the remaining species is more scarce. Abundance and biomass of long rough dab have oscillated since 1991 but they increased greatly in 2002 and 2003, reaching nearly 60 and 71.6 million individuals (2500 and 4000 tons, respectively). The abundance estimate for 2004 was lower, 43.4 million individuals, but apparently the fish were bigger since biomass was estimated to be 2700 tons (Figure 28C). The length range is from 5 to 43 cm, with frequency modes at 10–12, 15 and 30–32 cm between 1996 and 2002 (Engelstoft, 1996, 1997, 2002; Jørgensen and Carlsson, 1998; Storr-Paulsen and Jørgensen, 2005). Atlantic and spotted wolffish were much more abundant in 2003 than in previous years (Figure 28D and E). Wolffish ranged from 5- to 70-cm length during the period 1996–2002, but most individuals were <35 cm with higher abundance of fish between 8–12 and 25–30 cm (Engelstoft, 1996, 1997, 2002; Jørgensen and Carlsson, 1998).

Atlantic cod was mostly found in the southern part of the survey area. Cod was more abundant between 2002 and 2004 than in previous years, ranging between 4.1 and 6.5 million individuals and 1800–2400 tons. However, biomass in these years was very low compared with the biomass peak of 7000 tons in 1990, which was assumed to be due to the 1984–1985 year-classes recruited from Iceland (Figure 28F). Length range of cod was usually 15–60 cm except for 1996 when it was 25–42 cm. Frequency modes were at 23–27 and 34–38 cm (Engelstoft, 1997, 2002; Jørgensen and Carlsson, 1998; Storr-Paulsen and Jørgensen, 2005). Biomass and abundance trends of starry skate (Figure 28G) oscillated during the period 1988–2002 but peaked at 17 million individuals and nearly 4500 tons in 2003 (Pedersen and Kanneworff, 1995; Jørgensen and Carlsson, 1998). The species is found in all the survey area, but in 2004 it was most abundant between 70°37′N and 72°30′N (Storr-Paulsen and Jørgensen, 2005). Other bycatch species, such as Arctic cod (*B. saida*), sculpins (*Triglops* sp.) and sea snails are very abundant in number (Smidt, 1971; Pedersen and Kanneworff, 1995), but no data have been collected on these, probably because they have no commercial interest.

Rätz (1997a) recorded 66 bycatch fish species in annual groundfish surveys designed for cod assessment carried out in West and East Greenland during the period 1982–1996. The most abundant species in number were deep-sea redfish (*Sebastes mentella*, 21%), redfish (*Sebastes marinus*, 15.1%), Atlantic cod (14.5%) and long rough dab (4.7%). However, cod was most abundant in terms of biomass (45.7%), followed by redfish (27%), deep-sea redfish (14.6%), unidentified redfish and long rough dab, each of which contributed 2.4% (Figure 29). All these species represented 92% and 96% of the bycatch in abundance and biomass, respectively. There are some differences in abundance and biomass of bycatch species during the years 1988–1996 in West Greenland shown by Figures 28 and 29. This can probably be explained by the differences in the areas investigated during the different surveys. Bycatch data from the shrimp survey represent all Subarea 1

(Figure 17), whereas the groundfish survey data used by Rätz (1997b) had its northernmost limit at 67°N.

Fish were evenly distributed during the 1980s off West and East Greenland, until mean individual weight declined by 25–50% in the 1990s (Figure 30), indicating that the stocks were made up of juveniles (Rätz, 1997b). The stocks of spotted wolffish and starry skate in West Greenland were thought to be severely depleted considering their biomass trends since 1982 (Rätz, 1998b). It is known that shrimp grounds overlap with important nursery grounds of redfish, Greenland halibut, cod and Arctic cod (Smidt, 1969; Pedersen and Lehmann, 1989). Spawning of redfish is unknown in West Greenland. It is thought that they are recruited from breeding areas southwest of Iceland (Anonymous, 1984b). However, in northwest, west and southwest of Disko Island, there are very important nursery grounds for Greenland halibut. This spatial distribution of nursery grounds might be related to the hydrography of the area, food availability or predator–prey relationships among these species (Pedersen and Kanneworff, 1995). In fact, the stomach contents of redfish >20 cm and Greenland halibut >14 cm consist mostly of shrimp and juvenile

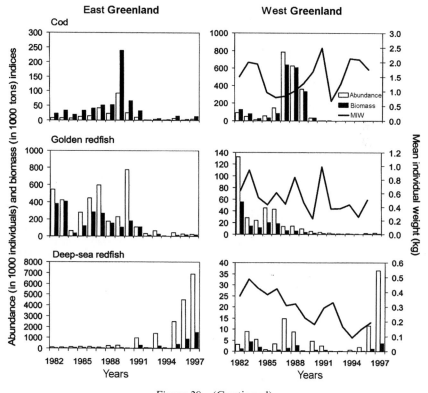

Figure 29   (Continued)

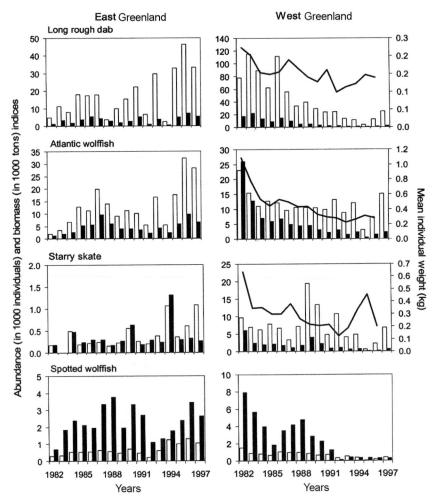

Figure 29 Abundance (in thousands of individuals, white bars) and biomass (in thousands of tons, black bars) indices of the most abundant demersal fish species in East (left column) and West (right column) Greenland estimated from the annual groundfish survey data. The black line represents mean individual weight (MIW, in kg) and refers to the right *y*-axis. These data were not available for samples from East Greenland. Data from Rätz (1998a).

redfish (Pedersen and Riget, 1993). Cod, redfish, long rough dab and starry skate also prey on shrimp (Pedersen, 1994).

Disagreement on bycatch data from the surveys and commercial fishing makes it difficult to estimate accurately the impact of the shrimp fishery. First

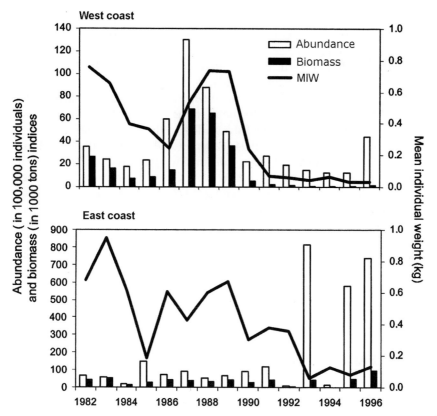

Figure 30   Indices for abundance, expressed as hundreds of thousands of individuals (white bars) and biomass, expressed as thousands of tons (black bars) of 66 demersal fish species recorded off the West and East coasts during the period 1982–1996. Mean individual weight (MIW, black line) is expressed in kg and refers to the right *y*-axis. Note that the left *y*-axis has different scales. Juvenile *Sebastes* spp., *Sebastes mentella* and *Sebastes marinus* represented 71.5% of the total abundance, followed by *G. morhua* (14.5%) and *H. platessoides* (4.7%). In terms of biomass, *G. morhua* represented 45.7% of the total, followed by *Sebastes* spp., *Sebastes mentella* *Sebastes marinus* (44%) and *H. platessoides* (2.4%). Data from Rätz (1997b).

bycatch estimations for redfish and Greenland halibut in the 1988 shrimp survey were 12% and 7%, whereas fishermen reported less than 1.6% (Pedersen and Lehmann, 1989). The introduction of observers in 1991 may have influenced the increase of bycatch reports from the commercial fishery from 1% in 1987 to 3% in 1998 (Hvingel, 2002a). Moreover, the use of sorting grids and other regulations (Table 6) have been considered effective in decreasing mortality of demersal fish in the shrimp fishery. Nevertheless, regular data on fish bycatch have been collected since the annual shrimp stock assessment survey started in 1988. Biomass estimates for demersal fish species obtained during the shrimp surveys are considered reliable and they provide an accurate estimate of the state of the demersal stocks. They are also supported by results from other studies on vertical distribution of shrimp and fish (Pedersen and Kanneworff, 1995; Storr-Paulsen, 2004).

The observed general decrease in abundance and biomass of the investigated demersal fish stocks has been attributed to the rising mortality, especially of juveniles, caused by the growing fishing effort in the offshore shrimp grounds (Rätz, 1992, 1997b; Pedersen, 1994; Pedersen and Kanneworff, 1995; Storr-Paulsen, 2004). The fact that mean individual weight started to increase again in East Greenland in 1994 but not in West Greenland suggests recruitment failure caused by the shrimp fishery (Rätz, 1998a). The changes in climate and surface currents recorded in Greenland during the last century that overlap during certain years with the observed changes in trends of recruitment and biomass of the main bycatch species may also have affected the non-marketable bycatch species (Smedstad and Torheim, 1990; Pedersen, 1994; Horsted, 2000; Buch et al., 2002, 2004). More information on climate variability can be found in Stein (2004), who investigated the climatic conditions off West Greenland from 1990 to 2001 and compared them with previous decades.

5. SVALBARD, THE BARENTS SEA AND JAN MAYEN

5.1. The fishery

The Norwegian shrimp fishery for *Pandalus borealis* started in inshore areas in southern Norway in 1897, and extended gradually northwards (Halmø, 1957). It reached Sør-Trøndelag in 1925, Nør-Trøndelag in 1927, Norland in 1929, Tromsø in 1931 and Finnmark in 1935 (Halmø, 1957). The inshore fishery became larger after the World War II, especially in northern Norway, from where it developed into the offshore fishery that started operating in Svalbard and the Barents Sea in 1970. However, offshore shrimp fishing had taken place in Norwegian waters as early as 1930 in the Skarregak (Ulltang, 1981). Unlike in Greenland and Iceland, 35% of the catch from the inshore fishery is sold locally on arrival to land either fresh or boiled on board (ICES, 2006).

5.1.1. Svalbard and Barents Sea area

The shrimp fishery started in the Barents Sea in 1973 with 30 Norwegian trawlers after a pilot study had mapped the shrimp resources in 1970 and 1971 (Strøm and Rasmussen, 1970; Strøm, 1971; Anonymous, 1976c).

However, other nations have traditionally taken part in the shrimp fishery in waters off Svalbard. Their catches are distributed according to historical rights as stipulated in the Svalbard Treaty and the regulation of a Fishery Protection Zone around the islands (Standal, 2003). It has been estimated that 69% of the distribution area of shrimp in the Svalbard zone lies within Norwegian waters, but on average 75.5% of the total catch is taken by Norway (Standal, 2003), 17.8% by Russia, which joined the fishery in 1974, and 6.7% by other countries (Anonymous, 2003b; Aschan *et al.*, 2004; ICES, 2006). The Faeroes fished most years during the period 1978–2001, whereas other countries belonging to the EU have participated sporadically (Anonymous, 1986b, 2004b; ICES, 2006) (Figure 31). It has been suggested that the rapid increase of catches in the Barents Sea could have been greatly influenced by the quota restrictions adopted within the Greenland economic zone in 1977 (Berenboim *et al.*, 1980).

Catches have shown great fluctuations after reaching a maximum of 128,000 tons in 1984 (Figure 31), due to variations in stock size caused by irregular recruitment and predation from cod, but also because of variations of effort provoked by market prices (Berenboim *et al.*, 1986a; Anonymous, 2003b; Standal, 2003; Aschan *et al.*, 2004; ICES, 2006). Thus, in 1975 there were no shrimp to be found off Svalbard and the fleet moved to new grounds that were found East of Bjørnøya and Hopen (Anonymous, 1976c). The lowest catch recorded was 25,000 tons in 1995 (Aschan *et al.*, 2004).

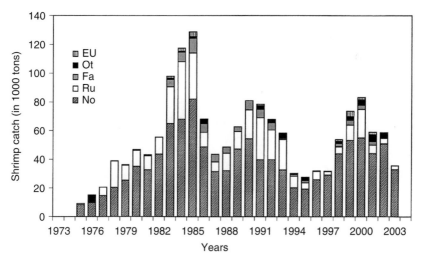

Figure 31 Shrimp catches of the international fleet in the Barents Sea and Svalbard area, expressed in thousands of tons. No, Norway; Ru: Russia; Fa, Faeroe Islands; EU, European Union countries; Ot, Others. Data from Anonymous (1986b, 1996c, 2004b).

Stock estimations from 1977 supposed that the areas of largest concentrations were confined within 15% of the shrimp distribution area; density was too low outside these grounds for a commercial fishery. The stock was estimated to be 15,500 tons in Svalbard, distributed over an area of 10,300 km²; and 13,300 tons in Hopen, over an area of 12,300 km² (Anonymous, 1978c). However, the fishery was constrained by sea ice conditions and the high variability in catches within the different areas. CPUE ranged during 1978 and 1979 from 24 to 105 kg hour^{-1} in Hopen, from 86 to 188 kg hour^{-1} in Bjørnøya and from 120 to 205 kg hour^{-1} in West Svalbard (Anonymous, 1979b). Nevertheless, catches of the Norwegian fleet increased from about 9700 to 82,000 tons from 1976 to 1985, with total catches rising to 128,000 tons in 1984 (Figure 31), partly due to the stock increase when the year-classes from 1977 and following years recruited to the stock. Migration of juvenile cod and haddock into the shrimp grounds prompted the closure of large areas within the shrimp grounds between 1983 and 1988. On the other hand, although shrimp abundance off Svalbard showed an increase from 8% to 19% during the period 1986–1988, the stock had been greatly reduced to about one-third of its size in 1984 (Figure 32). This decline was caused by poor recruitment, increased mortality by predation and input of cold water (below −1°C) in certain areas (Teigsmark and Øynes, 1983a; Anonymous,

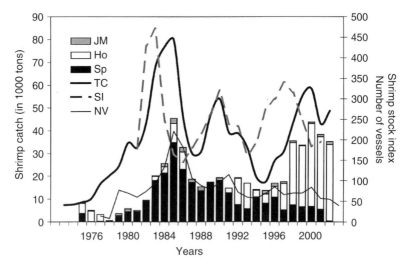

Figure 32 Norwegian shrimp catches from Jan Mayen, Hopendypet and Spitsbergen, expressed as thousands of tons and fleet size. JM, Jan Mayen; Ho, Hopendypet; Sp, Spitsbergen; TC, total catch for the Barents Sea north of 62°N; SI, shrimp stock index; NV, number of vessels. Catches during the years 1973–1974 amounted to 7300–7500 tons, most of it from the Barents Sea. A small percentage corresponds to the inshore fishery. Data from Anonymous (1976c, 1984c, 1986b, 2004b).

1984c, 1985b, 1986a, 1987c, 1988b; Tveranger and Øynes, 1985; Hylen, 1986; Hylen and Øynes, 1986; Berenboim et al., 1987). Moreover, fishing mortality (F) reached 0.4 after 1984–1985, and Berenboim et al. (1991) concluded that overfishing was partly responsible for the large variability of the shrimp stock size in the Barents Sea. They pointed out that shrimp stocks managed with TAC were more stable, and they suggested the implementation of a TAC for the Barents Sea, estimated with the multi-species VPA (MS VPA) as modified by Pope. As a result, catches in Svalbard decreased to 1980 levels (Teigsmark and Øynes, 1983a; Anonymous, 1984c, 1985b, 1986b, 1987c, 1988b; Tveranger and Øynes, 1985; Hylen, 1986; Hylen and Øynes, 1986; Berenboim et al., 1987).

Nevertheless, the recruitment of the strong year-class of 1983 in 1988 increased the stock to over 300,000 tons between 1987 and 1991 (Figure 32). This increase was detected by the Norwegian surveys, but not by the Russians, who estimated that the shrimp stock in the Barents Sea had declined by 40% in 1988 compared to that of 1987 (Berenboim et al., 1989). The large input of warmer Atlantic water and reduced predation due to the decline of the cod stock during these years might also have contributed to the biomass increase. The trend reversed again in 1991 and the stock declined to 50%, probably because of the increased effort since 1989, and the rise of the cod stock coinciding with the collapse of capelin and herring stocks, thus effecting heavier predation pressure on the shrimp (Hylen et al., 1987; Hylen and Øynes, 1988; Mukhin and Sheveleva, 1991; Aschan et al., 1993, 1994, 1995). Despite the declining biomass trend (Figure 33), the government allowed a renewal of licensed vessels in 2000, which led to a growth of the fleet beyond the capacity of the fishery (Standal, 2003) (Section 5.2). The subsequent increase in fishing effort and poor recruitment in 2000 caused the shrimp stock to decline further during the years 2001–2004 (Anonymous, 2004b; ICES, 2006). A map showing the distribution of shrimp in the Barents Sea in 2005 can be found in ICES (2006).

5.1.2. Jan Mayen

A pilot survey was carried out in 1974 and the fishery started in 1975–1976 with 15 Norwegian trawlers. Average catches were 150 kg hour^{-1}, but the grounds were small and ice conditions restricted the fishing operations to summer and autumn (Anonymous, 1974c, 1977d; Øynes, 1977). Annual catches showed great variability and depended not only on stock size, but also on accessibility to the area, fleet size and fishing conditions in Greenland, the Barents Sea, Svalbard and the Flemish Cap. Catches peaked in the 1970s, when they reached nearly 9000 tons. Most of this was taken by Norwegian vessels. The fleet consisted mostly of medium-sized vessels

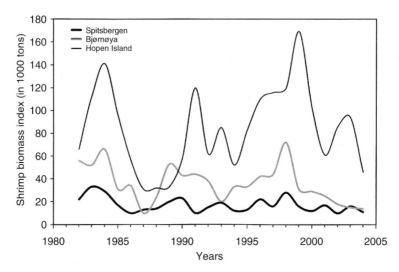

Figure 33 Shrimp biomass index (in thousands of tons) estimated from the Norwegian surveys for Spitsbergen, Bjørnøya and Hopen Island during the period 1984–2004. Data from Anonymous (2004b).

(1000–2000 HP), but the largest percentage of catches during the 1990s corresponded to those >2000 HP (Aschan *et al.*, 1996a).

5.2. Management

There is no TAC for the shrimp stock in Svalbard due to the lack of knowledge about stock structure and the problem of defining management units (Hansen and Aschan, 2000). The fishery is managed with a minimum carapace length (15 mm), fishing licenses, number of effective fishing days and maximum number of juvenile cod, haddock, redfish and Greenland halibut captured as bycatch (Aschan and Sunnanå, 1997; Aschan, 1999a,b; Hansen and Aschan, 2000; Table 9). Since 1983, fishing grounds are closed when the number of juvenile fish surpasses a pre-determined threshold. Thus, the Norwegian–Soviet Fisheries Commission imposed closure of grounds in 1983 when the number of juvenile haddock and cod was >300 individuals per ton of shrimp. This rule was also applied to juvenile Greenland halibut from 1992 onwards (Veim, 1999). This strategy is based on the future market price of the bycatch, which if higher than the value of the present shrimp catch, the fishing areas are closed. However, a good estimate of maximum allowable juvenile fish in the bycatch is very difficult to obtain. Accurate estimation of this parameter requires good information

Table 9   Management regulations for the shrimp fishery in Svalbard, Barents Sea area and Jan Mayen

Year	Regulation	References
1973	License mandatory for trawlers >50 GRT	Ulltang, 1981
1974	Mesh size in cod-end with stretched net must be at least 30 and 35 mm south and north of 65°N, respectively	Ulltang, 1981
1983	Joint Soviet–Norwegian Fisheries Commission decides to close shrimp grounds when bycatch of juvenile cod and haddock is >300 individuals per ton of shrimp	Veim, 1994
1989	Maximum catch of cod and haddock limited to eight individuals per 10 kg of shrimp	Aschan, 1999a
	Mesh size 35 mm (stretched mesh) north of 64°N and 40 mm in Jan Mayen	Aschan, 1999b
	Minimum CL established at 15 mm, maximum 10% of catch may be <15 mm	Veim, personal communication
1991	Sorting grid in Norwegian and Russian EEZ and in Jan Mayen, bar spacing in Barents Sea and Spitsbergen is 19 mm	ICES, 1998
1992	URSS and Norway agree on restrict bycatch of Greenland halibut to 300 individuals per metric ton	Reithe and Aschan, 2004
1993	Use of sorting grid (20-mm bar space) mandatory in Barents Sea to prevent catches of fish >18 cm	Anonymous, 2004b
1996	Restrictions on number of days at sea	Aschan and Sunnanå, 1997
	Restrictions on number of vessels per country	
1997	Closure of Svalbard area to nations without historical fishing rights	Standal, 2003

CL, carapace length.

on year-class strength, average yield per fish, cannibalism rates, knowledge of the magnitude of predation on the shrimp stock and a good prediction of market prices for bycatch fish species that will depend on supply and demand, among other factors (Aschan, 1999a; Veim, 1999).

The maximum allowable number of juvenile fish has been revised several times since the enforcement of the regulation, and especially after the obligatory use of sorting grids started in January 1993. The grid prevents catches of all fish under 18–20 cm, so the rules about closure of grounds were changed in 1995 to allow bycatches of up to 1000 cod or haddock per ton of shrimp (Aschan, 1999a,b; Veim, 1999). The maximum bycatch was recently limited to 8 cod and haddock, or 10 redfish or 3 Greenland halibut per 10 kg of shrimp (Anonymous, 2003b).

Despite the measures taken, the stocks of bycatch species have declined over the years. The five bycatch species of commercial value found in the

Barents Sea and Svalbard area, cod (*G. morhua*), redfish (*Sebastes mentella* and *Sebastes marinus*), Greenland halibut (*R. hippoglossoides*) and haddock (*Melanogrammus aeglefinus*) are all outside safe biological limits, meaning that regulating bycatch according to shrimp catches was not a good solution. In fact, the bio-economical model may in certain cases protect the shrimp stock rather than the bycatch species. As it is, the model estimates the bycatch criterion by current shrimp catch and future yield of fish. Therefore, it is most likely that closures are applied when shrimp stocks are small and fish stocks are large, as happened in 1993, rather than the other way round. The reason behind this is the lack of conservation factors in the model. These, however, are very difficult to incorporate in the short term due to lack of data. In the situation of low levels of the fish stocks, the best solution is probably to put aside economical considerations and apply a more conservative biological approach (Aschan, 2000b; Reithe and Aschan, 2004). Nevertheless, the main management problem the fishery faces currently is the large increase in capacity that the fleet has developed since the late 1980s, motivated by the lack of TAC. Until 1998, the Norwegian government limited the capacity of the fleet by controlling the renewal of licensed vessels. However, fishermen with 34-m (250 GRT) vessels fitted for cod and shrimp fishing wanted to be able to fish in sea ice to lengthen the fishing season. This required very powerful vessels that could not be built under the legal size restriction of 34 m and 250 GRT. Eventually the government rescinded, allowing vessel renewals. Before 1999, the shrimp fleet was on average made up by vessels 25 years old and 35 m in length. Between 1999 and 2000, 25 new trawlers were built, cod fishers with modern vessels applied for shrimp licenses and twin and triple trawls came into use. The fleet became very diverse, with numerous types represented in the range between small vessels and 65-m trawlers with triple trawls. The following statistics illustrate very well the scale of these changes (Standal, 2003):

1. The fleet consisted of 166 vessels in 1985, but 108 in 2001. However, annual catches in the Svalbard area increased by 50% between 1997 and 2000.
2. Catches from vessels 35–50 m have remained stable during the period 1990–2001 despite the fact that the number of vessels has declined by 40%.
3. The number of shrimp trawlers >50 m rose by 140%, whereas the number of shrimp trawlers <50 m declined by 138% from 1990 to 2002.
4. In 1996 only three trawlers had twin trawls, in 1997 there were 35 and in 2002 there were 40, among which there are two or four trawlers with triple trawls.

The development of the fleet has created a conflict of interest between fishermen, the government and society that has proved difficult to solve.

The new fleet requires a large income to keep going independently of the price of shrimp in the market. On the other hand, owners of small boats complain of the arrival of cod fishermen into the shrimp fishery, as they have increased the fleet capacity, but lowered profitability. Actually, the smaller boats in the fleet would benefit from a TAC, as the Institute of Marine Research has advised over the past few years. Stricter management would probably lead to improvement of the stock, which is in the long-term interest for society, although it would prove difficult to implement. It would also have an adverse effect on the Norwegian fishery, at least during the first years, because the stock would have to be divided among the countries involved (Standal, 2003).

5.3. Research

The Norwegian Directorate of Fisheries has supported surveys to search for shrimp fishing grounds in the Svalbard area since 1923 (see review by Strøm and Øynes, 1974). From 1980 to 1992, the Institute of Marine Research in Bergen carried out the surveys in the Barents Sea and Svalbard area (Aschan and Sunnanå, 1997). The former Polar Research Institute of Marine Fisheries and Oceanography (PINRO) in Murmansk, Union of Soviet Socialist Republics (USSR) also conducted its own Barents Sea surveys from 1984, until both countries agreed in 1992 on continuing the stock assessment with joint surveys and common methodology (Table 10), although they do not survey exactly the same area. Russia surveys the Kola coast and Goose Bank but does not survey the Island Trench, and has surveyed Bjørnøya, Storfjord Trench and Spitsbergen less often than Norway (ICES, 2005). Also in 1992, the Norwegian Institute of Fisheries and Aquaculture Ltd. (Fiskeriforskning) was appointed to carry out all research on shrimp in collaboration with the Norwegian College of Fisheries Science at the University of Tromsø (Aschan and Sunnanå, 1997). The main goals of the surveys are to estimate stock size and to increase knowledge of the population structure. The results from the Norwegian and the Russian surveys are well correlated (Figure 34) (Aschan et al., 2004). Moreover, the bycatch data can be used to estimate abundance of some fish species. The surveys cover the known shrimp grounds, or roughly 164,000 km^2 in the Barents Sea and 74,000 km^2 in Svalbard. These areas are divided into depth strata in Svalbard but into fishing grounds in the Barents Sea. Since their beginning, the surveys have undergone several modifications to improve accuracy. Thus, the original design of random stations within strata changed into a design with fixed stations in a regular grid within the strata, and later to fixed stations in a depth stratified system. These improvements may suggest that the indices obtained in the early period are not comparable to the more recent ones, but the fact is that the Russian survey has undergone very minor changes and the correlation among both surveys is

Table 10 Surveys carried out in Svalbard and Barents Sea area since 1970

Year	Survey	Main results	References
1970	Search	*Spitsbergen*: 140–150 kg hour^{-1} *Hopen*: 50 kg hour^{-1}, 120 kg of juvenile redfish, 20 kg of cod and 20 kg of juvenile haddock *Bjørnøya*: 24 kg hour^{-1}	Strøm and Rasmussen, 1970
1971	Search	*Bjørnøya*: Most hauls >50 kg hour^{-1}, redfish most abundant bycatch species, up to 150 kg	Strøm, 1971
1974	Mapping	Mapping of shrimp grounds in Svalbard and Barents Sea	Strøm and Øynes, 1974
1976	Search	Large shrimp in Kongsfjord, 170–180 individuals per kilogram	Strøm, 1976
1980–1990	Norwegian stock assessment surveys	Collection of catch data, bycatch and biological information to estimate stock size, abundance of juveniles of marketable species and investigate the structure of the shrimp population	Tavares and Øynes, 1980; Teigsmark and Øynes, 1981, 1982, 1983a,b; Hylen *et al.*, 1984, 1987, 1989; Hylen and Øynes, 1986, 1988; Tveranger and Øynes, 1986; Hylen and Ågotnes, 1990
1986–1991	Russian stock assessment surveys	Collection of catch data, bycatch and biological information to estimate stock size, abundance of juveniles of marketable species and investigate the structure of the shrimp population	Berenboim *et al.*, 1986a, 1987, 1990; Hylen and Ågotnes, 1990; Mukhin and Sheveleva, 1991
1987–2000	Bycatch	Estimation of cod taken as bycatch Bio-economic management of the shrimp fishery with bycatch of commercial species	Hylen and Jacobsen, 1987; Veim, 1994; Aschan, 1999a,b, 2000b
1992–1996	Joint Norwegian and Russian stock assessment surveys	Collection of catch and biological data, and information on bycatch to estimate stock size, abundance of juveniles of marketable species and investigate the structure of the shrimp population	Berenboim *et al.*, 1992; Aschan *et al.*, 1993, 1994, 1995, 1996b
1997	Evaluation of surveys	Assessment of the three approaches used during the period 1980–1997	Aschan and Sunnanå, 1997

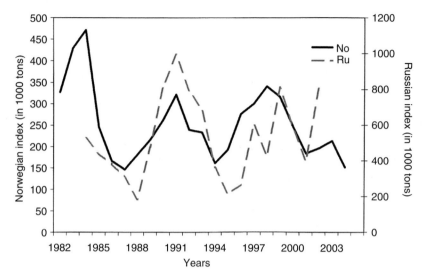

Figure 34 Trends of the biomass indices (in thousands of tons) estimated from the Norwegian (No) and Russian (Ru) surveys during the period 1982–2004 (ICES, 2006).

good. However, sampling of juvenile shrimp needs to be bettered to permit the estimation of recruitment indices. Nevertheless, the main criticism against the survey is that the results are not used in a model integrating survey and fisheries data, predation and other variables (Aschan *et al.*, 1993; Aschan and Sunnanå, 1997). Similarly, different models have been used to estimate stock size, a list of which can be found in Aschan *et al.* (2004). The main research projects carried out in the shrimp stock in the Barents Sea are summarised in Table 11.

The impact of cod on shrimp populations in the Barents Sea has also been investigated. Berenboim *et al.* (1986b) were the first to describe the predator–prey relationship between cod and shrimp in the Barents Sea. They found that shrimp biomass correlated very well with cod stock size ($r = -0.98$) and frequency of shrimp ocurrence in cod stomachs ($r = 0.83$). Later studies concluded that cod prey on shrimp of all age groups and the percentage of the shrimp stock eaten by cod ranged from 15% to 58% during the years 1981–1997. Cod is, therefore, one of the main factors influencing the variability of the shrimp stock size and cod predation was included in multi-species VPA and production models (Berenboim *et al.*, 1999, 2000). The bycatch of commercial fish species in the shrimp grounds and bio-economical management options for fishing areas with large catches of juvenile cod and haddock have also been investigated in the Barents Sea and Svalbard area (Hylen and Jacobsen, 1987; Veim, 1994; Aschan, 1999a,b, 2000b; Reithe and Aschan, 2004).

Table 11 Research carried out on the Barents Sea and Jan Mayen offshore shrimp stock

Barents Sea

Year	Topic	References
1903–	Shrimp biology	Wollebæk, 1903; Rasmussen, 1942; Lysy, 1978; Nilssen and Hopkins, 1991
	Growth, local differences in growth rate and age at first female maturity	Rasmussen, 1953; Teigsmark, 1983; Hansen and Aschan, 2000, 2001
	Spatial variation in length distribution depth-dependent	Aschan, 2000a
	Larval distribution and recruitment patterns	Aschan et al., 2000
	Cod-shrimp trophic relationship	Berenboim et al., 1986b, 1999, 2000
	Genetic variation among populations	Kartavtsev et al., 1991; Rasmussen et al., 1993; Aschan, 1997; Martinez et al., 1997; Drenstig et al., 2000
	Variation in egg size and reproductive output in relation to temperature	Clarke et al., 1991
	Larval distribution and dispersal	Pedersen et al., 2003
1968–	Improvement of fishing gear to decrease catches of juvenile shrimp and fish:	
	Vertical side panel	Isaksen, 1984
	Sorting grid	Isaksen et al., 1990, 1992; Valdemarsen and Misund, 2003
	Shrimp selectivity	Nilssen et al., 1986; Valdemarsen et al., 1993, 1996
	Use of different trawl designs	Rasmussen, 1973
	Mesh selection experiments	Rasmussen and Øynes, 1974; Thomassen and Ulltang, 1975
1970–	Mapping of shrimp fishing grounds	Strøm and Rasmussen, 1970; Strøm, 1971, 1976; Strøm and Øynes, 1974; Anonymous, 1976c; Aschan, 1997
1980–	Stock assessment surveys	(see also Table 10)
	Norwegian	Tavares and Øynes, 1980; Teigsmark and Øynes, 1981, 1982, 1983a, 1983b; Hylen et al., 1984, 1987, 1989; Tveranger and Øynes, 1985, 1986; Hylen and Øynes, 1986, 1988; Hylen and Jacobsen, 1987; Hylen and Ågotnes, 1990

Russian		
Joint	Survey evaluation and improvement, with complete list of surveys	Berenboim et al., 1980, 1986a, 1987, 1989, 1990; Mukhin and Sheveleva, 1991; Berenboim et al., 1992; Aschan et al., 1993, 1996b; Aschan and Sunnanå, 1997
1980–	Stock management	
	Variability of shrimp stock size related to predation and unregulated fishery, TAC suggested	Berenboim et al., 1980; Aschan et al., 2004; Berenboim et al., 1991
	Inclusion of predation by cod in multispecies model	Berenboim and Korzhev, 1997, Berenboim et al., 2000
	Use of models on survey data	Stolyarenko, 1986; Ivanov et al., 1988
	Use of geostatistic method to simulate the precision of abundance estimates	Harbitz and Aschan, 2003
1988	Shrimp catchability	Ivanov and Stolyarenko, 1988
1994–	Bycatch	
	Management: Use of sorting grid does not solve bycatch of juvenile redfish, Greenland halibut, haddock and cod.	Veim, 1994, 1999; Aschan, 1999a,b, 2000b; Reithe and Aschan, 2004
	Evaluation of bioeconomic model	
	Survival of gadoids caught as bycatch	Soldal, 1995; Soldal and Engaas, 1997
	Mortality of cod caught as bycatch	Ajiad et al., 2004
Jan Mayen		
1977	Experimental fishing	Torrisen, 1974; Strøm, 1976; Olsen, 1977; Øynes, 1977; Torheim, 1980
1995	Mapping of shrimp fishing grounds	Aschan, 1995
1996	Catch statistics and shrimp biology	Aschan et al., 1996b
	Bycatch	Torrisen, 1974; Strøm, 1976; Torheim, 1980; Aschan et al., 1996

Note: The early start of research on fishing gear improvement was due to the existence of an inshore shrimp fishery.

The first shrimp samples from Jan Mayen were obtained in 1950 (Rasmussen, 1953; Øynes, 1977), but search for shrimp grounds did not start until the 1970s (Anonymous, 1974c; Torrisen, 1974; Strøm, 1976; Olsen, 1977; Table 12). The distribution of the stock was mapped in 1979, and more surveys followed in 1980, 1981, 1994 and 1995. The Russians also surveyed the area intermittently between 1974 and 1990 (Aschan et al., 1996a). These Norwegian and Russian surveys are merely mentioned in Norwegian annual reports and the results of most of them have not been published. However, some of the Norwegian data have been compiled (Aschan et al., 1996b). The latest survey to map and assess the resources in the area was carried out in 1995 (Aschan, 1995).

Investigations on shrimp growth in the Barents Sea focused on local variation of length distribution (Aschan, 2000a), growth rates and age at maturity (Hansen and Aschan, 2000). Genetic investigations have been undertaken to identify shrimp sub-populations in the Barents Sea, Svalbard, Jan Mayen and the Norwegian coast, among other locations in the Northeast Atlantic. Genetic variation was found to be very high within locations and there are genetic gradients linked to water masses, but distinct sub-populations were not identified. However, it was possible to define 14 biologically homogeneous subareas on the basis of growth rates that can be used in production models and multi-species virtual population analysis (Kartavtsev et al., 1991; Rasmussen et al., 1993; Aschan, 1997; Martinez et al., 1997; Drengstig et al., 2000). Larval dispersal has also been investigated with the aim of identifying the source sink dynamics of Pandalus borealis, of interest for management. Larval settlement seems to depend on hydrodynamics of the Barents Sea, with most settlement taking place in the Polar Front area (Pedersen et al., 2003).

Research on selectivity of shrimp trawls must be mentioned here even though fishing experiments may not have been carried out within the study area. Mesh selection experiments commenced in 1968 and continued during the following years with comparisons of three different mesh sizes (Thomassen and Ulltang, 1975). The Institute of Fishery Technology Research conducted extensive research on the effectiveness of sorting panels in Norwegian shrimp trawls during the years 1975–1978. These panels were attached to the aft belly of the trawl with the purpose of directing fish towards the fish 'leading' channel. The channel consists of two small-meshed 'walls' that lead to release of fish through the lower belly. The upper part of the leading channel was a 60-mm mesh that allowed shrimp to go through and enter the cod-end. Results were very good in trawls used in the inshore fishery, but considered variable in the offshore shrimp trawls. The selectivity of the offshore shrimp trawls improved with the use of vertical side-sorting panels instead of the original horizontal panels (Isaksen, 1984). Nilssen et al. (1986) also investigated mesh selection by adding bags made of finer mesh at

Table 12 Surveys carried out in Jan Mayen since 1970

Year	Survey	Main results	References
1974	Search	Eight areas surveyed, two of them with catches >500 kg hour^{-1}, and three with 100–500 kg hour^{-1}	Anonymous, 1974c
		Four areas surveyed, best catches in NW ground, 60 tons in 69 hauls	Torrisen, 1974
1976	Search	Large shrimp found SE and SW of the island, 130–160 individuals per kilogram	Strøm, 1976
1977	Experimental fishing	Winter fishery is not profitable	Olsen, 1977
1979	Survey	Collection of catch data and biological information	Torheim, 1980
1995	Mapping	Larger shrimp found at depth >300 m, capelin, polar cod and long rough dab widely distributed, cod and Greenland halibut caught sporadically	Aschan, 1995

different places in the trawl. They confirmed that most selection occurs at the cod-end, but they also found that by capturing the shrimp that passed under the trawl, the catch increased by 27% per unit of swept volume.

The first trials to use a sorting device to minimise bycatch took place in 1970 and were performed on three different trawls, among them the 'Kodiak' design used in the Norwegian offshore shrimp fishery. The sorting device consisted of a net with mesh size 130 mm that would let the shrimp in, but would direct fish outside the trawl. Results with the experimental trawl were highly satisfactory, with a reduction in fish bycatch in the range of 75–98% (Rasmussen, 1973; Rasmussen and Øynes, 1974). Sorting grids have been successfully developed, and the newer designs allow fish to escape unharmed from the trawl without markedly reducing the shrimp catch (Isaksen et al., 1992; Valdemarsen et al., 1996). The Nordmøre grid is now mandatory in the shrimp fishery in Norwegian waters (Isaksen et al., 1990; Soldal, 1995; Soldal and Engaas, 1997; Valdemarsen and Misund, 2003; ICES, 2005). A comprehensive list of references regarding research on gear selectivity in Norway has been compiled in the ICES CM Report 1998/B:2 (ICES, 1998).

5.4. Bycatch in the Svalbard shrimp fishery

Bycatch data from the Norwegian fishery are scarce and comparisons are not straightforward because reports from different years present results in different units and the shrimp stock assessment survey has undergone several modifications since 1970.

5.4.1. Svalbard and the Barents Sea area

The first surveys searching for shrimp fishing grounds reported catches of Greenland halibut (*R. hippoglossoides*), polar cod (*B. saida*), haddock (*Melanogrammus aeglefinus*), wolffish (*Anarhichas lupus*), redfish (*Sebastes* spp.), cod (*G. morhua*), capelin (*Mallotus villosus*) and even one Greenland shark (*Somniosus microcephalus*). Close to Bjørnøya, grenadier (*Coryphaenoides rupestris*) and long rough dab (*H. platessoides*) were also found. These were small catches though, on average 143 kg of bycatch per haul in Jan Mayen and 40 kg in Bjørnøya, when all species were combined. Furthermore, most of the fish caught were juveniles (Strøm and Rasmussen, 1970). Despite the low figures recorded in 1970, Teigsmark and Øynes (1981) pointed out that species of commercial interest, such as cod, haddock and Greenland halibut, were more abundant in 1970. The records of both surveys are difficult to compare, especially because Strøm and Rasmussen (1970) did not record bycatch systematically and expressed their results in either biomass or numbers, often pooling the biomass of several species. The average abundance and biomass of bycatch estimated from the survey data included in their reports do not show the decline mentioned by Teigsmark and Øynes (1981) (Table 13).

Bycatch in the Norwegian offshore fishery was regularly recorded only during the period 1981–1986 (Table 14), whereafter it is not mentioned. Off West Svalbard, redfish were usually the most common bycatch species. In 1981, they accounted for 44% of bycatch, followed by polar cod (6.2%), cod (1.3%), Greenland halibut (1%) and capelin (1%). The remaining 47% of the bycatch was not identified. In 1982, cod was very scarce (0.5%) and 75% of the total was below the minimum landing size. Juvenile haddock was caught in the shallower stations south of Svalbard. Capelin was the most abundant species (76.5%), followed by polar cod (8.1%). Redfish (7.4%) was caught most often in tows and was most abundant at depths between 200 and 250 m and >350 m. The remaining species, Greenland halibut and long rough dab constituted 0.7% and 3.7% of the bycatch, respectively. Only 3% of the bycatch was not identified. Abundance of juvenile cod and haddock increased greatly in 1984, leading to the closure of areas where densities of these were high, with in some cases up to 4000 individuals caught per trawling hour. In 1986, polar cod was the most abundant species (37.2%), with cod remaining as abundant as in 1985 (3.8%) but with haddock (0.1%), redfish (34.8%), Greenland halibut (1.4%) and capelin (2.4%) having declined (Teigsmark and Øynes, 1981, 1983a; Hylen *et al.*, 1984; Hylen and Øynes, 1986). Other species present in the benthic assemblage were blue whiting (*Micromesistius poutassou*), saithe (*Pollachius virens*) and wolffish (*Anarhichas lupus*) (Godø and Nedreaas, 1986).

Table 13   Average of biomass and abundance of bycatch recorded during the 1970 and 1981 surveys in the Barents Sea and Svalbard area

Area	Cod	Haddock	Redfish	Gr Hal	Capelin	A cod	Others
Mean weight (kg) hour^{-1} trawled in 1970							
Barents	8.9	5.3	44.6	4.4		142.1	14.8
Bjørnøya			3.8	14.7			24.6
Spitsbergen		5.6	19.5	7.5		6.4	
Mean abundance (number of individuals) per hour trawled in 1981							
Barents	14.5	9.6	184.8	15.4	239.9	1354.5	310.9
Bjørnøya	19.9	1.0	104.1	11.5	18.0	259.9	412.3
Spitsbergen	17.1		576.2	14.8	12.1	176.4	640.5

Gr Hal; Greenland halibut, A cod; Arctic cod.

Note: The figures given here for the 1970 survey only include the hauls where weight of bycatch was recorded. Bycatch of capelin was expressed mostly in abundance but for some hauls it was recorded in weight. To convert these to abundance, the average weight of male and female capelin (ages 3 and 4 belonging to the Icelandic stock) was used.

Source: Data from Strøm and Rasmussen (1970), Teigsmark and Øynes (1981) and Vilhjálmsson (1994).

Table 14   Bycatch of commercial species in the shrimp fishery in Barents Sea and Svalbard area

Year	Co	Ha	Rf	Gh	Ca	Pc	Lrd	Ot	References
1981	3	0	39	2	25	94	55	16	Teigsmark and Øynes, 1981
1982	1	0	27	1	72	137	31	22	Teigsmark and Øynes, 1982
1983	5	3	302	13	684	74	56	115	Teigsmark and Øynes, 1983a,b
1984	31	14	519	25	23	41	84	92	Hylen et al., 1984
1986	26	4	215	6	15	250	43	55	Hylen and Øynes, 1986

Co, cod; Ha, haddock; Rf, redfish; Gh, Greenland halibut; Ca, capelin; Pc, polar cod; Lrd, long rough dab; Ot, others.

Note: Bycatch is expressed in average number of fish per haul and per 3 nautical miles trawled.

Hylen and Jacobsen (1987) compiled data from surveys and commercial fishery to estimate bycatch of cod in the shrimp fishery north of 69°N. Cod recruited to the fishery at age 1, and bycatch of cod aged 1–3 during the years 1982–1983 amounted to yield losses of 20,000 and 30,000 tons, respectively. Sorting grids were introduced in 1991 to minimise bycatch of fish species but seasonal catches of juvenile cod, haddock and redfish still remained a problem (Aschan, 1999a). The stocks of commercial bycatch species have decreased greatly, stressing the need for alternative bycatch management methods (Section 5.2).

5.4.2. Jan Mayen

Bycatch data from Jan Mayen are more limited than those from Svalbard. Very few surveys have been undertaken out there, and results were not always published. The first surveys recording bycatch were very limited: polar cod, some Greenland halibut, long rough dab and some capelin were reported (Torrisen, 1974; Strøm, 1976). During the period 1979–1995, blue whiting, cod and Arctic cod were also recorded in some years (Table 15). It is likely that changes in species composition are caused by the hydrographic conditions (Aschan et al., 1996b). Other bycatch fish species caught in Jan Mayen were snake blenny (*Lumpenus lampretaeformis*), spotted snake blenny (*Leptoclinus maculatus*), two horned sculpin (*Icelus bicornis*), pogge (*Agonus cataphractus*), Atlantic poacher (*Leptagonus decagonus*) and short horned sculpin (*Mioxocephalus scorpius*) (Torheim, 1980; Aschan, 1999b).

6. OVERVIEW

Catches of *Pandalus borealis* in the North Atlantic during the period 1982–2005 amounted to 9.07 million tons, of which 3.2 million tons or 35% correspond to the fisheries considered in this chapter (Table 16) (U. Skúladóttir, personal communication). Some of the increases observed in the different areas are due to development of the exploitation, especially during the first years of the fisheries, others to biomass increases caused by good recruitment (Hvingel, 1997). Table 17 gives an overview of the state of the main shrimp stocks in the North Atlantic.

Shrimp stock size is affected by water temperature, fishing mortality and groundfish predation, mainly from cod (Hvingel, 1997). Temperature seems

Table 15   Bycatch registered in Jan Mayen

Year	Gh	Lrd	Bw	Ca	Pc	Co	Ac
Mean weight (kg) per 3 nautical miles trawled							
1979	0.55	32.95	148.73	46.45	200.95	0.18	0.00
1980	3.82	10.98	1.40	0.33	2.83	0.83	0.00
1981	8.25	7.13	0.04	0.11	0.69	0.00	0.00
Mean abundance (number of individuals) per 3 nautical miles trawled							
1994	0.17	6.11	0.00	78.06	31.28	0.00	0.00
1995	0.09	10.09	0.00	1680.18	52.97	0.00	5.24

Gh, Greenland halibut; Lrd, long rough dab; BW, blue whiting; Ca, capelin; Pc, polar cod; Co, cod; Ac, Arctic cod.
Source: Data from the Norwegian surveys carried out in 1980, 1981, 1994 and 1995; Aschan *et al.* (1996a).

Table 16 Catch of *Pandalus borealis* (in tons) in the North Atlantic during the period 1982–2005

Area	Min		Max		2005	Total	Mean	No. y
	Year	Catch	Year	Catch				
Canada	1984	8547	2004	174,153	155,447	1565	65.2	24
Flemish Cap	2005	23,520	2003	62,180	23,520	519	40.0	13
Grand Bank	1989	11	2005	14,402	14,402	68	5.2	13
US (Maine)	2003	1209	1996	9524		81	3.5	23
BS, Sv	1995	25,220	1984	128,062	36,944	1499	62.4	24
Jan Mayen	1991	100	1985	2200		16	0.9	16
Iceland	2005	4469	1997	72,059	4469	852	35.5	24
Denmark Str	1983	5212	1988	15,306	8738	269	11.2	24
North S (IV)	1991	3810	1987	13,920	4564	157	6.8	23
Skarr (IIIa)	2001	6103	1998	11,546	6904	198	8.6	23
Davis Str*	1984	44,556	2005	133,801	133,801	1812	75.5	24
Greenl insh	1987	7613	2000	23,133	18,616	373	15.5	24

Min, year of minimum catch; Max, year of maximum catch; 2005, catch in 2005; Total, total catch in thousands of tons; Mean, mean annual catch in thousands of tons for the period 1982–2005; No. y, number of years fished in each country between 1982 and 2005; BS, Sv, Barents Sea and Svalbard; Denmark Str, Denmark Strait; North S (IV), North Sea ICES Division IV; Skarr (IIIa), Skarregak ICES Division (IIIa); Davis Str, Davis Strait; Greenl insh, Greenland inshore fishery. *92% of the catches from the Davis Strait correspond to Greenland. (U. Skúladóttir, personal communication).

to be the most important factor affecting shrimp populations (Anderson, 2000; Koeller, 2000). Overfishing and unfavourable environmental conditions, with lower water temperature, resulted in the collapse of the cod stock in Greenland and the subsequent increase of shrimp abundance in West Greenland (Smidt, 1969; Horsted, 2000). The variability in water temperature within the distribution area of *Pandalus borealis* is reflected in the diversity of the life cycles of populations from different locations, affecting survival, abundance and recruitment patterns (Nilssen and Hopkins, 1991; Aschan, 2000a; Koeller, 2000). In general, higher water temperatures lead to faster shrimp growth rates and to a decrease in mean size and age at sex change (Allen, 1959; Horsted and Smidt, 1965; Nilssen and Hopkins, 1991; Skúladóttir and Pétursson, 1999). Mean size at sex change is positively correlated to egg production and therefore to fecundity (Horsted and Smidt, 1965; Nilssen and Hopkins, 1991). Furthermore, water temperature may show large annual variability in the study area, depending on the amount of warmer, Atlantic water reaching the Barents Sea, North Iceland and West Greenland (Stefánsson, 1962; Aschan, 2000a; Buch, 2000). These local changes in temperature can affect shrimp populations. For example, annual variability in the distribution and recruitment of the shrimp stock in

Table 17 Comparison of the main *Pandalus borealis* fisheries in the North Atlantic

Area	Flemish Cap	Grand Bank	Davis Strait	DS/EG	BS/Sv*	Iceland**
Fishery began	1993	1993	Inshore in 1930s, offshore in 1969	1978 north of 65°N, 1993 south of 65°N	1970	1975
Data	Fishery, survey	Fishery, survey	Fishery, survey	Fishery, last survey in 1995	Fishery, survey	Fishery, survey
Cath trend	Oscillated during 1993–1995. Minimum in 1997 at 24,000 tons, rose to maximum in 2003. Sharp decline followed, with 9000 tons by autumn 2005	Below 1000 tons during 1994–2000, followed by oscillating increasing trend with catches between 5000 and 13,000 tons. Maximum catch 13,300 tons in 2005	Rather steady increasing trend since 1970, with small decreases between 1977 and 1979 and 1993–1997. Maximum catch in 2004 and 2005, 140,000 tons	Declined between 1980 and 1983, rose to maximum of 15,300 tons in 1988, declined again until 1997. Landings from south of 65°N peaked in 2001 at 11,600 tons	Increased until reaching peak at 128,000 tons in 1985, has oscillated between 27,000 and 83,000 since then	Increased steadily to maximum catch of 66,000 tons in 1995. Since 1996 declining, some increase in 2001–2002. Catch in 2004–2005 was 5400 tons
Min–Max CPUE (standard)	24–63 Oscillated without trend from 1994 to 1997 and increased afterwards	0.5–13 Increased during 2000–2003, then declined by 34% in large vessel fleet. Increased 86% to 750 kg hour^{-1} during 2003–2005 in small vessel fleet	1.2–140.5 Oscillating without trend between 1975 and 1987, decreased sharply during 1986–1990, increasing since then, very quickly after 1997	5–15 Decreased during 1987–1993, increased until late 1990s and oscillated thereafter with slightly increasing trend	5.5–128 Oscillating, declining trend since 1980, dropped by 30% between 2003 and 2004. Now at lowest level since 1987 according to Norwegian data	1–66 Oscillating trend, slightly decreasing. Maximum 199 kg hour^{-1} in 1974 and 1996. Declined during 1996–1999, rose to 140 kg hour^{-1} from 2000 to 2004. Historical low, 40 kg hour^{-1} in 2005

Recruit	Year-classes 2001–2002 good but 2003 the lowest since 1994	Year-classes 1997–2001 above average, 2002 and 2003 below	Good in 2006, but since 2002 age-2 shrimp below average, 2005 estimate is the lowest since 1992	Not available	Age-1 shrimp recruitment low but stable over past two years	Year-class 2002 below average, year-class 2003 is historical low
SSB	All indices rose in 1997 and have oscillated without trend since then	Rose during 1995–1997, established and has increased since 1999	Oscillating during 1987–1997, doubled since, maximum in 2003–2004	Not available	Has decreased by 32% since 2002	Historical low
Fishable biomass	May be predicted in future with abundance of age-2 shrimp	Increased from 1995 to 2001 and stable since then. Exploitation index increased during 2000–2001 and decreased after that	Stable from 1988 to 1997, has increased afterwards and reached maximum in 2003	Increased from 1993 to 2000, oscillated thereafter	Survey indices since 1985 indicate that it varies cyclically without trend	Oscillated with rising trend since 1988, peaked in 1996. Declining trend afterwards with small increase during 2000–2002. Historical low in 2004
State of stock	CPUE and biomass indices high, recruitment may decline in 2007. Biomass well above B_{LIM}	SSB and total biomass rising since 1999. Biomass well above B_{LIM}	Historical high but models of stock development and recruitment indices indicate possible declining trend in future. Stock well above B_{MSY}, mortality well below Z_{MSY}	Due to changes in fishing patterns in 2004 and 2005 it is uncertain if the biomass estimate is a good reflection of stock biomass	According to Norwegian survey, the biomass index is the lowest since 1987	Historical low since 2004
Rec	Mean annual catch of 48,000 tons since 1998 with no detectable effect on stock biomass	TAC should not be raised for years to monitor impact of fishery on stock	Catches over 130,000 tons in 2006 rise risk of exceding Z_{LIM} to over 20%. This risk will increase with time if such catch is maintained	Maintain TAC at 12,400, conduct survey, sampling of catches by observers	Implement TAC	Decrease effort while mortality due to predation remains high

(Continued)

Table 17 (Continued)

Area	Flemish Cap	Grand Bank	Davis Strait	DS/EG	BS/Sv*	Iceland**
TAC (tons)	48,000	22,000	130,000	12,400	40,000	10,000
Reference points	B_{LIM} = index of stock size that is 85% smaller than the maximum observed index	B_{LIM} = index of stock size that is 85% smaller than the maximum observed index	B_{LIM} = 30% of B_{MSY}, $Z_{LIM} = Z_{MSY}$	Not available	Not available	Not available
Man	TAC (2000), sorting grid mandatory	TAC (1999), sorting grid (2000)***	Mesh size (1977), no discarding (1979), TAC (1981), logbooks (1986), ITQ (1991), sorting grid (2000)	Mesh size (1979), TAC (1981), logbooks (1987)	Minimum carapace length (1985), licenses, fishing days, bycatch limits, sorting grids (1993)	Logbooks, licenses and mesh size (1970), TAC (1987), ITQ (1990), sorting grid (1995)
Other		Fishing restricted to 3L with sorting grid (22-mm bar spacing). Canada gets 83% of the TAC since the stock became regulated	Stock development depending on cod abundance, that has increased lately	The apparent TAC increase after 2003 was due to data revision, not to increase in stock production. Access to fishing grounds depends on ice conditions	This stock has no explicit management objectives. Stock very influenced by cod predation	The observed decrease is due to both cod predation and low recruitment levels

Davis Strait includes NAFO Subareas 0 and 1; DS/EG, Denmark Strait and East Greenland (Greenland EEZ); BS/Sv, Barents Sea and Svalbard. Min–Max, minimum and maximum catches in tons since the beginning of each fishery; CPUE is standardised but expressed in different indexes; Recruit, expected recruitment; SSB, spawning stock biomass; Rec, recommendations from scientific committees; Man, management; kg hour^{-1}, kg per hour. Data from NAFO (2006), *ICES (2006) and **Anonymous (2006). *** D. Orr, personal communication.

the Barents Sea seems to be related to the location of the Polar Front (Hylen et al., 1984; Aschan et al., 2000). Water temperature increased in the mid-1990s in West Greenland, leading to an increase of stock biomass. However, the subsequent year-classes were considerably smaller than year-classes produced in colder waters as a result of the reduced length at sex change (Wieland, 2004a,c). Thus, environmental conditions are an important factor to consider for management of shrimp stocks, and fishing effort should be reduced when they worsen. The Maine shrimp fishery collapsed in the 1970s due to the combination of overfishing and unfavourable environmental conditions (Parsons and Fréchette, 1989; Clark et al., 2000).

One of the goals of stock assessment is the estimation of the appropriate fishing mortality to ensure the sustainability of the fishery. This requires an accurate estimate of stock biomass, which is particularly difficult to obtain in the case of shrimp stocks for the following reasons (Hvingel, 1997):

- The variability of shrimp growth and age at sex change result in approximate ageing that limits greatly the use of virtual and sequential population analysis (VPA and SPA).
- Natural mortality is probably high for all ages, making yield/recruit (Y/R) reference points unfeasible.
- Natural mortality may occasionally be higher than fishing mortality, a situation not considered in most fishery models.
- The vertical and horizontal range of distribution of shrimp renders it impossible to calculate absolute estimates of abundance or biomass.
- Difficulty to estimate recruitment (Wieland, 2004c). It is known that in the Barents Sea, there is a strong correlation between egg production, number of females and age-1 shrimp. However, larval shrimp transport may vary among years due to hydrographical conditions and this may affect recruitment (Pedersen et al., 2002). These processes need to be better understood.

The difficulty in estimating accurate figures for shrimp stock size and fishing mortality calls for a precautionary approach and conservative management, yet managers rarely accept the TAC recommended by scientists and landings often surpass the quota set by managers. Fleet size is another problem in limiting effort. This has become apparent in Norway after the great increase of the offshore shrimp fleet allowed by the government in 1999 led to the use of double and triple trawls in the Barents Sea (Standal, 2003) and has undoubtely contributed to the decline of stock biomass and landings observed since 2000 (Figures 31 and 33). The shrimp stocks in the Barents Sea and off North Iceland have coped well with high fishing effort in years of good recruitment (Berenboim et al., 1992; Anonymous, 1994b), but landings decrease promptly if recruitment is poor several years in a row, as is currently the case (Tveranger and Øynes, 1985; Hylen, 1986; Anonymous, 2006; ICES, 2006).

Within the study area considered in this chapter, low shrimp market prices, ice conditions and fuel price seem to be the most effective factors to keep fishing effort low (Anonymous, 1979b, 2006; Siegstad and Hvingel, 2005; Skúladóttir and Pétursson, 2005; NAFO, 2006). Fishing effort should also decrease if cod abundance is thought to have a major impact on the shrimp population, since the fishery targets the larger shrimp, which are all female, and low levels of spawning stock biomass produce fewer eggs, leading to poor recruitment (Clark *et al.*, 2000).

Although it was known that groundfish prey on shrimp (Horsted and Smidt, 1956), it was not until the late 1980s that the cod–shrimp interaction was first investigated in some detail (Berenboim *et al.*, 1986b). The Icelandic and Norwegian offshore shrimp stocks seem to depend largely on the cod stock size (Figures 35 and 36), although shrimp are not the main prey species of cod, and consumption of shrimp by cod is highly correlated with predator size. Although cod >20 cm seem to prefer fish and it is mostly young cod that prey on shrimp (Pálsson, 1994), shrimp was found to represent up to 15% of stomach content weight in cod between 20- and 45-cm length (Pálsson, 1997). Stefánsson *et al.* (1998) found a highly significant negative correlation between juvenile cod abundance and shrimp biomass and recruitment indices. Research on cod stomach contents in Iceland revealed that shrimp was mostly consumed by cod of age 3 or sizes 15–19 cm and

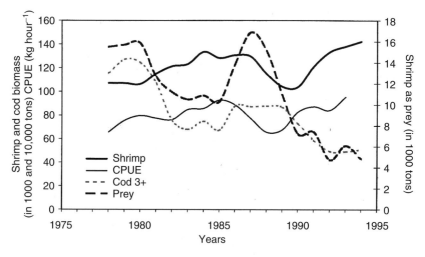

Figure 35 Trends of shrimp and cod 3+ (age 3 and older) biomass (in thousands and tens of thousands of tons, respectively) and CPUE (in kg hour^{-1}), in left y-axis, and shrimp consumed by cod (in thousands of tons, right y-axis) in Icelandic waters during the period 1978–1994. Data from Stefánsson *et al.* (1994) and Anonymous (2003a).

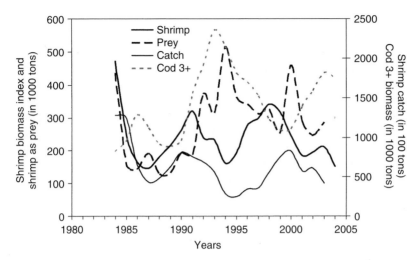

Figure 36 The Barents Sea and Svalbard area. Trends of shrimp biomass index and shrimp consumed by cod (in thousands of tons, left *y*-axis), cod 3+ (age 3 and older) biomass (in thousands of tons) and shrimp catch (in hundreds of tons), both in right *y*-axis, in the Barents Sea during the period 1984–2004. Data from ICES (2006).

20–24 cm, but even within this age and size range shrimp only constitutes a small part of the diet (Magnússon and Pálsson, 1989, 1991). Shrimp consumption is also very variable among seasons and years (Pálsson, 1983). Magnússon and Pálsson (1989) estimated that shrimp represented between 0.8% and 19% of the cod diet (ages 2–8 years) in March 1980–1986, but 3.4–32.4% in October–November 1979–1983 and 1985. A stock-production model using survey data estimated shrimp consumption by cod in the offshore grounds off North Iceland to be within the range of 4700 to 15,800 tons (3.4–14.9% of the shrimp stock biomass) during the period 1978–1997, with an average annual consumption of 11,560 thousand tons (9.4%) (Stefánsson *et al.*, 1994).

In the Barents Sea, shrimp biomass follows trends opposite to those of biomass of cod age 3+ and predation on shrimp, and the trend of shrimp catch is very similar to that of shrimp biomass, as in Iceland (Figure 36). The amount of shrimp eaten by cod has been estimated to be between 129,000 and 516,000 tons, or 282,000 tons on average per year. However, it has been pointed out that these figures may have been overestimated, and in fact the figures for biomass of shrimp consumed by cod are in some years higher than the shrimp index. Shrimp landings have ranged from 28,300 to 128,100 tons, or 55,200 tons on average (ICES, 2005, 2006; Johannesen and Aschan, 2005). Cod biomass in Greenland was between 1.2 and 1.8 million tons from 1955 to 1967, when it decreased by one order of magnitude. There was a further decrease below 10,000 tons during 1983–1986, and the stock has rarely gone above 1000 tons between 1991 and 2000. However, an increase has been detected since 2001, and in 2005 the stock was estimated to be nearly 40,000 tons, an eightfold increase compared to 2004 (Kingsley and Hvingel, 2005). Predation was estimated to range from 1000 to 85,000 tons during 1989–1992, when the cod stock was between 8500 and 191,600 thousand tons. Further increases of the cod stock could increase natural mortality significantly and force a TAC reduction, and it has been estimated that the presence of two large year-classes of cod could increase predation within 3 or 4 years to 88,000–163,000 tons (Kingsley and Hvingel, 2005; NAFO, 2006).

Another problem for management of the shrimp stocks is the bycatch of commercial fish species. Based on FAO data from 1980, Andrew and Pepperell (1992) gave a rough estimate of the bycatch:shrimp ratio of 5:1 for temperate areas. Furthermore, extrapolating FAO data from 1986, they estimated bycatch in the world shrimp fisheries at 16.5 million tons year^{-1}, or 20.5% of the total world catch for all species. The most recent FAO estimates of discards conclude that *Pandalus* fisheries have a discard rate of 11.6% that decreases to 5.4% with the use of sorting grids (Kelleher, 2005). Within the study area of this chapter, sorting grids are widely used and have been apparently successful in preventing the capture of fish larger than 18 or 20 cm, yet abundance and biomass trends of many bycatch species are declining and the commercial fish species are below safe biological limits in the Barents Sea. A detailed discussion on the bycatch problem is out of the scope of this chapter, but it should be mentioned that research and literature on bycatch in the Northeast Atlantic shrimp fisheries is astonishingly scarce, considering how little is known about their impact on demersal communities within the study area and the economical status of the countries involved. The use of sorting grids has reduced the landings of fish bycatch by 99% in some shrimp fisheries, and it has been estimated that bycatch of cod and haddock in the Barents Sea is about 1% (for each species) of the shrimp landings (Ajiad *et al.*, 2004). However, in shrimp fisheries where sorting grids are not used, such as in the Norwegian Deeps, bycatch of commercial fish species ranges from 13% to 45% of the shrimp catch (ICES, 2005). Interestingly enough, most of the available literature on bycatch within the study area corresponds to Greenland, which has more limited resources than Iceland and Norway. In addition, bycatch data from the commercial fleet are extremely scarce in the literature, indicating a need for systematic sampling of bycatch on board the commercial fleet. In Iceland, for example, observers went on board the shrimp vessels occasionally before the implementation of the sorting grids, however the sampling focused on juvenile redfish. Bycatch data from the commercial fleet are limited to landings of marketable species above the legal minimum landing size, but there are no data on discards of undersized bycatch species or non-marketable species. The impact of shrimp trawling in the study area should also be investigated. Even if the use of sorting grids has decreased significantly the amount of bycatch, habitat destruction and indirect mortality may inflict important mortality rates on fish species whose distributions overlap with shrimp fishing grounds.

The large fluctuations of environmental factors due to alternation of cold and warm periods related to large-scale atmospheric circulation over the North Atlantic also play a major role in fish and shrimp stock abundance in Greenland, Iceland and the Barents Sea. During cold periods, cold Arctic water masses predominate over warm and also more saline Atlantic water, leading to increased stratification of surface layers, shorter duration of phytoplankton blooms and therefore to a decline of zooplankton resources that affects pelagic species and their predators. This was the situation at the beginning of the twentieth century until 1920, and after the climate deterioration period that started in the mid-1960s and culminated with the 'Great Salinity Anomaly' during the 1970s. These climate changes influenced the following events, among others (Jakobsson, 1992; Vilhjálmsson, 1997; Jakobsson and Stefánsson, 1998; Buch, 2000; Hamilton et al., 2000):

1. Onset of the cod fishery in Greenland at the beginning of a warm period, which peaked at 130,000 tons in 1936 and collapsed at the end of the 1960s when water temperature decreased.
2. Onset of the shrimp fishery, coincident with the beginning of the cooling period in the late 1960s and early 1970s and the collapse of the cod fishery.
3. Change in the distribution of capelin in West Greenland, which reached 76°N in 1936.
4. Drift of cod between Iceland and Greenland, which took place intermittently during the period 1920–1970 and in 1984–1985, when cold Arctic water blocked the Atlantic water carried by the Irminger current.
5. Collapse of the Icelandic capelin and herring fisheries in the 1980s.
6. Collapse of the capelin and herring stocks in the Barents Sea in the mid-1980s, which significantly decreased the food availability for cod.
7. Increased migration of cod to North and East Iceland coinciding with the warming up of the ocean since 1997 has led to an important reduction of shrimp biomass.

Changes in environmental conditions, therefore, influence many of the variables used in stock assessment such as growth rates, recruitment, migrations and predator–prey relationships. As illustrated by shrimp in this chapter, good knowledge of the environmental characteristics of each ecosystem and understanding of the impact of both environmental variables and fishing activities on marine resources is essential for adequate management strategies.

ACKNOWLEDGEMENTS

This study has been funded by the Nordic Council of Ministers and it is part of the project 'Bottom trawling and scallop dredging in the Arctic', developed within the frame of the Nordic Action Plan. Special thanks are due to U. Skúladóttir for sharing her knowledge, time, data and library with me, and for her comments on an earlier version of the chapter. The author also wishes to thank Sigurlína Gunnarsdóttir, librarian at the MRI, for her invaluable help. G. S. Bragason, S. Brynjólfsson, M. Aschan, G. Søvik, R. Frandsen, H. Siegstad and A. K. Veim are thanked for answering many questions and providing useful information. I am very thankful to C. Hvingel and M. Storr-Paulsen who supplied the fishing effort maps for the Greenland section. B. Isaksen and R. Larsen supplied the illustration of the Nordmøre grid, and B. Bergström allowed me the use of his distribution map for *Pandalus borealis* and the illustration of changes in pleopods. H. Siegstad, O. S. Tendal, S. A. Steingrímsson, S. A. Ragnarsson and H. Eiríksson read earlier versions of the chapter. Thanks are also due to A. J. Southward, D. W. Sims and an anonymous referee, whose comments and suggestions improved the chapter significantly.

REFERENCES

Ajiad, A., Ageln, A. and Nedreaas, K. (2004). Cod by-catch mortality from the Barents Sea shrimp fishery 1983–2002. ICES Arctic Fisheries Working Group, Working Document 24, p. 37.

Allen, J. A. (1959). On the biology of *Pandalus borealis* Krøyer with reference to a population off the Northumberland coast. *Journal of the Marine Biological Association of the United Kingdom* **38**, 189–220.

Andersen, M. (1994). The small vessel shrimp fishery in West Greenland. *NAFO SCR Doc. 94/89. Serial No. N2476*, p. 6.

Andersen, M., Carlsson, D. and Kanneworff, P. (1993a). Stratified-random trawl survey for shrimp (*Pandalus borealis*) offshore in NAFO Subareas 0 and 1, in 1993. *NAFO SCR Doc. 93/132. Serial No. N2344*, p. 19.

Andersen, M., Carlsson, D. and Kanneworff, P. (1993b). Stratified random trawl survey for shrimp (*Pandalus borealis*) in Disko Bay, West Greenland, 1993. *NAFO SCR Doc. 93/129. Serial No. N2481*, p. 13.

Andersen, M., Carlsson, D. and Kanneworff, P. (1994a). Stratified-random trawl survey for shrimp (*Pandalus borealis*) in Disko Bay, West Greenland, 1994. *NAFO SCR Doc. 94/94. Serial No. N2481*, p. 10.

Andersen, M., Carlsson, D. and Kanneworff, P. (1994b). Trawl survey for shrimp (*Pandalus borealis*) in Denmark Strait, 1994. *NAFO SCR Doc. 94/90. Serial No. N2477*, p. 13.

Anderson, P. J. (2000). Pandalid shrimp as indicators of ecosystem regime shift. *Journal of Northwest Atlantic Fishery Science* **27**, 1–10.

Andrew, N. and Pepperell, J. (1992). The by-catch of shrimp trawl fisheries. *Ocean-ography and Marine Biology* **30**, 527–565.
Anonymous (1971). Rækjuleit 1970. *Hafrannsóknir* **3**, 117–126.
Anonymous (1973). Rækjurannsóknir. *Hafrannsóknir* **4**, 19–21.
Anonymous (1974a). Rækjurannsóknir. *Hafrannsóknir* **5**, 16–18.
Anonymous (1974b). Veiðarfærarannsóknir. *Hafrannsóknir* **5**, 32–33.
Anonymous (1974c). Rapport fra forsøksfiske etter reker ved Jan Mayen med M/S "Alvenes" T-95-LK i tiden 1/7–4/8 1974. Skipper Otto Godtlibsen. Rapport Fiskerinæringens Forsøksfond Fiskeridirektoratet p. 35.
Anonymous (1975a). Rækjurannsóknir. *Hafrannsóknir* **6**, 17–19.
Anonymous (1975b). Veiðarfærarannsóknir. *Hafrannsóknir* **6**, 40–41.
Anonymous (1976a). Rækja. *Hafrannsóknir* **9**, 18–19.
Anonymous (1976b). Veiðarfærarannsóknir. *Hafrannsóknir* **9**, 47–48.
Anonymous (1976c). Reker. *Fisket og Havet.*, Ressursoversikt for 1976, pp. 96–99.
Anonymous (1977a). Rækja. *Hafrannsóknir* **11**, 14–18.
Anonymous (1977b). Veiðarfærarannsóknir. *Hafrannsóknir* **11**, 46–48.
Anonymous (1977c). International Convention of Northwest Atlantic Fisheries (ICNAF). Redbook, pp. 13–18.
Anonymous (1977d). Reker. *Fisket og Havet.* Ressursoversikt for 1977, pp. 84–86.
Anonymous (1978a). Rækja. *Hafrannsóknir* **14**, 18–21.
Anonymous (1978b). Veiðarfærarannsóknir. *Hafrannsóknir* **14**, 46–47.
Anonymous (1978c). Reker. *Fisket og Havet.*, Ressursoversikt for 1978, pp. 7–15.
Anonymous (1979a). Rækja. *Hafrannsóknir* **17**, 32–35.
Anonymous (1979b). Reker. *Fisket og Havet.*, Ressursoversikt for 1979, pp. 98–106.
Anonymous (1980a). Rækja. *Hafrannsóknir* **21**, 17–18.
Anonymous (1980b). Rækjuleit. *Hafrannsóknir* **21**, 18–19.
Anonymous (1981a). Rækja. *Hafrannsóknir* **22**, 43–48.
Anonymous (1981b). Rækja. *Hafrannsóknir* **23**, 17–19.
Anonymous (1981c). Rækja. Ástand nytjastofna á Íslandsmiðum og aflahorfur 1981. *Hafrannsóknir* **22**, 43–48.
Anonymous (1982a). Rækja. *Hafrannsóknir* **24**, 46–49.
Anonymous (1982b). Northwest Atlantic Fisheries Organization. Scientific Council Report 1982, p. 35.
Anonymous (1983). Northwest Atlantic Fisheries Organization. Scientific Council Reports 1983, pp. 13–16.
Anonymous (1984a). Rækja. *Hafrannsóknir* **29**, 17–19.
Anonymous (1984b). Report on the joint NAFO/ICES study group in biological relationships of the West Greenland and Irminger Sea redfish stocks. *International Council for the Exploration of the Sea C.M. 1984/G:3*, p. 13.
Anonymous (1984c). Reker. *Fisket og Havet.* Ressursoversikt for 1984, pp. 59–65.
Anonymous (1985a). Rækja. *Hafrannsóknir* **31**, 50–58.
Anonymous (1985b). Reker. *Fisket og Havet.* Ressursoversikt for 1985, pp. 64–71.
Anonymous (1986a). Rækja. *Hafrannsóknir* **33**, 44–48.
Anonymous (1986b). Reker. *Fisket og Havet.* Ressursoversikt for 1986, pp. 60–66.
Anonymous (1987a). Rækja. *Hafrannsóknir* **37**, 13–15.
Anonymous (1987b). Rækja. *Hafrannsóknastofnun Fjölrit* Nr. **11**, 29, 59.
Anonymous (1987c). Reker. *Fisket og Havet.* Ressursoversikt for 1987, pp. 66–73.
Anonymous (1988a). Ástand húmar-og úthafsrækjustofna 1988. *Hafrannsóknastof-nun Fjölrit* Nr. **15**, 5–8.
Anonymous (1988b). Reker. *Fisket og Havet.* Ressursoversikt for 1988, pp. 74–79.
Anonymous (1989). Rækja. *Hafrannsóknir* **40**, 22–24.

Anonymous (1990). Rækja. *Hafrannsóknastofnun Fjölrit* Nr. **21**, 59–65.
Anonymous (1991). Rækja. *Hafrannsóknastofnun Fjölrit* Nr. **25**, 53–58.
Anonymous (1993). Rækja. *Hafrannsóknastofnunin Fjölrit* Nr. **34**, 60–65, 135–137.
Anonymous (1994a). Rækja. *Hafrannsóknir* **46**, 28–30.
Anonymous (1994b). Rækja. *Hafrannsóknastofnun Fjölrit* Nr. **37**, 64–69.
Anonymous (1994c). Nytjastofnar sjávar 1993/1994. Aflahorfur fiskveiðiárið 1994/
    1995. State of marine stocks in Icelandic waters 1993/1994. Prospects for the quota
    year 1994/1995. *Hafrannsóknastofnun Fjölrit* **37**, 150.
Anonymous (1995). Nytjastofnar sjávar 1994/1995. Aflahorfur fiskveiðiárið 1995/1996.
    State of marine stocks in Icelandic waters 1994/1995. Prospects for the quota year
    1995/1996. *Hafrannsóknastofnun Fjölrit* **43**, 172.
Anonymous (1996a). Rækja. *Hafrannsóknastofnun Fjölrit* Nr. **46**, 79–85, 161.
Anonymous (1996b). Nytjastofnar sjávar 1995/1996. Aflahorfur fiskveiðiárið 1996/
    1997. State of marine stocks in Icelandic waters1995/1996. Prospects for the quota
    year 1996/1997. *Hafrannsóknastofnun Fjölrit* **46**, 175.
Anonymous (1996c). Reker nord for 62°N. *Fisket og Havet.*, Ressursoversikt for
    1996, pp. 42–45.
Anonymous (1997a). Rækja. *Hafrannsóknir* **52**, 28–30.
Anonymous (1997b). Rækja. *Hafrannsóknastofnun Fjölrit* Nr. **56**, 77–82.
Anonymous (1998). Rækja. *Hafrannsóknir* **53**, 30–33.
Anonymous (1999). Assessment methods for shrimp stocks in the North-Eastern
    Atlantic. *TemaNord* **533**, 64.
Anonymous (2003a). Nytjastofnar sjávar 2002/2003. Aflahorfur fiskveiðiárið 2003/
    2004. State of marine stocks in Icelandic waters 2002/2003. Prospects for the quota
    year 2003/2004. *Hafrannsóknastofnun Fjölrit* Nr. **97**, 73–79, 156–159.
Anonymous (2003b). Report of the Arctic Fisheries Working Group. ICES Advisory
    Committee on Fisheries Management. *International Council for the Exploration of
    the Sea C.M. 2003*, pp. 369–385.
Anonymous (2004a). Nytjastofnar sjávar 2003/2004. Aflahorfur fiskveiðiárið 2004/
    2005. State of marine stocks in Icelandic waters 2003/2004. Prospects for the quota
    year 2004/2005. *Hafrannsóknastofnun Fjölrit* Nr. **102**, 74–80, 159–161.
Anonymous (2004b). Reker. *Havets ressurser*, 45–48.
Anonymous (2005). Nytjastofnar sjávar 2004/2005. Aflahorfur fiskveiðiárið 2005/
    2006. State of marine stocks in Icelandic waters 2004/2005. Prospects for the
    quota year 2005/2006. *Hafrannsóknastofnun Fjölrit* Nr. **121**, 74–81, 99.
Anonymous (2006). Nytjastofnar sjávar 2005/2006. Aflahorfur fiskveiðiárið 2006/
    2007. State of marine stocks in Icelandic waters 2005/2006. Prospects for the
    quota year 2006/2007. *Hafrannsóknastofnun Fjölrit* Nr. **126**, 79–85.
Arnason, R. and Friis, P. (1995). The Greenland fisheries. Developing a modern fish-
    ing industry. *In* "The North Atlantic Fisheries: Successes, Failures and Challenges"
    (R. Arnason and L. Felt, eds), pp. 171–196. Island Studies Press, Charlottetown.
Aschan, M. (1995). Kartlegging av rekebestanden i Jan Mayen området. Fiskeri-
    forskning, Tromsø, p. 26.
Aschan, M. (1997). Kartlegging av bestandsstrukturen til reke (*Pandalus borealis*)
    i Nordøst-Atlanteren. Sluttrapport til Norges Forskningsråd, Programmet for
    Marin Ressursforvaltning, p. 12.
Aschan, M. (1999a). Bioeconomic analyses of by-catch of juvenile fish in the shrimp
    fisheries. Report 24/1999. Fiskeriforskning, Tromsø, p. 62.
Aschan, M. (1999b). Yngelinnblanding in rekefisket i Barentshavet og Svalbardsonen
    i perioden 1995–1998. Rappport 12/1999. Fiskeriforskning, Tromsø, p. 77.

Aschan, M. (2000a). Spatial variability in length frequency and growth of shrimp (*Pandalus borealis* Krøyer, 1838) in the Barents Sea. *Journal of Northwest Atlantic Fishery Science* **27**, 93–105.

Aschan, M. (2000b). Working document on by-catch in the shrimp fishery in the Barents Sea. Preliminary report. Fiskeriforskning, Tromsø, p. 48.

Aschan, M. and Sunnanå, K. (1997). Evaluation of the Norwegian shrimp surveys conducted in the Barents Sea and the Svalbard area 1980–1997. *International Council for the Exploration of the Sea C.M. 1997/Y:07*, p. 24 (mimeo).

Aschan, M., Berenboim, B., Mukhin, S. and Sunnanå, K. (1993). Results of Norwegian and Russian investigations of shrimp (*Pandalus borealis*) in the Barents Sea and Svalbard area in 1992. *International Council for the Exploration of the Sea C.M. 1993/K:9*, p. 22 (mimeo).

Aschan, M., Berenboim, B. and Mukhin, S. (1994). Results of Norwegian and Russian investigations of shrimp (*Pandalus borealis*) in the Barents Sea and Svalbard area in 1992. *International Council for the Exploration of the Sea C.M. 1994/K:37*, p. 14 (mimeo).

Aschan, M., Berenboim, B. and Mukhin, S. (1995). Results of Norwegian and Russian investigations of shrimp (*Pandalus borealis*) in the Barents Sea and Svalbard area in 1994. *International Council for the Exploration of the Sea C.M. 1995/K:11*, p. 17 (mimeo).

Aschan, M., Nilssen, E., Ofstad, L. and Torheim, S. (1996a). Catch statistics and life history of shrimp, *Pandalus borealis*, in the Jan Mayen area. *International Council for the Exploration of the Sea C.M. 1996/K:11*, p. 26 (mimeo).

Aschan, M., Berenboim, B. and Mukhin, S. (1996b). Results of Norwegian and Russian investigations of shrimp (*Pandalus borealis*) in the Barents Sea and Svalbard area 1995. *International Council for the Exploration of the Sea C.M. 1996/K:6*, p. 26 (mimeo).

Aschan, M., Adlansvik, B. and Tjelmeland, S. (2000). Spatial and temporal patterns in recruitment of shrimp *Pandalus borealis* in the Barents Sea. *International Council for the Exploration of the Sea C.M. 2000/N:32 Theme session on the Spatial and Temporal Patterns in Recruitment Processes*, p. 19 (mimeo).

Aschan, M., Bakenev, S., Berenboim, B. and Sunnanå, K. (2004). Management of the shrimp fishery (*Pandalus borealis*) in the Barents Sea and Spitsbergen area. *In* "Proceedings of the 10th Norwegian–Russian Symposium, Bergen, Norway, 27–29 August 2003" (Å. Bjordal, H. Gjøsæter and S. Mehl, eds), pp. 94–103. IMR/PINRO Joint Report Series Nr. 1

Barr, L. (1970). Diel vertical migration of *Pandalus borealis* in Kachemak Bay, Alaska. *Journal of the Fisheries Research Board of Canada* **27**, 669–676.

Begley, J. and Howell, D. (2004). An overview of Gadget, the Globally applicable Area-Dissaggregated General Ecosystem Toolbox. *International Council for the Exploration of the Sea C.M. 2004/FF:13*, p. 16 (mimeo).

Berenboim, B. and Korzhev, V. A. (1997). On possibility to apply the Stefansson stock production model for assessment of shrimp (*Pandalus borealis*) stock in the Barents Sea. *International Council for the Exploration of the Sea Council Meeting Papers 1991/DD:1*, p. 16 (mimeo).

Berenboim, B., Lysy, A. and Serebrov, L. (1980). On distribution, stock state and regulation measures of shrimp (*Pandalus borealis* Krøyer) fishery in the Barents Sea. *International Council for the Exploration of the Sea C.M. 1980/K:15*, p. 18 (mimeo).

Berenboim, B., Lysy, A. and Salmov, V. (1986a). Soviet investigations on shrimp (*Pandalus borealis*) in the Barents Sea and Spitsbergen area in May 1985. *International Council for the Exploration of the Sea C.M. 1986/K:1*, p. 19 (mimeo).

Berenboim, B., Ponomarenko, I. Ya. and Yaragina, N. A. (1986b). On "predator prey" relationship between cod and shrimp *Pandalus borealis* in the Barents Sea. *International Council for the Exploration of the Sea C.M. 1986/G:21*, p. 11 (mimeo).

Berenboim, B., Mukhin, S. and Sheveleva, G. (1987). URSS investigations of shrimp (*Pandalus borealis*) in the Barents Sea and in the Spitsbergen area in 1986. *International Council for the Exploration of the Sea C.M. 1987/K:28*, p. 15 (mimeo).

Berenboim, B., Mukhin, S. and Sheveleva, G. (1989). Soviet investigations of shrimp *Pandalus borealis* in the Barents Sea and off the Spitsbergen in 1988. *International Council for the Exploration of the Sea C.M. 1989/K:14*, p. 19 (mimeo).

Berenboim, B., Mukhin, S. and Sheveleva, G. (1990). Soviet investigations of shrimp (*Pandalus borealis*) in the Barents Sea and off the Spitsbergen in 1989. *International Council for the Exploration of the Sea C.M. 1990/K:4*, p. 18 (mimeo).

Berenboim, B., Korzhev, V. A., Tretjak, V. L. and Sheveleva, G. K. (1991). On methods of stock assessment and evaluation of TAC for shrimp *Pandalus borealis* in the Barents Sea. *International Council for the Exploration of the Sea C.M. 1991/K:15*, p. 22 (mimeo).

Berenboim, B., Mukhin, S. and Sunnanå, K. (1992). Results of Norwegian and Russian investigations of shrimp (*Pandalus borealis*) in the Barents Sea and Svalbard area in 1991. *International Council for the Exploration of the Sea C.M. 1992/K:39*, p. 20 (mimeo).

Berenboim, B., Dolgov, A., Korzhev, V. and Yaragina, N. (1999). Impact of cod on dynamics of shrimp stock, *Pandalus borealis*, in the Barents Sea and the use of this factor in multispecies models. *NAFO SCR Doc 99/87. Serial No. N4159*, p. 12.

Berenboim, B., Dolgov, A., Korzhev, V. and Yaragina, N. (2000). The impact of cod on dynamics of Barents Sea shrimp (*Pandalus borealis*) as determined by multispecies models. *Journal of Northwest Atlantic Fishery Science* **27**, 69–75.

Bergman, M. J. N. and Santbrink, J. W. V. (2000). Fishing mortality of populations of megafauna in sandy sediments. In "The Effects of Fishing on Non-Target Species and Habitats. Biological, Conservation and Socio-Economic Issues" (M. J. Kaiser and S. J. Groot, eds), pp. 49–68. Blackwell Sciences, Oxford.

Bergström, B. (1997). Do protandric pandalid shrimp have environmental sex determination? *Marine Biology* **128**, 397–407.

Bergström, B. (2000). The biology of *Pandalus*. *Advances in Marine Biology* **38**, 55–256.

Berkeley, A. A. (1930). The post-embryonic development of the common pandalids of British Columbia. *Contributions to Canadian Biology and Fisheries, New Series* **6**, 79–163.

Bertrand, J., Battaglia, A. and Derible, P. (1988a). Catch, effort and biological data of shrimp (*Pandalus borealis*) in the French fisheries off Greenland in 1986 and 1987. *NAFO SCR Doc. 88/50. Serial No. N1490*, p. 17.

Bertrand, J., Maucorps, A. and Poulard, J. (1988b). Shrimp abundance indices from the French fisheries off East and West Greenland. *NAFO SCR Doc. 88/55. Serial No. N1495*, p. 5.

Biseau, A., Fontaine, B. and Forest, A. (1984). Catch, effort and biological data of shrimp (*Pandalus borealis*) in the French fishery off East Greenland in 1984. *NAFO SCR Doc. 84/I/7. Serial No. N776*, p. 17.

Boje, J. and Lehmann, K. (1994). Selectivity in shrimp trawl codend from an alternate haul experiment. *International Council for the Exploration of the Sea C.M. 1994/B:12, Ref. K*, p. 9 (mimeo).

Bowering, W. R., Parsons, D. G. and Lilly, G. R. (1984). Predation on shrimp (*Pandalus borealis*) by Greenland halibut (*Reinhardtius hippoglossoides*) and Atlantic cod (*Gadus morhua*) off Labrador. *International Council for the Exploration of the Sea C.M. 1984/G:54*, p. 30 (mimeo).

Buch, E. (2000). A monograph on the physical oceanography of the Greenland waters. *Danish Metereological Institute, Scientific Report* **00–12**, 405.

Buch, E., Nielsen, M. and Pedersen, S. (2002). Ecosystem variability and regime shift in West Greenland waters. *NAFO SCR Doc. 02/16. Serial No. N4617*, p. 19.

Buch, E., Pedersen, S. and Ribergaard, M. H. (2004). Ecosystem variability in West Greenland waters. *Journal of Northwest Atlantic Fishery Science* **34**, 13–28.

Butler, T. H. (1971). A review of the biology of the pink shrimp *Pandalus borealis*. *In* "Proceedings of the Conference on the Canadian Shrimp Fishery, Saint John, New Brunswick, October 27–29, 1970". *Canadian Fisheries Report* **17**, pp. 17–24.

Caddy, J. F. (1973). Underwater observations on tracks of dredges and trawls and some effects of dredging on a scallop ground. *Journal of the Fisheries Research Board of Canada* **30**, 173–180.

Cadrin, X. and Skúladóttir, U. (1998). Surplus production analysis of shrimp in the Denmark Strait, 1977–1998. *NAFO SCR Doc. 98/117. Serial No. 4026*, p. 21.

Carlisle, D. B. (1959). On the sexual biology of *Pandalus borealis* (Crustacea Decapoda). I. Histological of incretory elements. *Journal of the Marine Biological Association of the United Kingdom* **38**, 381–394.

Carlsson, D. (1976). Research and management of the shrimp *Pandalus borealis* in Greenland waters. *ICES C.M. 1976/Shellfish stocks nr. 28*, p. 10 (mimeo).

Carlsson, D. (1980a). Data on the Greenland shrimp fishery in NAFO Subarea 0 + 1 in 1980 compared to earlier years. *NAFO SCR Doc. 80/XI/174. Serial No. N261*, p. 46.

Carlsson, D. (1980b). Observation on the shrimp fishery at East Greenland in 1980. *NAFO SCR Doc. 80/XI/164. Serial No. N251*, p. 10.

Carlsson, D. (1981a). Data on the shrimp fishery in NAFO Subarea 0 + 1 in 1980 and 1981. *NAFO SCR Doc. 81/XI/151. Serial No. N458*, p. 14.

Carlsson, D. (1981b). Data on the shrimp fishery at East Greenland in 1980 and 1981. *NAFO SCR Doc. 80/XI/165. Serial No. N473*, p. 12.

Carlsson, D. (1983a). Data on the shrimp fishery at East Greenland, 1980–1982. *NAFO SCR Doc. 83/I/9. Serial No. N647*, p. 16.

Carlsson, D. (1983b). Data on the shrimp fishery in NAFO Subareas 0 and 1 in 1981 and 1982. *NAFO SCR Doc 83/I/8. Serial No. N646*, p. 20.

Carlsson, D. (1984a). Data on the shrimp fishery in NAFO Subarea 1 in 1982 and 1983. *NAFO SCR Doc 84/I/9. Serial No. N778*, p. 37.

Carlsson, D. (1984b). Data on the shrimp fishery at East Greenland in 1983 compared to earlier years. *NAFO SCR Doc. 84/I/5. Serial No. N774*, p. 12.

Carlsson, D. (1985). Data on the shrimp fishery at East Greenland in 1984 compared to earlier years. *NAFO SCR Doc. 85/I/12. Serial No. N946*, p. 13.

Carlsson, D. (1986). Data on the shrimp fishery at East Greenland in 1985 compared to earlier years. *NAFO SCR Doc. 86/5. Serial No. N1103*, p. 14.

Carlsson, D. (1988). The commercial shrimp fishery in the Denmark Strait in 1987. *NAFO SCR Doc. 88/57. Serial No. N1497*, p. 12.

Carlsson, D. (1990a). Data and preliminary assessment of shrimp in Denmark Strait. *NAFO SCR Doc. 90/64. Serial No. N1786*, p. 7.

Carlsson, D. (1990b). Data and preliminary assessment of shrimp in Subareas 0 + 1. *NAFO SCR Doc. 90/65. Serial No. N1787*, p. 8.

Carlsson, D. (1991). Data and preliminary assessment of shrimp in Subareas 0 + 1. *NAFO SCR Doc. 91/74. Serial No. N1958*, p. 4.

Carlsson, D. (1993). The shrimp fishery in NAFO Subarea 1 January to October 1993. *NAFO SCR Doc. 93/130. Serial No. N2342*, p. 29.

Carlsson, D. (1996). Trawl survey for shrimp (*Pandalus borealis*) in Denmark Strait, 1996. *NAFO SCR Doc. 96/116. Serial No. N2813*, p. 8.

Carlsson, D. (1997a). A new interpretation of the age-at-length key for shrimp (*Pandalus borealis*) in the Disko area (Disko Bay and Vaigat) in West Greenland (NAFO Subarea 1). *NAFO SCR Doc. 97/104. Serial No. N2961*, p. 12.

Carlsson, D. (1997b). A first report on a special study on variations in catch of shrimp (*Pandalus borealis*) by depth in West Greenland (NAFO Subarea 1) in 1997. *NAFO SCR Doc. 97/108. Serial No. N2965*, p. 4.

Carlsson, D. (2000). Assessment data for northern shrimp in Denmark Strait in 2000. *NAFO SCR Doc. 00/76. Serial No. N4333*, p. 9.

Carlsson, D. and Hvingel, C. (2000). The fishery for northern shrimp (*Pandalus borealis*) off East Greenland in 1999 and 2000. *NAFO SCR Doc. 00/75. Serial No. N4332*, p. 24.

Carlsson, D. and Kanneworff, P. (1987a). Problems with bottom photography as a method for estimating biomass of shrimp (*Pandalus borealis*) off West Greenland. *NAFO SCR Doc. 887/05. Serial No. N1273*, p. 11.

Carlsson, D. and Kanneworff, P. (1987b). The commercial shrimp fishery in the Denmark Strait in 1985 and 1986. *NAFO SCR Doc. 87/09. Serial No. N1277*, p. 26.

Carlsson, D. and Kanneworff, P. (1989a). The shrimp fishery in NAFO Subarea 1 in 1988. *NAFO SCR Doc. 89/53. Serial No. N1633*, p. 30.

Carlsson, D. and Kanneworff, P. (1989b). Report on a stratified-random trawl survey for shrimp (*Pandalus borealis*) in NAFO Subareas 0 + 1 in July 1988. *NAFO SCR Doc. 89/40. Serial No. N1617*, p. 16.

Carlsson, D. and Kanneworff, P. (1989c). The commercial shrimp fishery in Denmark Strait in 1988. *NAFO SCR Doc. 89/70. Serial No. N1650*, p. 21.

Carlsson, D. and Kanneworff, P. (1990). The commercial shrimp fishery in Denmark Strait in 1989 and early 1990. *NAFO SCR Doc. 90/42. Serial No. N1759*, p. 22.

Carlsson, D. and Kanneworff, P. (1991a). The shrimp fishery in NAFO Subarea 1 in 1990. *NAFO SCR Doc. 91/69. Serial No. N1953*, p. 19.

Carlsson, D. and Kanneworff, P. (1991b). Report on stratified-random trawl surveys for shrimp (*Pandalus borealis*) in NAFO Subarea 0 + 1 in July-August 1990, and a comparison with earlier surveys. *NAFO SCR Doc. 91/70. Serial No. N1954*, p. 34.

Carlsson, D. and Kanneworff, P. (1991c). The commercial shrimp fishery in Denmark Strait in 1990 and early 1991. *NAFO SCR Doc. 91/53. Serial No. N1936*, p. 27.

Carlsson, D. and Kanneworff, P. (1992a). The shrimp fishery in NAFO Subarea 1 in 1991. *NAFO SCR Doc. 92/65. Serial No. N2119*, p. 21.

Carlsson, D. and Kanneworff, P. (1992b). Report on a stratified-random trawl survey for shrimp (*Pandalus borealis*) in NAFO Subareas 0 + 1 in July-September 1991, and a comparison with earlier surveys. *NAFO SCR Doc. 92/67. Serial No. N2121*, p. 27.

Carlsson, D. and Kanneworff, P. (1992c). The commercial shrimp fishery in Denmark Strait in 1991 and early 1992. *NAFO SCR Doc. 92/64. Serial No. N2118*, p. 26.

Carlsson, D. and Kanneworff, P. (1993a). The shrimp fishery in NAFO Subarea 1 in 1992 and early 1993. *NAFO SCR Doc. 93/64. Serial No. N2248*, p. 32.

Carlsson, D. and Kanneworff, P. (1993b). Stratified-random trawl survey for shrimp (*Pandalus borealis*) in inshore areas at West Greenland, NAFO Subarea 1, in 1992. *NAFO SCR Doc. 93/72. Serial No. N2256*, p. 13.

Carlsson, D. and Kanneworff, P. (1993c). The commercial shrimp fishery in Denmark Strait in 1992 and early 1993. *NAFO SCR Doc. 93/60. Serial No. N2243*, p. 23.

Carlsson, D. and Kanneworff, P. (1993d). Stratified-random trawl survey for shrimp (*Pandalus borealis*) in Denmark Strait in 1992. *NAFO SCR Doc. 93/66. Serial No. N2250*, p. 13.

Carlsson, D. and Kanneworff, P. (1995). Trawl survey for shrimp (*Pandalus borealis*) in Denmark Strait, 1995. *NAFO SCR Doc. 95/109. Serial No. N2648*, p. 14.

Carlsson, D. and Kanneworff, P. (1997a). Stratified random trawl survey for shrimp (*Pandalus borealis*) in Disko Bay and Vaigat, inshore West Greenland. *NAFO SCR Doc. 97/99. Serial No. N2956*, p. 10.

Carlsson, D. and Kanneworff, P. (1997b). Evaluation of year-class strength in samples from a commercial shrimp fishery, a theoretical approach. *NAFO SCR Doc. 97/105. Serial No. N2962*, p. 8.

Carlsson, D. and Kanneworff, P. (1997c). Offshore stratified random trawl survey for shrimp (*Pandalus borealis*) in NAFO Subarea 0 + 1, in 1997. *NAFO SCR Doc. 97/101. Serial No. N2958*, p. 19.

Carlsson, D. and Kanneworff, P. (1997d). A trawl survey for small shrimp (*Pandalus borealis*) in shallow waters at West Greenland (NAFO Subarea 1) in 1991. *NAFO SCR Doc. 97/107. Serial No. N2964*, p. 8.

Carlsson, D. and Kanneworff, P. (1998a). Stratified random trawl survey for shrimp (*Pandalus borealis*) in Disko Bay and Vaigat, inshore West Greenland. *NAFO SCR Doc. 98/115. Serial No. N4024*, p. 13.

Carlsson, D. and Kanneworff, P. (1998b). Occurrence of various species taken as by-catch in stratified random trawl surveys from shrimp (*Pandalus borealis*) in NAFO Subarea 0 + 1, 1988–98. *NAFO Working Doc. 98/79*, p. 7.

Carlsson, D. and Kanneworff, P. (1999). Bottom temperatures and possible effect on growth and size at sex change of northern shrimp in West Greenland. *NAFO SCR Doc. 99/110. Serial No. N4190*, p. 6.

Carlsson, D. and Kanneworff, P. (2000). Stratified random trawl survey for Northern shrimp (*Pandalus borealis*) in NAFO Subareas 0 + 1, in 2000. *NAFO SCR Doc. 00/78. Serial No. N4335*, p. 27.

Carlsson, D. and Lassen, H. (1991). A catch-rate index for large shrimp in the Greenland shrimp fishery in NAFO Division 1B. *NAFO SCR Doc. 91/57. Serial No. N1941*, p. 14.

Carlsson, D. and Smidt, E. (1976). *Pandalus borealis* stocks at Greenland. Biology, exploitation and possible protective measures. *International Convention of Northwest Atlantic Fisheries. Research Document 76/VI/16, Serial No. 3796*, p. 20.

Carlsson, D. and Smidt, E. (1978). Shrimp, *Pandalus borealis* Krøyer, stocks off Greenland: Biology, exploitation and possible protective measures. *Selected Papers. International Convention of Northwest Atlantic Fisheries* **4**, 7–14.

Carlsson, D., Kanneworff, P. and Lehmann, K. (1990). Report on a stratified-random trawl survey for shrimp (*Pandalus borealis*) in NAFO Subareas 0 + 1 in July-August 1989. *NAFO SCR Doc. 90/46. Serial No. N1763*, p. 16.

Carlsson, D., Kanneworff, P. and Nygaard, K. (1992). Report on a stratified-random trawl survey for shrimp (*Pandalus borealis*) in inshore areas West Greenland, NAFO subarea 1, in 1991. *NAFO SCR Doc. 92/55. Serial No. N2108*, p. 18.

Carlsson, D., Andersen, M. and Kanneworff, P. (1993a). Assessment of shrimp in Davis Strait (Subareas 0 + 1). *NAFO SCR Doc. 93/81. Serial No. N2266*, p. 19.

Carlsson, D., Kanneworff, P. and Parsons, D. (1993b). Stratified-random survey for shrimp (*Pandalus borealis*) in NAFO Subarea 0 + 1, in 1992. *NAFO SCR Doc. 93/70. Serial No. N2254*, p. 13.

Carlsson, D., Folmer, O., Hvingel, C. and Kanneworff, P. (1995a). Offshore trawl survey for shrimp (*Pandalus borealis*) in NAFO Subareas 0 and 1, in 1995. *NAFO SCR Doc. 95/113. Serial No. N2652*, p. 23.

Carlsson, D., Folmer, O., Hvingel, C. and Kanneworff, P. (1995b). Stratified random trawl survey for shrimp (*Pandalus borealis*) in Disko Bay and Vaigat, inshore West Greenland 1995. *NAFO SCR Doc. 95/111. Serial No. N2650*, p. 11.

Carlsson, D., Folmer, O., Hvingel, C., Kanneworff, P, Pennington, M. and Siegstad, H. (1998). A review of the trawl survey of the shrimp stock off West Greenland. *NAFO SCR Doc. 98/114. Serial No. N4023*, p. 20.

Carlsson, D., Kanneworff, P. and Kingsley, M. (1999). Stratified random trawl survey for shrimp (*Pandalus borealis*) in NAFO Subarea 0 + 1, in 1999. *NAFO SCR Doc. 99/109. Serial No. N4189*, p. 27.

Carlsson, D., Kanneworff, P., Folmer, O., Kingsley, M. and Pennington, M. (2000). Improving the West Greenland survey for shrimp (*Pandalus borealis*). *Journal of Northwest Atlantic Fishery Science* **27**, 151–160.

Christensen, S. (1990). Selection in shrimp trawl. *NAFO SCR Doc. 90/56. Serial No. N1777*, p. 4.

Clark, S. H., Cadrin, S. X., Schick, D. F., Diodati, P. J., Armstrong, M. P. and McCarron, D. (2000). The Gulf of Maine northern shrimp (*Pandalus* borealis) fishery: A review of the record. *Journal of Northwest Atlantic Fishery Science* **27**, 193–226.

Clarke, A., Hopkins, C. C. E. and Nilssen, E. M. (1991). Egg size and reproductive output in the deep-water prawn *Pandalus borealis. Functional Ecology* **5**, 724–730.

Collie, J. S., Escanero, G. A. and Valentine, P. C. (1997). Effects of bottom fishing on the benthic megafauna of Georges Bank. *Marine Ecology Progress Series* **155**, 159–172.

Collie, J. S., Hall, S. J., Kaiser, M. J. and Poiner, I. R. (2000). A quantitative analysis of fishing impacts on shelf-sea benthos. *Journal of Animal Ecology* **69**, 785–798.

Currie, D. R. and Parry, G. D. (1996). Effects of scallop dredging on a soft sediment community: A large-scale experimental study. *Marine Ecology Progress Series* **134**, 131–150.

Currie, D. R. and Parry, G. D. (1999). Impacts and efficiency of scallop dredging on different soft substrates. *Canadian Journal of Fisheries and Aquatic Sciences* **56**, 539–550.

Derible, P., Dupouy, H. and Minet, J. (1980). Catch, effort and biological characteristics of shrimp (*Pandalus borealis*) in the French fishery off West Greenland, 1980. *NAFO SCR Doc. 80/XI/159. Serial No. N246*, p. 18.

Drengstig, A., Fevolden, S., Galand, P. and Aschan, M. (2000). Population structuring of the deep sea shrimp (*Pandalus borealis*) in the NE Atlantic based on allozymic differentiation. *Aquatic Living Resources* **13**, 1–9.

Dupouy, H. and Fontaine, B. (1983). Catch, effort, and biological characteristics of shrimp (*Pandalus borealis*) in the French fishery off West Greenland, 1982. *NAFO SCR Doc 83/I/3. Serial No. N641*, p. 16.

Dupouy, H. and Fréchette, J. (1979). Biomass estimate of the northern deepwater shrimp, *Pandalus borealis*, in NAFO divisions 1B and 0B -R/V *Thalassa* survey, September-October 1979. *NAFO SCR Doc. 79/XI/6. Serial No. N017*, p. 17.

Dupouy, H., Minet, J. and Derible, P. (1981a). Catch, effort, and biological characteristics of shrimp (*Pandalus borealis*) in the French fishery off West Greenland, 1981. *NAFO SCR Doc. 81/XI/147. Serial No. N456*, p. 15.

Dupouy, H., Minet, J. and Derible, P. (1981b). Catch, effort and biological characteristics of shrimp (*Pandalus borealis*) in the French fishery off East Greenland in 1981. *NAFO SCR Doc. 81/XI/157. Serial No. N465*, p. 15.

Dupouy, H., Derible, P. and Biseau, A. (1983). Catch, effort and biological data of shrimp (*Pandalus borealis*) in the French fishery off East Greenland in 1983. *NAFO SCR Doc. 83/I/4. Serial No. N642*, p. 21.

Einarsson, S. (1975). Rækjuleit fyrir Austur-og Norðurlandi. *Ægir* **68**, 238–239.

Eiríksson, H. (1968). Rækjuleit við Austurland. *Ægir* **61**, 399–401, 405.

Eiríksson, H. (1971). Rækjuleit r.s. Hafþórs við Vestur-og Norðvesturland í maí 1971. *Ægir* **64**, 308–310.

Eiríksson, H., Hallgrímsson, I. and Bragason, G. S. (1974). Experiments with *Pandalus* sorting machines at Iceland. *International Council for the Exploration of the Sea C.M. 1974/K:27*, p. 6 (mimeo).

Eleftheriou, A. and Robertson, M. R. (1992). The effects of experimental scallop dredging on the fauna and physical environment of a shallow sandy community. *Netherlands Journal of Sea Research* **30**, 289–299.

Engelstoft, J. (1996). By-catches in the shrimp fishery at West Greenland. *NAFO SCR Doc. 96/36. Serial No. N2711*, p. 11.

Engelstoft, J. (1997). Biomass and abundance of demersal fish stocks off West Greenland estimated from the Greenland trawl survey, 1988–1996. *NAFO SCR Doc. 97/39. Serial No. 2871*, p. 17.

Engelstoft, J. (2002). Biomass and abundance of demersal fish stocks off West Greenland estimated from the Greenland shrimp survey, 1998–2001. *NAFO SCR Doc. 02/48. Serial No. 4660*, p. 24.

Fisheries and Oceans Canada (2003). Integrated Fisheries Management Plan. Northern shrimp Northeast Newfoundland, Labrador Coast and Davis Strait, p. 41. Report published online: www.dfo-mpo.gc.ca.

Folmer, O. (1996). Occurrence of striped shrimp (*Pandalus montagui*) along the west coast of Greenland from 1988 to 1996. *NAFO SCR Doc. 96/113. Serial No. 2810*, p. 4.

Folmer, O. and Pennington, M. (2000). A statistical evaluation of the design and precision of the shrimp trawl survey off West Greenland. *Fisheries Research* **49**, 165–178.

Folmer, O., Carlsson, D., Hvingel, C. and Kanneworff, P. (1996a). Offshore trawl survey for shrimp (*Pandalus borealis*) in NAFO Subareas 0 and 1, in 1996. *NAFO SCR Doc. 96/114. Serial No. N2811*, p. 20.

Folmer, O., Carlsson, D., Hvingel, C. and Kanneworff, P. (1996b). Stratified random trawl survey for shrimp (*Pandalus borealis*) in Disko Bay and Vaigat, inshore West Greenland. *NAFO SCR Doc. 96/112. Serial No. N2809*, p. 12.

Fréchette, J. and Dupouy, H. (1979). Preliminary biological data on the shrimp stocks of Davis Strait. *NAFO SCR Doc. 79/XI/8. Serial No. N019*, p. 15.

Gislason, H. (1994). Ecosystem effects of fishing activities in the North Sea. *Marine Pollution Bulletin* **29**, 520–527.

Godø, O. and Nedreaas, K. (1986). Preliminary report of the Norwegian groundfish survey at Bear Island and West-Spitzbergen in the autumn 1985. *International Council for the Exploration of the Sea C.M. 1986/G:81*, p. 30 (mimeo).

Greenstreet, S. and Rogers, S. (2000). Effects of fishing on non-target fish species. *In* "Effects of Fishing on Non-Target Species and Habitats" (M. J. Kaiser and S. J. Groot, eds), pp. 217–234. Blackwell Science, Oxford.

Hall-Spencer, J. and Moore, P. (2000). Scallop dredging has profound, long-term impacts on maerl habitats. *ICES Journal of Marine Science* **557**, 1407–1415.

Hallgrímsson, I. (1977). On the regulation of the Icelandic shrimp fisheries. *International Council for the Exploration of the Sea C.M. 1977/K:37*, p. 7 (mimeo).

Hallgrímsson, I. (1980). Preliminary information on the Icelandic shrimp fishery in the Icelandic–East Greenland area. Special Meeting of Scientific Council. *NAFO SC Working Paper 80/XI/53*, p. 16.

Hallgrímsson, I. (1981). Preliminary information on the Icelandic shrimp (*Pandalus borealis*) fishery in the Denmark Strait in 1981. *NAFO SCR Doc. 81/XI/167. Serial No. N475*, p. 2.

Hallgrímsson, I. (1993a). Upphaf rækjuveiða við Ísland. *Ægir* **12**, 524–529.

Hallgrímsson, I. (1993b). Rækjuleit á djúpslóð við Ísland. *Hafrannsóknastofnun Fjölrit* **33**, 63.

Hallgrímsson, I. and Einarsson, S. (1978). Rækjuleit og rækjuveiðar á djúpslóð. *Ægir* **78**, 411–418.

Hallgrímsson, I. and Skúladóttir, U. (1979). The history of research and management of the Icelandic shrimp fisheries. *In* "Proceedings of the International Pandalid Symposium, Kodiak, Alaska, February 13–15, 1979, Kodiak" (T. Frady, ed.), pp. 81–86. U.S. Sea Grant Report, No. 81–83.

Hallgrímsson, I. and Skúladóttir, U. (1984). Some data on the Icelandic catch of shrimp in the Denmark Strait area in 1983. *NAFO SCR Doc. 84/I/8. Serial No. N777*, p. 4.

Hallgrímsson, I. and Skúladóttir, U. (1985). The Icelandic shrimp (*Pandalus borealis*) fishery in Denmark Strait in 1984. *NAFO SCR Doc. 85/I/11. Serial No. N945*, p. 13.

Hallgrímsson, I. and Skúladóttir, U. (1988). The Icelandic shrimp (*Pandalus borealis*) fishery in Denmark Strait in 1987. *NAFO SCR Doc. 88/64. Serial No. N1506*, p. 10.

Halmø, K. (1957). Praktiske fiskeforsøk 1954 og 1955. *Årsberetning vedkommende Norges Fiskerier 1955*. **9**, 100–115.

Hamilton, L., Lyster, P. and Otterstad, O. (2000). Social change, ecology and climate in 20th-century Greenland. *Climatic Change* **47**, 193–211.

Hansen, H. and Aschan, M. (2000). Growth, size- and age-at-maturity of shrimp, *Pandalus borealis*, at Svalbard related to environmental parameters. *Journal of Northwest Atlantic Fishery Science* **27**, 83–91.

Hansen, H. and Aschan, M. (2001). Maturity stages of shrimp *Pandalus borealis* Krøyer 1838. Method for classification and description of characteristics. Fiskeriforskning. Report 8/2001, p. 9.

Harbitz, A. and Aschan, M. (2003). A two-dimensional geostatistic method to simulate the precision of abundance estimates. *Canadian Journal of Fisheries and Aquatic Sciences* **60**, 1539–1551.

Hassager, T. (1992). Do length-at-age vary with depth? *NAFO SCR Doc. 92/47. Serial No. N2098*, p. 7.

Haynes, E. (1983). Distribution and abundance of larvae of king crab, *Paralithodes camtschatica*, and pandalid shrimp in the Kachemak Bay, Alaska, 1972 and 1976. *U.S. Department of Commerce, National Ocean and Atmospheric Administration Technical Report NMFS SSRF-765*, p. 64.

Haynes, E. B. and Wigley, A. L. (1969). Biology of the Northern shrimp, *Pandalus borealis* in the Gulf of Maine. *Transactions of the American Fisheries Society* **98**, 60–76.

Hermansson, H. (1980). Upphaf og þróun rækjuveiða og–vinnslu á Íslandi. *Ægir* **73**, 67–85.

Hill, B. and Wassenberg, T. (2000). The probable fate of discards from prawn trawlers fishing near coral reefs: A study in the northern Great Barrier Reef, Australia. *Fisheries Research* **48**, 277–286.

Hoffman, D. L. (1973). Observed acts of copulation in the protandric shrimp *Pandalus platyceros* Brandt (Decapoda, Pandalidae). *Crustaceana* **24**, 242–244.

Horsted, S. A. (1969a). Rejeforekomsterne i Davisstraedet. *Tidsskriftet Grønland* **5**, 129–144.

Horsted, S. A. (1969b). Tagging of deep sea prawn (*Pandalus borealis* Kr.) in Greenland waters. *International Council for the Exploration of the Sea C.M. 1969/K:26*, p. 4 (mimeo).

Horsted, S. A. (1978). Rejerne ved Grønland. *Fisk og Hav-78. Skrifter fra Danmarks Fiskeri-og Havundersøgelser* **37**, 17–25.

Horsted, S. A. (1991). Biological advice for and management of some of the major fisheries resources in Greenland waters. *NAFO Scientific Council Studies* **16**, 79–94.

Horsted, S. A. (2000). A review of the cod fisheries at Greenland, 1910–1995. *Journal of Northwest Atlantic Science* **28**, 121.

Horsted, S. A. and Smidt, E. (1956). The deep sea prawn (*Pandalus borealis* Kr.) in Greenland waters. *Meddelelser fra Danmarks Fiskeri og Havundersøgelser*, Bind I, 1952–1956: Nr. 11, p. 118.

Horsted, S. A. and Smidt, E. (1965). Influence of cold water on fish and prawn stocks in West Greenland. *International Convention of Northwest Atlantic Fisheries Special Publications* **6**, 199–207.

Hoydal, K. (1980). Observations on the Faroese prawn fishery in East Greenland, March to June 1980. *NAFO SCR Doc. 80/XI/172. Serial No. N259*, p. 11.

Humborstad, O.-B., Nøttestad, L., Løkkeborg, S. and Rapp, H. T. (2004). RoxAnn bottom classification system, sidescan sonar and video-sledge: Spatial resolution and their use in assessing trawling impacts. *ICES Journal of Marine Science* **61**, 53–63.

Hvingel, C. (1996). Geographical changes in the fishing pattern of Greenlandic shrimp trawlers in the Davis Strait, 1987–1996. *NAFO SCR Doc. 96/110. Serial No. N2807*, p. 5.

Hvingel, C. (1997). Northern shrimp research in the North Atlantic: State of the art and future research strategy. *TemaNord* **592**, 64.

Hvingel, C. (1999). The fishery for northern shrimp (Pandalus borealis) off East Greenland, Greenlandic zone, 1987–1999. *NAFO SCR Doc. 99/108. Serial No. N4188*, p. 14.

Hvingel, C. (2000). The fishery for northern shrimp (*Pandalus borealis*) off West Greenland, 1970–2000. *NAFO SCR Doc. 00/81. Serial No. N4338*, p. 27.

Hvingel, C. (2001). Data for the assessment of the shrimp (*Pandalus borealis*) stock in Denmark Strait/off East Greenland, 2001. *NAFO SCR Doc. 01/174. Serial No. N4519*, p. 27.

Hvingel, C. (2002a). The fishery for northern shrimp (*Pandalus borealis*) off West Greenland, 1970–2002. *NAFO SCR Doc. 02/151. Serial No. N4780*, p. 19.

Hvingel, C. (2002b). Assessment, prediction and risk analysis of stock development: Shrimp off West Greenland, 2002. *NAFO SCR Doc. 02/157. Serial No. N4786*, p. 10.

Hvingel, C. (2003a). Correction of reported past catches of northern shrimp within the Greenland EEZ to conform to a revision of reporting practices. *NAFO SCR Doc 03/74. Serial No. N4913*, p. 3.

Hvingel, C. (2003b). Assessment, prediction and risk analysis of stock development: Shrimp off West Greenland, 2002. *NAFO SCR Doc 03/73. Serial No. N4912*, p. 14.

Hvingel, C. (2003c). Data for the assessment of the shrimp (*Pandalus borealis*) stock in Denmark Strait/off East Greenland, 2003. *NAFO SCR Doc 03/77. Serial No. N4918*, p. 25.

Hvingel, C. (2004a). The fishery for northern shrimp (*Pandalus borealis*) off West Greenland, 1970–2004. *NAFO SCR Doc. 04/75. Serial No. N5045*, p. 28.

Hvingel, C. (2004b). An assessment of the shrimp stock in Denmark Strait/off East Greenland. *NAFO SCR Doc. 94/81. Serial No. N5051*, p. 25.

Hvingel, C. and Folmer, O. (1998). The Greenlandic fishery for Northern shrimp (*Pandalus borealis*) off West Greenland, 1970–1998. *NAFO SCR Doc. 98/111. Serial No. N4020*, p. 24.

Hvingel, C. and Kingsley, M. (1998). Jack-knifing a logistic model of biomass dynamics of the West Greenland shrimp stock. *NAFO SCR Doc. 98/116. Serial No. N4025*, p. 4.

Hvingel, C. and Kingsley, M. (2002). A framework for the development of management advice on a shrimp stock using a Bayesian approach. *NAFO SCR Doc. 02/158. Serial No. N4787*, p. 28.

Hvingel, C., Lassen, H. and Parsons, D. (1996). A biomass index for Northern shrimp (*Pandalus borealis*) in Davis Strait based on multiplicative modelling of commercial catch-per-unit-effort data (1976–1995). *NAFO SCR Doc. 96/111. Serial No. N2808*, p. 19.

Hvingel, C., Folmer, O. and Siegstad, H. (1997). The Greenlandic fishery for Northern shrimp (*Pandalus borealis*) off West Greenland, 1970–1997. *NAFO SCR Doc. 97/98. Serial No. N2955*, p. 24.

Hvingel, C., Lassen, H. and Parsons, D. (1998). A biomass index for Northern shrimp (*Pandalus borealis*) in Davis Strait based on multiplicative modelling of commercial catch-per-unit-effort data (1976–1997). *NAFO SCR Doc. 98/113. Serial No. N4022*, p. 17.

Hvingel, C., Lassen, H. and Parsons, D. (2000). A biomass index for Northern shrimp (*Pandalus borealis*) in Davis Strait based on multiplicative modelling of commercial catch-per-unit-effort data (1976–1997). *Journal of Northwest Atlantic Fishery Science* **26**, 25–36.

Hylen, A. (1986). Bestands-og forvaltningshistorikk for torsk, hyse, uer og reke i 1970- og 1980-åra. *In* "Seminar on Barentshavets ressurser, 6–7 mai 1986" (O. Hemsett and N. Olsen, eds), pp. 44–64. Scandic Hotel, Trondheim.

Hylen, A. and Ågotnes, P. (1990). Results of stratified shrimp trawl surveys for shrimps (*Pandalus borealis*) in the Barents Sea and in the Svalbard region in 1986. *Survey report*, p. 8.

Hylen, A. and Jacobsen, J. (1987). Estimation of cod taken as by-catch in the Norwegian fishery for shrimp north of 69°N. *International Council for the Exploration of the Sea C.M. 1987/G:34*, p. 21 (mimeo).

Hylen, A. and Øynes, P. (1986). Results of stratified bottom trawl surveys for shrimps (*Pandalus borealis*) in the Barents Sea and the Svalbard region in 1986. *International Council for the Exploration of the Sea C.M. 1986/K:34*, p. 25 (mimeo).

Hylen, A. and Øynes, P. (1988). Results of stratified bottom trawl surveys for shrimps (*Pandalus borealis*) in the Barents Sea and the Svalbard region in 1988. *International Council for the Exploration of the Sea C.M. 1988/K:18*, p. 12 (mimeo).

Hylen, A., Tveranger, B. and Øynes, P. (1984). Norwegian investigations on the deep sea shrimp (*Pandalus borealis*) in the Barents Sea in April-May 1984 and in the Spitsbergen area in July-August 1984. *International Council for the Exploration of the Sea C.M. 1984/K:21*, p. 25 (mimeo).

Hylen, A., Jacobsen, J. and Øynes, P. (1987). Results of stratified bottom trawl surveys for shrimps (*Pandalus borealis*) in the Barents Sea and the Svalbard region in 1987. *International Council for the Exploration of the Sea C.M. 1987/K:39*, p. 13 (mimeo).

Hylen, A., Sunnanå, K. and Øynes, P. (1989). Results of stratified bottom trawl surveys for shrimps (*Pandalus borealis*) in the Barents Sea and the Svalbard region in 1989. *International Council for the Exploration of the Sea C.M. 1989/K:26*, p. 12 (mimeo).

ICES (1998). Report of the study group on grid (grate) sorting systems in trawls, beam trawls and seine nets. *International Council for the Exploration of the Sea C.M. 1998/B:2*, p. 62.

ICES (2005). *Pandalus* Assessment Working Group Report (WGPAND 2004). ICES Advisory Committee on Fisheries Management. *International Council for the Exploration of the Sea C.M. 2005/ACFM:05*, p. 80.

ICES (2006). *Pandalus* Assessment Working Group Report (WGPAND 2005). ICES Advisory Committee on Fisheries Management. *International Council for the Exploration of the Sea C.M. 2006/ACFM:10 Ref. G*, p. 76.

Isaksen, B. (1984). Experiments with vertical side-sorting panels in Norwegian shrimp trawls 1982–1983. *International Council for the Exploration of the Sea C.M. 1984/B:10*, p. 17 (mimeo).

Isaksen, B., Valdemarsen, J. and Larsen, R. (1990). Reduction of fish by-catch in shrimp trawl using a solid separator grid in the aft belly. *International Council for the Exploration of the Sea C.M. 1990/B:13*, p. 13 (mimeo).

Isaksen, B., Valdemarsen, J., Larsen, R. and Karlsen, L. (1992). Reduction of fish by-catch in shrimp trawl using a rigid separator grid in the aft belly. *Fisheries Research* **13**, 335–352.

Ivanov, B. G. and Stolyarenko, D. A. (1988). Relation between currents and orientation of shrimps and its effect on trawl catchability: A new hypothesis. *International Council for the Exploration of the Sea C.M. 1988/K:14*, p. 9 (mimeo).

Ivanov, B. G., Stolyarenko, D. A. and Berenboim, B. (1988). The stock assessment of the deep water shrimp (*Pandalus borealis*) off Spitsbergen and Bear Island with a usage of the spline survey designer software system. *International Council for the Exploration of the Sea C.M. 1988/K:13*, p. 19 (mimeo).

Jakobsen, T. (1980). The Norwegian trial fishery for shrimp, *Pandalus borealis*, at East Greenland in 1980. *NAFO SC Working Paper 80/XI/52*, p. 1.

Jakobsen, T. and Torheim, S. (1980). Norwegian investigations on shrimps, *Pandalus borealis*, off West Greenland in 1980. *NAFO SCR Doc. 80/XI/163. Serial No. N250*, p. 10.

Jakobsen, T. and Torheim, S. (1981a). The Norwegian fishery for shrimp, *Pandalus borealis*, off West Greenland in 1981. *NAFO SCR Doc. 81/XI/156. Serial No. N464*, p. 4.

Jakobsen, T. and Torheim, S. (1981b). Norwegian investigations on shrimp, *Pandalus borealis*, in East Greenland waters in 1981. *NAFO SCR Doc. 81/XI/158. Serial No. N466*, p. 9.

Jakobsen, T. and Torheim, S. (1983a). Norwegian investigations on shrimp, *Pandalus borealis*, off West Greenland in 1982. *NAFO SCR Doc 83/I/5. Serial No. N643*, p. 8.

Jakobsen, T. and Torheim, S. (1983b). Norwegian investigations on shrimp, *Pandalus borealis*, in East Greenland waters in 1982. *NAFO SCR Doc. 83/I/6. Serial No. N644*, p. 7.

Jakobsson, J. (1992). Recent variability in the fisheries of the North Atlantic. *International Council for the Exploration of the Sea Marine Science Symposia* **195**, 291–315.

Jakobsson, J. and Stefánsson, G. (1998). Rational harvesting of the cod-capelin-shrimp complex in the Icelandic marine ecosystem. *Fisheries Research* **37**, 7–21.

Jennings, S. and Kaiser, M. (1998). The effects of fishing on marine ecosystems. *Advances in Marine Biology* **34**, 201–252.

Johannesen, E. and Aschan, M. (2005). How much shrimp does the cod really eat?—Five years later. *NAFO SCR Doc. 05/97. Serial No. N5203*, p. 10.

Jónsdóttir, O. D. B., Imsland, A. K. and Nævdal, G. (1998). Population genetic studies of northern shrimp, *Pandalus borealis*, in Icelandic waters and the Denmark Strait. *Canadian Journal of Fisheries and Aquatic Sciences* **55**, 770–780.

Jónsson, E. and Hallgrímsson, I. (1981). The Icelandic shrimp (*Pandalus borealis*) fishery in the Denmark Strait. *International Council for the Exploration of the Sea* C.M. 1981/K:7, p. 16 (mimeo).

Jónsson, J. (1990). Hafrannsóknir við Ísland. II. Eftir 1937 Bókaútgafa Menningarsjóðs, Reykjavík.

Jørgensen, A. and Kanneworff, P. (1980). Biomass of shrimp (*Pandalus borealis*) in NAFO Subarea 1 estimated by means of bottom photography. *NAFO SCR Doc.* 80/XI/169. Serial No. N256, p. 14.

Jørgensen, O. and Carlsson, D. (1998). An estimate of by-catch of fish in the West Greenland shrimp fishery based on survey data. *NAFO SCR Doc.* 98/41. Serial No. 3030, p. 20.

Kaiser, M. J. and Ramsay, K. (1997). Opportunistic feeding by dabs within areas of trawl disturbance: Possible implications for increased survival. *Marine Ecology Progress Series* **152**, 307–310.

Kaiser, M. J. and Spencer, B. E. (1994). Fish scavenging behaviour in recently trawled areas. *Marine Ecology Progress Series* **112**, 41–49.

Kaiser, M. J., Ramsay, K., Richarson, C., Spence, F. and Brand, A. (2000). Chronic fishing disturbance has changed shelf sea benthic community structure. *Journal of Animal Ecology* **69**, 494–503.

Kanneworff, P. (1979a). Stock biomass 1979 of shrimp (*Pandalus borealis*) in NAFO Subarea 1 estimated by means of bottom photography. *NAFO SCR Doc.* 79/XI/9. Serial No. N020, p. 6.

Kanneworff, P. (1979b). Density of shrimp (*Pandalus borealis*) in Greenland water observed by means of photography. *Rapport et Procès-verbaux des Rèunions. Conseil Permanent International pour l'Exploration de la Mer* **175**, 134–138.

Kanneworff, P. (1981). Biomass of shrimp (*Pandalus borealis*) in NAFO Subarea 1 in 1977–1981 estimated by means of bottom photography. *NAFO SCR Doc.* 81/XI/155. Serial No. N463, p. 19.

Kanneworff, P. (1983). Biomass of shrimp (*Pandalus borealis*) in NAFO Subarea 1 in 1977–1982 estimated by means of bottom photography. *NAFO SCR Doc.* 83/I/1. Serial No. N639, p. 24.

Kanneworff, P. (1984). Biomass of shrimp (*Pandalus borealis*) in NAFO Subarea 1 in 1978–1983 estimated by means of bottom photography. *NAFO SCR Doc* 84/I/6. Serial No. N775, p. 24.

Kanneworff, P. (1986). Biomass of shrimp (*Pandalus borealis*) in NAFO SA1 in 1981–1985 estimated by means of bottom photography. *NAFO SCR Doc.* 86/3. Serial No. N1101, p. 41.

Kanneworff, P. (2003). Occurrence of *Pandalus montagui* in trawl survey samples from NAFO Subareas 0 + 1. *NAFO SCR Doc.* 03/70. Serial No. N4909, p. 4.

Kanneworff, P. and Pedersen, S. (1992). Survey biomass of Greenland halibut (*Reinhardtius hippoglossoides*) off West Greenland (NAFO Subarea 0 + 1), July-August 1988–1991. *NAFO SCR Doc.* 92/41. Serial No. N2092, p. 15.

Kanneworff, P. and Wieland, K. (2001). Stratified random trawl survey for Northern shrimp (*Pandalus borealis*) in NAFO Subareas 0 + 1 in 2001. *NAFO SCR Doc* 01/175. Serial No. N4520, p. 23.

Kanneworff, P. and Wieland, K. (2002). Stratified random trawl survey for Northern shrimp (*Pandalus borealis*) in NAFO Subareas 0 + 1 in 2002. *NAFO SCR Doc* 02/148. Serial No. N4777, p. 25.

Kanneworff, P. and Wieland, K. (2003). Calculating a TAC for northern shrimp (*Pandalus borealis*) in West Greenland waters (NAFO Subareas 0 + 1). *NAFO SCR Doc.* 03/86. Serial No. N4928, p. 5.

Kartavtsev, V., Berenboim, B. and Zugurovsky, K. (1991). Population genetic differenciation on the pink shrimp *Pandalus borealis* Krøyer 1838, from the Barents and Bering Seas. *Journal of Shellfish Research* **10**, 333–339.

Kelleher, K. (2005). Discards in the world's marine fisheries. An update. *FAO Fisheries Technical Paper* **470**, 131.

Kingsley, M. (2000). Effects of fixing stations in the 2000 West Greenland shrimp survey on the detection of changes in biomass. *NAFO SCR Doc. 00/86. Serial No. N4343*, p. 4.

Kingsley, M. (2001). Effects in 2001 of recent modifications to the design of the West Greenland shrimp survey. *NAFO SCR Doc. 01/176. Serial No. N4565*, p. 4.

Kingsley, M. and Carlsson, D. (1998). An experimental investigation on spatial and depth variation in catch of shrimp, Greenland halibut and redfish. *NAFO SCR Doc. 98/119. Serial No. N4028*, p. 6.

Kingsley, M. and Hvingel, C. (2005). The fishery for northern shrimp (*Pandalus borealis*) off West Greenland, 1970–2005. *NAFO SCR Doc. 05/83. Serial No. N5188*, p. 19.

Kingsley, M. and Siegstad, H. (2001). A preliminary analysis of size distributions of northern shrimp taken in the commercial fishery in East Greenland. *NAFO SCR Doc. 01/180. Serial No. N4569*, p. 5.

Kingsley, M., Kanneworff, P. and Carlsson, D. (1999a). Modifications to the design of the trawl survey for *Pandalus borealis* in West Greenland waters: Effects on bias and precision. *NAFO SCR Doc. 99/105. Serial No. N4148*, p. 12.

Kingsley, M., Kanneworff, P. and Carlsson, D. (1999b). By-catches of fish in the West Greenland shrimp survey: An initial analysis. *NAFO SCR Doc. 99/11.*

Koeller, P. A. (2000). Relative importance of abiotic and biotic factors to the management of the northern shrimp (*Pandalus borealis*) fishery of the Scotian Shelf. *Journal of Northwest Atlantic Fishery Science* **27**, 21–34.

Lassen, H. and Carlsson, D. (1990). A catch-rate index for the Greenland Shrimp fishery in NAFO Subarea 1. *NAFO SCR Doc. 90/90. Serial No. N1817*, p. 14.

Lehmann, K. (1989). Report on commercial trial fishery for shrimp at East Greenland in 1987. *NAFO SCR Doc. 89/39. Serial No. N1616*, p. 10.

Lehmann, K. (1990). Report on a commercial trial fishery for shrimp (*Pandalus borealis*) off Southwest, Southeast and East Greenland. *NAFO SCR Doc. 90/62. Serial No. N1784*, p. 6.

Lehmann, K., Valdemarsen, J. and Riget, F. (1993). Selectivity in shrimp trawl codends tested in a fishery in Greenland. *International Council for the Exploration of the Sea Marine Science Symposia* **196**, 80–85.

Lilliendahl, K. and Solmundsson, J. (1997). An estimate of summer food consumption of six seabird species in Iceland. *ICES Journal of Marine Science* **54**, 624–630.

Lilly, G., Parsons, D. and Kulka, D. (2000). Was the increase in shrimp biomass on the Northeast Newfoundland shelf a consequence of a release in predation pressure from cod? *Journal of Northwest Atlantic Fisheries Science* **27**, 45–61.

Lindeboom, H. and De Groot, S. J. (1998). The effects of different types of fisheries on the North Sea and Irish Sea benthic ecosystems. *Netherlands Institute for Sea Research*, NIOZ Rapport 1998–1–RIVLO-DLO Report C003/98, p. 404.

Lund, H. (1988a). On environment and reproduction of the West Greenland shrimp stock (*Pandalus borealis* Kr.) north of 71°N (NAFO Division 1A). *NAFO SCR Doc. 88/59. Serial No. N1499*, p. 15.

Lund, H. (1988b). Greenland fishery for shrimp (*Pandalus borealis* Kr.) in NAFO Division 1A, (Greenland management areas NV1 and NV2) in 1986 and 1987. *NAFO SCR Doc. 88/58. Serial No. N1498*, p. 15.

Lund, H. (1989a). Greenland fishery for shrimp (*Pandalus borealis* Kr.) in NAFO Division 1A, (Greenland management areas NV1 and NV2) in 1986 and 1987. *NAFO SCR Doc. 89/38. Serial No. N1615*, p. 19.

Lund, H. (1989b). Report on a stratified–random trawl survey for shrimp (*Pandalus borealis*) in NAFO subareas 0 + 1 in July 1988. *NAFO SCR Doc. 89/40. Serial No. N1617*, p. 16.

Lund, H. (1990a). Fecundity of shrimp (*Pandalus borealis*) sampled on fishing grounds at North West Greenland and West Greenland. *NAFO SCR Doc 90/63. Serial No. N1785*, p. 7.

Lund, H. (1990b). Greenland fishery for shrimp (*Pandalus borealis*) at North West Greenland from 1985 to 1989. *NAFO SCR Doc. 90/44. Serial No. N1761*, p. 12.

Lysy, A. (1978). Investigations on deep-water shrimp (*Pandalus borealis* Kr.) in the Norwegian and Barents Seas in 1978. *Annales Biology* **35**, 253–256.

Løkkeborg, S. (2005). Impacts of trawling and scallop dredging on benthic habitats and communities. *FAO Fisheries Technical Paper* **472**, 57.

Magnússon, K. G. and Pálsson, O. K. (1989). Trophic ecological relationships of Icelandic cod. *Rapport et Procès-Verbaux des Rèunions. Conseil International pour l'Exploration de la Mer* **188**, 206–224.

Magnússon, K. G. and Pálsson, O. K. (1991). The predatory impact of cod on shrimps in Icelandic waters. *International Council for the Exploration of the Sea C.M. 1991/K:31 Shellfish Committee*, p. 12 (mimeo).

Martinez, I., Skjeldal, T., Dreyer, B. and Aljanabi, S. (1997). Genetic structuring of *Pandalus borealis* in the NE Atlantic. II. RAPD analysis. *International Council for the Exploration of the Sea C.M. 1997/T:24*, p. 12 (mimeo).

McCrary, J. (1971). Sternal spines as a characteristic for differentiating between females of some Pandalidae. *Journal of the Fisheries Research Board of Canada* **28**, 98–100.

Minet, J. (1979). Data on catches, CPUE and biomass of shrimp (*Pandalus borealis*) from the French fishery off West Greenland in 1979. *NAFO SCR Doc. 79/XI/5. Serial No. N016*, p. 11.

Minet, J., Dupouy, H. and Derible, P. (1980). Information on shrimp, *Pandalus borealis*, off East Greenland in 1980. *NAFO SCR Doc. 80/XI/260. Serial No. N260*, p. 5.

Mistakidis, M. N. (1957). The biology of *Pandalus montagui* Leach. *Fishery Investigations Series II* **XXI**, p. 52.

Mukhin, S. and Sheveleva, G. (1991). Soviet investigations on shrimp in the Barents Sea and off the Spitsbergen in 1990. *International Council for the Exploration of the Sea C.M. 1991/K:14*, p. 17 (mimeo).

NAFO (2004). Scientific Council Meeting–October/November 2004. *NAFO SCR Doc. 04/20. Serial No. N5061*, p. 14.

NAFO (2006). *Scientific Council Reports 2005*, p. 373.

Nilssen, E. and Hopkins, C. (1991). Population parameters and life histories of the deep-water prawn *Pandalus borealis* from different regions. *International Council for the Exploration of the Sea C.M. 1991/K:2*, p. 27 (mimeo).

Nilssen, E., Larsen, R. and Hopkins, C. (1986). Catch and size-selection of *Pandalus borealis* in a bottom trawl and implications for population dynamics analyses. *International Council for the Exploration of the Sea C.M. 1986/K:4*, p. 20 (mimeo).

Olsen, H. (1977). Rapport fra prøvefiske etter reker på Jan Mayenfeltet. M/S "Kjelløy" T-97-T fra 31/1–11/2 1977. Rapport Fiskerinæringens Forsøksfond. Fiskeridirektoratet.

Øynes, A. (1977). Rekefisket ved Jan Mayen. Rapport Fiskerinæringens Forsøksfond. Fiskeridirektoratet .

Pálsson, O. K. (1976). Um líffræði fiskungviðis í Ísafjarðardjúpi. *Hafrannsóknir* **8**, 5–42(English summary).

Pálsson, O. K. (1983). The feeding habits of demersal fish species in Icelandic waters. *Rit Fiskideildar* **VII**(1), 60.

Pálsson, O. K. (1994). A review of the trophic interactions of cod stocks in the North Atlantic. *International Council for the Exploration of the Sea Marine Science Symposia* **198**, 553–575.

Pálsson, O. K. (1997). Fæðunám þorsks. *Hafrannsóknastofnun Fjölrit* **57**, 177–192.

Pálsson, O. K. and Thorsteinsson, G. (1985). The management of juvenile fish by-catch in an Icelandic shrimp fishery. *International Council for the Exploration of the Sea C. M.1985/K:47Shelfish Committee Ref: Demersal Fish Committee/Sess:Z*, 19 (mimeo).

Parsons, D. (1979). Canadian research efforts for shrimp (*Pandalus borealis*) in Division 0A and Subarea 1 in 1979. *NAFO SCR Doc. 79/XI/7. Serial No. N018*, p. 16.

Parsons, D. (1980). Canadian observations on the shrimp fishery at West Greenland in 1980. *NAFO SCR Doc. 80/XI/165. Serial No. N252*, p. 8.

Parsons, D. and Fréchette, J. (1989). Fisheries for northern shrimp (*Pandalus borealis*) in the Northwest Atlantic from Greenland to the Gulf of Maine. *In* "Marine Invertebrate Fisheries: Their Assessment and Management" (J. F. Caddy, ed.), pp. 63–85. Wiley Interscience Publications, New York.

Parsons, D. and Khan, R. A. (1986). Microsporidiosis in the northern shrimp, *Pandalus borealis*. *Journal of Invertebrate Pathology* **47**, 74–81.

Parsons, D., Mercer, V. and Veitch, P. (1989). Comparison of the growth of northern shrimp (*Pandalus borealis*) from four regions of the Northwest Atlantic. *Journal of Northwest Atlantic Fisheries Science* **9**, 123–132.

Parsons, D., Savard, L., Carlsson, D., Kanneworff, P. and Siegstad, H. (1992). Assessment of shrimp in Davis Strait Subareas 0 + 1. *NAFO SCR Doc. 92/83. Serial No. N2140*, p. 14.

Pauly, D., Christensen, V., Dalsgaard, J., Froese, R. and Torres, F. (1998). Fishing down marine webs. *Science* **279**, 860–863.

Pedersen, S. A. (1994). Multispecies interactions on the offshore West Greenland shrimp grounds. *International Council for the Exploration of the Sea C.M.1994/P:2 Theme session Ref. K*, p. 26 (mimeo).

Pedersen, S. A. (1998). Distribution and lipid composition of *Pandalus* shrimp larvae in relation to hydrography in West Greenland waters. *Journal of Northwest Atlantic Fishery Science* **24**, 39–60.

Pedersen, O. P., Aschan, M., Hvingel, C. and Gulliksen, B. (2000). Påviker reketråling diversitet og biomasse i arktiske bunndyrsamfunn? Rapport fra en studie av arktiske bunndyrsamfunn på reketrålfelt ved Grønland og Svalbard med forslag til videre undersøkelser. Akvaplan-niva rapport nr APN-421.1553, p. 32.

Pedersen, O. P., Aschan, M., Rasmussen, T., Tande, K. S. and Slagstad, D. (2003). Larval dispersal and mother populations of *Pandalus borealis* investigated by a Lagrangian particle-tracking model. *Fisheries Research* **65**, 173–190.

Pedersen, S. A. and Kanneworff, P. (1992). Survey biomass of redfish (*Sebastes* spp.) off West Greenland (NAFO Subarea 0 + 1), July-August 1988–91. *NAFO SCR Doc. 92/44. Serial No. N2095*, p. 11.

Pedersen, S. A. and Kanneworff, P. (1995). Fish on the West Greenland shrimp grounds, 1988–1992. *ICES Journal of Marine Science* **52**, 165–182.

Pedersen, S. A. and Lehmann, K. (1989). By-catch of redfish and Greenland halibut in the shrimp fishery off West Greenland, 1988. *NAFO SCR Doc. 89/41. Serial No. N1618*, p. 12.

Pedersen, S. A. and Riget, F. (1993). Feeding habits of redfish (*Sebastes* spp.) and Greenland halibut (*Reinhardtius hippoglossoides*) in West Greenland waters. *ICES Journal of Marine Science* **50**, 445–459.

Pedersen, S. A. and Storm, L. (2002). Northern shrimp (*Pandalus borealis*) recruitment in West Greenland waters. Part II. Lipid classes and fatty acids in *Pandalus* shrimp larvae: Implications for survival expectations and trophic relationships. *Journal of Northwest Atlantic Fisheries Science* **30**, 47–60.

Pedersen, S. A., Storm, L. and Simonsen, C. (2002). Northern shrimp (*Pandalus borealis*) recruitment in West Greenland waters. Part I. Distribution of *Pandalus* shrimp larvae in relation to hydrography and plankton. *Journal of Northwest Atlantic Fisheries Science* **30**, 19–46.

Petersen, C. (1892). Fiskens biologiske forhold i Holboek Fjord, 1890–91. *Beretning fra den Danske Biologiske Station* **1**, 121–183.

Poulard, J. and Fontaine, B. (1985). Catch, effort and biological data of shrimp (*Pandalus borealis*) in the French fishery off East Greenland in 1984. *NAFO SCR Doc. 85/I/10. Serial No. N944*, p. 12.

Poulard, J., Fontaine, B., Battaglia, A. and Derible, P. (1986). Catch, effort and biological data of shrimp (*Pandalus borealis*) in the French fishery off East Greenland in 1985. *NAFO SCR Doc. 86/6. Serial No. N1104*, p. 17.

Pranovi, F., Raicevich, S., Franceschini, G., Torricelli, P. and Giovanardi, O. (2001). Discard analysis and damage to non-target species in the "rapido" trawl fishery. *Marine Biology* **139**, 863–875.

Prena, J., Schwinghamer, P., Rowell, T. W., Gordon, J. D. C., Gilkinson, K. D., Wass, W. P. and McKeown, D. L. (1999). Experimental otter trawling on a sandy bottom ecosystem of the Grand Banks of Newfoundland: Analysis of trawl by-catch and effects on epifauna. *Marine Ecology Progress Series* **181**, 107–124.

Ramseier, R., Garrity, C., Parsons, D. and Koeller, P. (2000). Influence of particulate organic carbon sedimentation within the seasonal sea-ice regime on the catch distribution of northern shrimp (*Pandalus borealis*). *Journal of Northwest Atlantic Fisheries Science* **27**, 35–44.

Rasmussen, B. (1942). Om dypvannsreken ved Spitsbergen. *Fiskeridirektoratet Skrifter Serie Havundersøkelser* **VII**, 43.

Rasmussen, B. (1953). On the geographical variation in growth and sexual development of the deep sea prawn (*Pandalus borealis* Kr.). *Reports on Norwegian Fishery and Marine Investigations* **X**, 159.

Rasmussen, B. (1973). Fishing experiments with selective shrimp trawl in Norway 1970 to 1973. Report of the Expert Consultation on Selective Shrimp Trawls. *FAO Fisheries Reports* **139**, 50–56.

Rasmussen, B. and Øynes, A. (1974). Forsøk med reketrål som sorterer bort fisk og fiskeyngel. Fiskeridirektoratet. Fiskerinæringens Forsoksfond, Rapport **4**, 3–15.

Rasmussen, T., Thollesson, M. and Nilssen, E. (1993). Preliminary investigations on the population genetic differenciation of the deep water prawn, *Pandalus borealis* Krøyer 1838, from Northern Norway and the Barents Sea. *International Council for the Exploration of the Sea C.M. 1993/K:11*, p. 5 (mimeo).

Rätz, H. (1992). Decrease in fish biomass off West Greenland (Subdivisions 1B-1F) continued. *NAFO SCR Doc. 92/40. Serial No. N2088*, p. 8.

Rätz, H. (1997a). Biomass indices and geographical distribution patterns of survey catches for shrimp (*Pandalus borealis*) off West and East Greenland, 1982–1996. *NAFO SCR Doc. 97/96. Serial No. N2953*, p. 8.

Rätz, H. (1997b). Structures and changes of the demersal fish assemblage off Greenland, 1982–1996. *NAFO Scientific Council Studies* 32, 1–15.

Rätz, H. (1998a). Abundance, biomass and size composition of dominant demersal fish stocks and trend in near bottom temperature off West and East Greenland, 1982–97. *NAFO SCR Doc. 98/21. Serial No. N3005*, p. 22.

Rätz, H. (1998b). Assessment of other finfish in NAFO Subarea 1. *NAFO SCR Doc. 98/45. Serial No. N3036*, p. 5.

Reithe, S. and Aschan, M. (2004). Bioeconomic analysis of by-catch of juvenile fish in the shrimp fisheries—an evaluation of management procedures in the Barents Sea. *Environmental Resources Economics* 28, 55–72.

Savard, L., Parsons, D. G. and Carlsson, D. M. (1994). Estimation of age and growth of northern shrimp (*Pandalus borealis*) in Davis Strait (NAFO Subareas 0 + 1) using cluster and modal analyses. *Journal of Northwest Atlantic Fisheries Science* 16, 63–74.

Schwinghamer, P., Gordon, D. C., Jr., Rowell, T. W., Prena, J., McKeown, D. L., Sonnichsen, G. and Guigné, J. Y. (1998). Effects of experimental otter trawling on surficial sediment properties of a sandy-bottom ecosystem of the Grand Banks of Newfoundland. *Conservation Biology* 12, 1215–1222.

Shumway, S., Perkins, H., Schick, D. and Stickney, A. (1985). Synopsis of biological data on the pink shrimp, *Pandalus borealis* Krøyer, 1838. *National Oceanic and Atmospheric Administration Technical Report*, NMFS 30 (*FAO Fisheries Synopsis* No. 144), p. 57.

Siegstad, H. (1994a). The shrimp fishery in NAFO Subarea 1 in 1993 and January-October 1994. *NAFO SCR Doc. 94/93. Serial No. N2480*, p. 45.

Siegstad, H. (1994b). The commercial shrimp fishery in Denmark Strait in 1993 and January-October 1994. *NAFO SCR Doc. 94/91. Serial No. N2478*, p. 34.

Siegstad, H. (1996). Preliminary assessment of shrimp (*Pandalus borealis*) in Davis Strait, 1996 (Subareas 0 + 1). *NAFO SCR Doc. 96/115. Serial No. N2812*, p. 20.

Siegstad, H. (1998). Preliminary assessment of shrimp (*Pandalus borealis*) in Davis Strait 1998 (Subareas 0 + 1). *NAFO SCR Doc. 98/123. Serial No. N4032*, p. 19.

Siegstad, H. (1999). Preliminary assessment of shrimp (*Pandalus borealis*) in Davis Strait 1999 (Subareas 0 + 1). *NAFO SCR Doc. 99/113. Serial No. N4193*, p. 19.

Siegstad, H. and Hvingel, C. (2005). An assessment of the shrimp stock in Denmark Strait/off East Greenland–2005. *NAFO SCR Doc. 05/93. Serial No. N5198*, p. 19.

Siegstad, H., Hvingel, C. and Folmer, O. (1995). The Greenland fishery for Northern shrimp (*Pandalus borealis*) in Davis Strait in 1994 and January-October 1995. *NAFO SCR Doc. 95/110. Serial No. N2649*, p. 34.

Skúladóttir, U. (1969). Rækjurannsóknir. *Hafrannsóknir* 1, 50–65.

Skúladóttir, U. (1970). Rækjurannsóknir 1969. *Hafrannsóknir* 2, 45–55.

Skúladóttir, U. (1971). Rækjurannsóknir 1970. *Hafrannsóknir* 3, 45–55.

Skúladóttir, U. (1981). The deviation method. A simple method for detecting year-classes of a population of *Pandalus borealis* from length distributions. *In* "Proceedings of the International Pandalid Shrimp Symposium, February 13–15, 1979" (T. Frady, ed.), pp. 181–196. Kodiak, Alaska.

Skúladóttir, U. (1985a). Tagging and recapture results of *Pandalus borealis* (Krøyer) in Icelandic waters. *International Council for the Exploration of the Sea C.M. 1985/ K:42*, p. 12 (mimeo).

Skúladóttir, U. (1985b). The sustainable yield of *Pandalus borealis* in the Denmark Strait area. *NAFO SCR Doc. 85/I/15. Serial No. N949*, p. 4.

Skúladóttir, U. (1989a). Ástand og hörfur í úthafsrækju 1988 og 1989. *Ægir* **82**, 2–7.

Skúladóttir, U. (1989b). The Icelandic shrimp fishery (*Pandalus borealis*) in Denmark Strait. *NAFO SCR Doc. 89/50. Serial No. N1628*, p. 16.

Skúladóttir, U. (1989c). A review of the shrimp fishery, *Pandalus borealis* in Denmark Strait. *NAFO SCR Doc. 89/36. Serial No. N1613*, p. 14.

Skúladóttir, U. (1990a). The sustainable yield of *Pandalus borealis* in the Denmark Strait area based on data for the years 1980–89. *NAFO SCR Doc. 90/91. Serial No. N1818*, p. 7.

Skúladóttir, U. (1990b). A review of the shrimp fishery (*Pandalus borealis*) in the Denmark Strait, in the years 1978–89. *NAFO SCR Doc. 90/82. Serial No. N1804*, p. 14.

Skúladóttir, U. (1991a). The Icelandic shrimp fishery (*Pandalus borealis*) in Denmark Strait in 1989 and 1990. *NAFO SCR Doc. 91/72. Serial No. N1956*, p. 16.

Skúladóttir, U. (1991b). The catch statistics of the shrimp fishery (*Pandalus borealis*) in the Denmark Strait in the years 1989–1990. *NAFO SCR Doc. 91/58. Serial No. N1946*, p. 16.

Skúladóttir, U. (1992a). The catch statistics of the shrimp fishery (*Pandalus borealis*) in the Denmark Strait in the years 1980–1991. *NAFO SCR Doc. 92/49. Serial No. N2101*, p. 9.

Skúladóttir, U. (1992b). The Icelandic shrimp fishery (*Pandalus borealis*) in Denmark Strait in 1991 and early 1992. *NAFO SCR Doc. 92/62. Serial No. N2116*, p. 7.

Skúladóttir, U. (1993a). The sexual maturity of female shrimp (*Pandalus borealis*) in the Denmark Strait in the years 1985–1992 and a comparison to the nearest Icelandic shrimp stocks in 1992. *NAFO SCR Doc 93/65. Serial No. N2249*, p. 7.

Skúladóttir, U. (1993b). The Icelandic shrimp fishery (*Pandalus borealis*) in the Denmark Strait in 1992 and 1993. *NAFO SCR Doc. 93/135. Serial No. N2347*, p. 2.

Skúladóttir, U. (1993c). The catch statistics of the shrimp fishery (*Pandalus borealis*) in the Denmark Strait in the years 1980–1992. *NAFO SCR Doc. 93/63. Serial No. N2246*, p. 12.

Skúladóttir, U. (1994a). The catch statistics of the shrimp fishery (*Pandalus borealis*) in the Denmark Strait in the years 1980–1994. *NAFO SCR Doc. 94/92. Serial No. N2479*, p. 17.

Skúladóttir, U. (1994b). Preliminary assessment of shrimp in the Denmark Strait. *NAFO SCR Doc. 94/96. Serial No. N2483*, p. 12.

Skúladóttir, U. (1995a). Size at sexual maturity of female Northern shrimp (*Pandalus borealis* Kr.) in the Denmark Strait 1985–1993 and a comparison to the nearest Icelandic shrimp populations. *Journal of Northwest Atlantic Fishery Science* **24**, 27–37.

Skúladóttir, U. (1995b). The Icelandic shrimp fishery (*Pandalus borealis* Kr.) in the Denmark Strait in 1994–1995 and some reflection on age groups in the years 1991–1995. *NAFO SCR Doc. 95/108. Serial No. N2647*, p. 7.

Skúladóttir, U. (1995c). The catch statistics of the shrimp fishery (*Pandalus borealis*) in the Denmark Strait in the years 1980–1995. *NAFO SCR Doc. 95/114. Serial No. N2653*, p. 18.

Skúladóttir, U. (1995d). Assessment of shrimp in the Denmark Strait in 1994–1995. *NAFO SCR Doc. 95/115. Serial No. N2654*, p. 7.

Skúladóttir, U. (1996a). The Icelandic shrimp fishery (*Pandalus borealis* Kr.) in the Denmark Strait in 1995–1996 and some reflection on age groups in the years 1991–1996. *NAFO SCR Doc 96/108. Serial No. N2805*, p. 8.

Skúladóttir, U. (1996b). The catch statistics of the shrimp fishery (*Pandalus borealis*) in the Denmark Strait in the years 1980–1996. *NAFO SCR Doc. 96/107. Serial No. N2804*, p. 17.

Skúladóttir, U. (1996c). Preliminary assessment of shrimp in the Denmark Strait in 1996. *NAFO SCR Doc 96/118. Serial No. N2815*, p. 16.

Skúladóttir, U. (1997a). The Icelandic shrimp fishery (*Pandalus borealis* Kr.) in the Denmark Strait in 1996–1997 and some reflection on age groups in the years 1991–1996. *NAFO SCR Doc 97/103. Serial No. N2960*, p. 8.

Skúladóttir, U. (1997b). The catch statistics of the shrimp fishery (*Pandalus borealis*) in the Denmark Strait in the years 1980–1997. *NAFO SCR Doc. 97/102. Serial No. N2959*, p. 19.

Skúladóttir, U. (1997c). Preliminary assessment of shrimp in the Denmark Strait in 1997. *NAFO SCR Doc 97/106. Serial No. N2936*, p. 13.

Skúladóttir, U. (1998a). Size at sexual maturity of female Northern shrimp (*Pandalus borealis* Krøyer) in the Denmark Strait 1985–93 and a comparison with the nearest Icelandic shrimp populations. *Journal of Northwest Atlantic Fishery Science* **24**, 27–37.

Skúladóttir, U. (1998b). The Icelandic shrimp fishery (*Pandalus borealis*) in the Denmark Strait in 1997–1998 and some reflection on age groups in the years 1991–1997. *NAFO SCR Doc 98/120. Serial No. N4029*, p. 12.

Skúladóttir, U. (1998c). The catch statistics of the shrimp fishery (*Pandalus borealis*) in the Denmark Strait in the years 1980–1998. *NAFO SCR Doc. 98/122. Serial No. N4031*, p. 21.

Skúladóttir, U. (1998d). Preliminary assessment of shrimp in the Denmark Strait in 1998. *NAFO SCR Doc 98/121. Serial No. N4030*, p. 16.

Skúladóttir, U. (1999). The Icelandic shrimp fishery (*Pandalus borealis*) in the Denmark Strait in 1998–1999. *NAFO SCR Doc 99/115. Serial No. N4196*, p. 10.

Skúladóttir, U. (2004). The Icelandic shrimp fishery (*Pandalus borealis* Kr.) in the Denmark Strait in 2004. *NAFO SCR Doc 04/83. Serial No. N5053*, p. 4.

Skúladóttir, U. and Eiríksson, H. (1970). Rækju-og skelfiskleit í Faxaflóa og Hafnarleir. *Ægir* **1**, 7–11.

Skúladóttir, U. and Hallgrímsson, I. (1986). Sustainable yield of shrimp (*Pandalus borealis*) in the Denmark Strait area, 1978 to 1984. *NAFO SCR Doc. 86/2. Serial No. N1100*, p. 4.

Skúladóttir, U. and Pétursson, G. (1999). Defining populations of northern shrimp, *Pandalus borealis* (Krøyer 1838), in Icelandic waters using the maximum length and maturity ogive of females. *Rit Fiskideildar* **16**, 247–262.

Skúladóttir, U. and Pétursson, G. (2005). Assessment of the international fishery for shrimp (*Pandalus borealis*) in Division 3M (Flemish Cap), 1993–2005. *NAFO SCR Doc. 05/89. Serial No. N5194*, p. 21.

Skúladóttir, U. and Stefánsson, G. (1985). Bivariate measurements with regard to detection of age-classes in *Pandalus borealis* (Kröyer). *International Council for the Exploration of the Sea C.M. 1985/K:41* (mimeo).

Skúladóttir, U., Pálsson, J., Bragason, G. S. and Brynjólfsson, S. (1991a). The variation in size and age at change of sex, maximum length and length of ovigerous periods of the shrimp, *Pandalus borealis*, at different temperatures in Icelandic waters. *International Council for the Exploration of the Sea C.M. 1991/K:5*, p. 14 (mimeo).

Skúladóttir, U., Savard, L., Lehmann, K., Parsons, D. and Carlsson, D. (1991b). Preliminary assessment of shrimp in Denmark Strait. *NAFO SCR Doc. 91/96. Serial No. N1985*, p. 24.

Skúladóttir, U., Andersen, M., Carlsson, D. M., Kanneworf, P., Parsons, D. G. and Siegstad, H. (1993). Assessment of shrimp in the Denmark Strait. *NAFO SCR Doc. 93/84. Serial No. N2269*, p. 10.

Skúladóttir, U., Guðmundsdóttir, Á., Bragason, G., Einarsson, S. and Brynjólfsson, S. (1995). Stofnmæling úthafsrækju 1988–1994. *Ægir* **88**, 18–24.

Smedstad, O. (1985). Preliminary report of a cruise with M/T "Masi" to East Greenland waters in September 1984. *NAFO SCR Doc. 85/I/5. Serial No. N939*, p. 6.

Smedstad, O. (1988). Preliminary report of a cruise with M/T "Masi" to East Greenland waters in September 1987. *NAFO SCR Doc. 88/48. Serial No. N1488*, p. 10.

Smedstad, O. (1989). Preliminary report of a cruise with M/T "Håkøy-II" to East Greenland waters in September 1988. *NAFO SCR Doc. 88/19. Serial No. N1595*, p. 11.

Smedstad, O. (1992). Norwegian investigations on shrimp (*Pandalus borealis*) in East Greenland waters in 1991. *NAFO SCR Doc. 92/17. Serial No. N2062*, p. 8.

Smedstad, O. and Torheim, S. (1984a). Norwegian investigations on shrimp, *Pandalus borealis*, in West Greenland waters in 1983. *NAFO SCR Doc 84/I/2. Serial No. N771*, p. 7.

Smedstad, O. and Torheim, S. (1984b). Norwegian investigations on shrimp, *Pandalus borealis*, in East Greenland waters in 1983. *NAFO SCR Doc. 84/I/1. Serial No. N770*, p. 9.

Smedstad, O. and Torheim, S. (1985). Norwegian investigations on shrimp (*Pandalus borealis*) in East Greenland waters in 1984. *NAFO SCR Doc. 85/I/7. Serial No. N941*, p. 6.

Smedstad, O. and Torheim, S. (1986). Investigations on shrimp (*Pandalus borealis*) in the Norwegian fishery off East Greenland in 1985. *NAFO SCR Doc. 86/9. Serial No. N1107*, p. 7.

Smedstad, O. and Torheim, S. (1987). Norwegian investigations on shrimp (*Pandalus borealis*) in East Greenland waters in 1986. *NAFO SCR Doc. 87/03. Serial No. N1271*, p. 9.

Smedstad, O. and Torheim, S. (1989). Norwegian investigations on shrimp (*Pandalus borealis*) in East Greenland waters in 1988. *NAFO SCR Doc. 89/18. Serial No. N1594*, p. 12.

Smedstad, O. and Torheim, S. (1990). Norwegian investigations on shrimp (*Pandalus borealis*) in East Greenland waters in 1989. *NAFO SCR Doc. 90/11. Serial No. N1723*, p. 13.

Smedstad, O. and Torheim, S. (1991). Norwegian investigations on shrimp (*Pandalus borealis*) in East Greenland waters in 1990. *NAFO SCR Doc. 91/20. Serial No. N1897*, p. 12.

Smidt, E. (1960). *Pandalus borealis* Kr. in Greenland waters. *International Council for the Exploration of the Sea C.M. 160/2*, p. 3 (mimeo).

Smidt, E. (1965). Deep-sea prawns and the prawn fishery in Greenland waters. *Rapports et Procès-verbaux des Réunions. Conseil Permanent International pour l'Exploration de la Mer* **156**, 100–104.

Smidt, E. (1969). *Pandalus borealis* in Greenland waters: Its fishery and biology. *FAO Fisheries Reports*, No. 57, 893–901.

Smidt, E. (1970). The *Pandalus* fisheries of the ICES area. *International Council for the Exploration of the Sea C.M. 1970/K:20*, p. 13 (mimeo).

Smidt, E. (1971). Fishery for *Pandalus borealis*. ICES Cooperative Research Report Series A 27, p. 26.

Soldal, A. (1995). Survival of one year cod (*Gadus morhua*), haddock (*Melanogrammus aeglefinus*) and whiting (*Merlangius merlangius*) escaping a shrimp trawl with a metal grid sorting device. *Fisken og Havet* **24**, 22.

Soldal, A. and Engaas, A. (1997). Survival of young gadoids excluded from a shrimp trawl by a rigid deflecting grid. *ICES Journal of Marine Science* **54**, 117–124.

Squires, H. J. (1992). Recognition of *Pandalus eous* Makarov, 1935, as a Pacific species not a variety of the Atlantic *Pandalus borealis* Krøyer, 1838 (Decapoda, Caridea). *Crustaceana* **63**, 257–262.

Standal, D. (2003). Fishing the last frontier-controversies in the regulations of shrimp trawling in the high Arctic. *Marine Policy* **27**, 375–388.

Stefánsson, G. and Pálsson, O. K. (1997). BORMICON. A Boreal migration and consumption model. *Hafrannsóknastofnun Fjölrit* **57**, 223.

Stefánsson, G. and Pálsson, O. K. (1998). A framework for multispecies modelling of Boreal systems. *Reviews in Fish biology and fisheries* **8**, 101–104.

Stefánsson, G., Skúladóttir, U. and Pétursson, G. (1994). The use of a stock production type model in evaluating the offshore *Pandalus borealis* stock of North Icelandic waters, including the predation of northern shrimp by cod. *International Council for the Exploration of the Sea C.M. 1994/K:25, Ref. D, G*, p. 13 (mimeo).

Stefánsson, G., Skúladóttir, U. and Steinarsson, B. Æ. (1998). Aspects of the ecology of a Boreal system. *ICES Journal of Marine Science* **55**, 859–862.

Stefánsson, U. (1962). North Icelandic Waters. *Rit Fiskideildar*, 3. Atvinnudeild Háskólans–Fiskideild, p. 269.

Stein, M. (2004). Climatic overview of NAFO Subarea 1, 1991–2000. *Journal of Northwest Atlantic Fisheries Science* **34**, 29–41.

Stickney, A. P. (1978). A previously unreported peridinian parasite in the eggs of the northern shrimp, *Pandalus borealis*. *Journal of Invertebrate Pathology* **32**, 212–215.

Stickney, A. P. and Perkins, H. C. (1977). Environmental physiology of commercial shrimp, *Pandalus borealis*. Project 3–202-R Completion Report, February 1, 1974 to January 31, 1977. Department of Marine Resources, W. Boothbay Harbour, Mayne, p. 78.

Stickney, A. P. and Perkins, H. C. (1981). Observations on the food of the larvae of the northern shrimp, *Pandalus borealis* Krøyer (Decapoda, Caridea). *Crustaceana* **40**, 36–49.

Stolyarenko, D. A. (1986). Data analysis of trawl shrimp survey with spline approximation density. *International Council for the Exploration of the Sea C.M. 1986/K:25*, p. 15 (mimeo).

Storm, L. and Pedersen, S. A. (2003). Development and drift of northern shrimp larvae (*Pandalus borealis*) at West Greenland. *Marine Biology* **143**, 1083–1093.

Storr-Paulsen, M. (2004). Biomass and abundance of demersal fish stocks off West Greenland estimated from the Greenland shrimp survey, 1988–2003. *NAFO SCR Doc. 04/18. Serial No. N4966*, p. 28.

Storr-Paulsen, M. and Jørgensen, O. A. (2005). Biomass and abundance of demersal fish stocks off West Greenland estimated from the Greenland shrimp survey, 1988–2004. *NAFO SCR Doc. 05/39. Serial No. N5125*, p. 27.

Strøm, A. (1971). Rapport on rekeforsøk i området Bjørnøja og Nordsjøkysten með M/S "Halvarson". *Fiskets og Gang* **33**, 608–610.

Strøm, A. (1976). Rapport fra forsøksfiske etter reker ved Jan Mayen og Spitsbergen, 2/8–29/8–76, med M/S "Feiebas". Rapport Fiskerinæringens Forsøksfond. Fiskeridirektoratet, pp. 63–66.

Strøm, A. and Øynes, P. (1974). Rekefelter Langs Norskekysten, Barentshavet og Svalbard. Fiskerinæringens Forsøksfond Rapporter. Fiskeridirektoratet, Bergen, p. 29.

Strøm, A. and Rasmussen, B. (1970). Forsøksfiske etter reker i Barentshavet og Svalbardområdet Juni-August 1970. *Fiskets Gang* **50**, 912–917.

Sund, O. (1930). The renewal of fish population studied by means of measurement of commercial catches. Example: The Arcto Norwegian cod stock. *Rapport et Procès-Verbaux des Rèunions. Conseil International pour l'Exploration de la Mer* **65**, 10–17.

Tavares, A. and Øynes, P. (1980). Results of a stratified bottom trawl survey for shrimps (*Pandalus borealis*) in the Barents Sea and the Spitsbergen area in May-June 1980. *International Council for the Exploration of the Sea C.M. 1980/K:22*, p. 14 (mimeo).

Taylor, L. and Stefánsson, G. (2004). Gadget models of cod-capelin-shrimp interactions in Icelandic waters. *International Council for the Exploration of the Sea C.M. 2004/FF:29*, p. 30 (mimeo).

Teigsmark, G. (1983). Populations of the deep-sea shrimp (*Pandalus borealis* Krøyer) in the Barents Sea. *Fiskeridirektoratet Skrifter Serie Havundersøkelser* **17**, 377–430.

Teigsmark, G. and Øynes, P. (1981). Results of a stratified bottom trawl survey for shrimps (*Pandalus borealis*) in the Barents Sea in May-June 1981. *International Council for the Exploration of the Sea C.M. 1981/K:21*, p. 19 (mimeo).

Teigsmark, G. and Øynes, P. (1982). Norwegian investigations on the deep sea shrimp (*Pandalus borealis*) in the Barents Sea in 1982. *International Council for the Exploration of the Sea C.M. 1982/K:12*, p. 18 (mimeo).

Teigsmark, G. and Øynes, P. (1983a). Norwegian investigations on the deep sea shrimp (*Pandalus borealis*) in the Barents Sea in April-May 1983 and in the Spitsbergen area in July 1983. *International Council for the Exploration of the Sea C.M. 1983/K:46*, p. 18 (mimeo).

Teigsmark, G. and Øynes, P. (1983b). Results of a stratified bottom trawl survey for shrimps (*Pandalus borealis*) in the Spitsbergen area in July 1982. *International Council for the Exploration of the Sea C.M. 1983/K:17*, p. 11 (mimeo).

Thomassen, T. and Ulltang, Ø. (1975). Report from mesh selection experiments on *Pandalus borealis* in Norwegian waters. *International Council for the Exploration of the Sea C.M. 1975/K:51. Shellfish and Benthos Committee. Ref: Gear and Behaviour Committee*, p. 16 (mimeo).

Thorsteinsson, G. (1973). Selective prawn-trawl experiments in Icelandic waters. Report of the Expert Consultation on Selective Shrimp Trawls. *FAO Fisheries Reports* **139**, 57–63.

Thorsteinsson, G. (1974). On the influence of increased mesh size on the size distribution of *Pandalus borealis* in Icelandic waters. *International Council for the Exploration of the Sea, Shellfish and Benthos Committee Ref: Gear and Behaviour Committee, C.M. 1974/K:4*, p. 9 (mimeo).

Thorsteinsson, G. (1980). On the influence of mesh size and hanging ratio on the escape of *Pandalus borealis* through the side panels of a A4-seam bottom trawl. *International Council for the Exploration of the Sea, Fish Capture Committee, Ref: Shellfish Committee, C.M. 1980/B:4*, p. 13 (mimeo).

Thorsteinsson, G. (1981). The effect of net slack in the side panels of shrimp trawls on the size distribution of the catch. *International Council for the Exploration of the Sea, C.M. 1981/B:5*, p. 9 (mimeo).

Thorsteinsson, G. (1989). Icelandic experiments with square mesh netting in the shrimp fishery. *International Council for the Exploration of the Sea C.M. 1989/B:49*, p. 11 (mimeo).

Thorsteinsson, G. (1992). The use of square mesh codends in the Icelandic shrimp (*Pandalus borealis*) fishery. *Fisheries Research* **13**, 255–266.

Thorsteinsson, G. (1995). Survival of shrimp and small fish in the inshore shrimp fishery in Iceland. *International Council for the Exploration of the Sea Study Group on Unaccounted Fishing Mortality in Fisheries. Aberdeen*, 17–18 April, 1995. p. 13 (mimeo).

Torheim, S. (1980). Rekeundersøkelser ved Jan Mayen i oktober 1979. *Fisken og Havet* **1980**, 1–9.

Torrisen, H. (1974). Rapport fra forsøksfiske etter reker ved Jan Mayen med M/S "Langskjær" T-95-LK fra 1/7–4/10 1974. *Rapport* Fiskerinæringens Forsøksfond. Fiskeridirektoratet, pp. 16–20.

Tuck, I. D., Hall, S. J., Robertson, M. R., Armstrong, E. and Basford, D. J. (1998). Effects of physical trawling disturbance in a previously unfished sheltered Scottish sea Loch. *Marine Ecology Progress Series* **162**, 227–242.

Tveranger, B. and Øynes, A. (1985). Results of stratified trawl surveys for shrimps (*Pandalus borealis*) in the Barents Sea in May and in the Svalbard region in July-August 1985. *International Council for the Exploration of the Sea C.M. 1985/K:50*, p. 25 (mimeo).

Tveranger, B. and Øynes, P. (1986). Results of stratified trawl survey for shrimps (*Pandalus borealis*) in the Barents Sea in May and in the Svalbard region in July-August. *Council Meeting of the International Council for the Exploration of the Sea* 1986 (G:76) p. 10 (mimeo).

Ulltang, Ø. (1981). The Norwegian shrimp fisheries. History of research and management. *In* "Proceedings of the International Pandalid Symposium, February 13–15, 1979" (T. Frady, ed.), p. 87. Kodiak, Alaska.

Ulltang, Ø. and Torheim, S. (1979). Norwegian investigations on shrimp, *Pandalus borealis*, off West Greenland in 1979. *NAFO SCR Doc. 79/XI/2. Serial No. N013*, p. 9.

Valdemarsen, J. and Misund, R. (2003). Forsøk med 19 og 22 mm spileavstand i sorteringsrist i fisket etter rognreke i Nordsjøen våren 2002. Havforskningsinstituttet og Fiskeridirektoratet, p. 8.

Valdemarsen, J., Lehmann, K., Riget, F. and Boje, J. (1993). Grid devices to select shrimp size in trawls. *International Council for the Exploration of the Sea C.M. 1993/B:35, Ref. K*, p. 12 (mimeo).

Valdemarsen, J., Thorsteinnsson, G., Boje, J., Lehmann, K. and Jakobsen, J. (1996). Seleksjon i reketrål. *TemaNord* **520**, p. 216.

van den Hoonaard, W. C. (1992). "Reluctant Pioneers: Constraints and Opportunities in an Icelandic Fishing Community". Peter Lang, New York.

Veim, A. K. (1994). By-catch of juvenile fish in the shrimp fishery-management based on bio-economic criteria. *International Council for the Exploration of the Sea C.M. 1994/T:14*, p. 14 (mimeo).

Veim, A. K. (1999). When should shrimp fisheries be closed? A bio-economic analysis of by-catch regulation, p. 12. Unpublished report.

Vilhjálmsson, H. (1994). The Icelandic capelin stock. *Rit Fiskideildar* **XIII**, 281.

Vilhjálmsson, H. (1997). Climatic variations and some examples of their effects on the marine ecology of Icelandic and Greenland waters, in particular during the present century. *Rit Fiskideildar* **XV**, 9–32.

Watling, L. and Norse, E. A. (1998). Disturbance of the sea bed by mobile fishing gear: A comparison to forest clearcutting. *Conservation Biology* **12**, 1180–1197.

Wieland, K. (2001). Precision and consistency of size measurements for northern shrimp (*Pandalus borealis*) in the West Greenland bottom survey. *NAFO SCR Doc. 01/179. Serial No. N4568*, p. 5.

Wieland, K. (2002a). The use of fine-meshed bags for sampling juvenile northern shrimp (*Pandalus borealis*) in the West Greenland bottom trawl survey. *NAFO SCR Doc. 02/145. Serial No. N4774*, p. 11.

Wieland, K. (2002b). Abundance indices for juvenile (age 1 and 2) northern shrimp (*Pandalus borealis*) off West Greenland (NAFO subareas 0 + 1), 1993–2002. *NAFO SCR Doc. 02/146. Serial No. N4775*, p. 20.

Wieland, K. (2003). Abundance of young (age 1, 2 and 3) northern shrimp (*Pandalus borealis*) off West Greenland (NAFO subareas 0 + 1) in 1993–2003, and changes in mean size at age related to temperature and stock size. *NAFO SCR Doc. 03/76. Serial No. N4915*, p. 22.

Wieland, K. (2004a). Length at sex transition in northern shrimp (*Pandalus borealis*) off West Greenland in relation to changes in temperature and stock size. *Fisheries Research* **69**, 49–56.

Wieland, K. (2004b). Abundance and mean size at age of northern shrimp (*Pandalus borealis*) juveniles and males off West Greenland (NAFO Subarea 1 and Div. 0A) in 1993–2004. *NAFO SCR Doc. 04/73. Serial No. N5043*, p. 26.

Wieland, K. (2004c). Recruitment of northern shrimp (*Pandalus borealis*) off West Greenland in relation to spawning stock size and environmental variation, 1993–2004. *NAFO SCR Doc. 04/74. Serial No. N5044*, p. 9.

Wieland, K., Kanneworff, P. and Bergström, B. (2004). Results of the Greenland bottom trawl survey for northern shrimp (*Pandalus borealis*) off West Greenland (NAFO Subarea 1 and Division 0A, 1988–2004). *NAFO SCR Doc. 04/72. Serial No. N5042*, p. 31.

Wienberg, R. W. (1981). On the food and feeding habits of *Pandalus borealis* Krøyer 1838. *Archiv fuer Fischereiwissenschaft* **31**, 123–137.

Wollebæk, A. (1903). Ræker of rækefiske. Årsberetning Vedkommende Norges Fiskerier for 1903. Bergen, 167–229.

Yamada, H., Okada, K. and Jøgersen, O. (1988). West Greenland groundfish biomass estimated from a stratified-random trawl survey in 1987. *NAFO SCR Doc 88/31. Serial No. N1469*, p. 6.

Þorsteinsson, G. (1969). Veiðarfærarannsóknir. *Hafrannsóknir* **1**, 110–117.

Þorsteinsson, G. (1970). Rækjuleit r/s Hafþórs fyrir Norðurlandi í desember 1969 og febrúar 1970. *Ægir* **63**, 197–202.

Þorsteinsson, G. (1976). Veiðiaðferðir og veiðarfæri við rækjuveiðar. *Hafrannsóknir* **8**, 65–110.

Þorsteinsson, G. (1980). Er hægt að losna við smárækju úr trollinu? *Sjávarfréttir* **8**, 25–30.

Þorsteinsson, G. (1981). Af tilraunum til að draga úr smárækjuveiði. *Ægir* **74**, 478–481.

Þorsteinsson, G. (1991). Rækjuveiðar með fiskafælu. *Fiskifréttir*, 26.júlí, p. 8.

Þorsteinsson, G. (1993). Tilraunir með seiðaskiljur í rækjuvörpu. *Morgunblaðið Sjávarfréttir*, 11.júní.

Þorsteinsson, G. (1994a). Afföl af rækju við rækjuveiðar. *Fiskifréttir*, 21.október, p. 8.

Þorsteinsson, G. (1994b). Nýjar tilraunir með smáfiskaskilju við Ísland. *Fiskifréttir*, 2.júlí, pp. 8–9.

Þorsteinsson, G. (1994c). Veiðiútbúnaður við seiðaskilju. *Fiskifréttir*, 9.september, p. 8.

Þorsteinsson, G. (1995). Seiðaskiljur við rækjuveiðar á Íslandsmiðum. *Fiskifréttir*, 10. febrúar, p. 8.

Þorsteinsson, G. and Eiríksson, H. (1971). Rækjuleit 1970. *Hafrannsóknir* **3**, 117–126.

Protein Metabolism in Marine Animals: The Underlying Mechanism of Growth

Keiron P. P. Fraser* and Alex D. Rogers*,[1]

*British Antarctic Survey, Natural Environment Research Council, Cambridge CB3 OET, United Kingdom

[1] Present address: Zoological Society of London, Institute of Zoology, Regents Park, London NW1 4RY, United Kingdom.

ADVANCES IN MARINE BIOLOGY VOL 52
© 2007 Elsevier Ltd. All rights reserved

0065-2881/07 $35.00
DOI: 10.1016/S0065-2881(06)52003-6

Growth is a fundamental process within all marine organisms. In soft tissues, growth is primarily achieved by the synthesis and retention of proteins as protein growth. The protein pool (all the protein within the organism) is highly dynamic, with proteins constantly entering the pool via protein synthesis or being removed from the pool via protein degradation. Any net change in the size of the protein pool, positive or negative, is termed protein growth. The three inter-related processes of protein synthesis, degradation and growth are together termed protein metabolism. Measurement of protein metabolism is vital in helping us understand how biotic and abiotic factors affect growth and growth efficiency in marine animals. Recently, the developing fields of transcriptomics and proteomics have started to offer us a means of greatly increasing our knowledge of the underlying molecular control of protein metabolism. Transcriptomics may also allow us to detect subtle changes in gene expression associated with protein synthesis and degradation, which cannot be detected using classical methods. A large literature exists on protein metabolism in animals; however, this chapter concentrates on what we know of marine ectotherms; data from non-marine ectotherms and endotherms are only discussed when the data are of particular relevance. We first consider the techniques available to measure protein metabolism, their problems and what validation is required. Protein metabolism in marine organisms is highly sensitive to a wide variety of factors, including temperature, pollution, seasonality, nutrition, developmental stage, genetics, sexual maturation and moulting. We examine how these abiotic and biotic factors affect protein metabolism at the level of whole-animal (adult and larval), tissue and cellular protein metabolism. Available gene expression data, which help us understand the underlying control of protein metabolism, are also discussed. As protein metabolism appears to comprise a significant proportion of overall metabolic costs in marine organisms, accurate estimates of the energetic cost per unit of synthesised protein are important. Measured costs of protein metabolism are reviewed, and the very high variability in reported costs highlighted. Two major determinants of protein synthesis rates are the tissue concentration of RNA, often expressed as the RNA to protein ratio, and the RNA activity (k_{RNA}). The effects of temperature, nutrition and developmental stage on

RNA concentration and activity are considered. This chapter highlights our complete lack of knowledge of protein metabolism in many groups of marine organisms, and the fact we currently have only limited data for animals held under a narrow range of experimental conditions. The potential assistance that genomic methods may provide in increasing our understanding of protein metabolism is described.

1. INTRODUCTION

1.1. Overview

Growth is a fundamental process in all animals. Individuals need to achieve a threshold body size before they are able to reproduce; the ultimate purpose of an individual being to pass on its genes (Jobling, 2002). It has also been suggested that growth rate is positively correlated with animal fitness (Yamahira and Conover, 2002). In many marine animals, the risk of mortality resulting from predation is considerably higher in larvae and juveniles than adults (Laurence, 1975; Peterson and Wroblewski, 1984); therefore, the rate at which animals grow can affect the degree of predation that populations are exposed to over a generation (Peterson and Wroblewski, 1984). Not only is the growth rate of animals important, but also the energetic efficiency with which that animal grows. In particular, growth efficiencies of commercially important marine species are of great interest to fisheries researchers and aquaculturists. A considerable literature now exists, examining a wide range of factors affecting growth in marine organisms, for example feeding regime (Mente *et al.*, 2001; Bolliet *et al.*, 2004; Schofield, 2004), temperature (Laurence, 1975; Jobling, 1995), nutrition (Cowey, 1992; Hewitt, 1992; Paul *et al.*, 1994) and social interactions (Carter *et al.*, 1992; McCarthy *et al.*, 1992).

There is considerable current interest in how global changes in the climate will affect marine organisms. Mean global air temperatures have increased by $0.6 \pm 0.2\,°C$ in the twentieth century, probably as a result of anthropogenic increases in greenhouse gases, with some geographic regions, including the Antarctic Peninsula, showing temperature increases as high as $2\,°C$ over the last 40–50 years (King and Harangozo, 1998; Houghton *et al.*, 2001; Quayle *et al.*, 2002). In concert with increasing air temperatures, sea-water temperatures are also rising and predicted to rise further, for example the third climate model produced by the Hadley Centre predicts a further $2\,°C$ increase in global sea-water temperatures (Levitus *et al.*, 2000; Gille, 2002; Meridith and King, 2005). Understanding how current and predicted increases in sea water temperatures will affect protein metabolism and hence growth in marine organisms is therefore vital in assessing the likely impacts of global climate change at both the organism and ecosystem scale. This is particularly true of protein metabolism, as this process is likely to have an important role in allowing organisms to adapt to environmental changes, including increasing water temperatures (Hawkins, 1991; Reid et al., 1998). Increases in the expression and synthesis of stress-related proteins, such as heat-shock proteins (HSPs), have a vital role in allowing animals to survive short-term exposures to temperatures approaching their upper or lower thermal limits (Hofmann et al., 2002; Ali et al., 2003; Hofmann, 2005).

An animal effectively produces soft tissue growth by synthesising protein and retaining a proportion of the synthesised protein as protein growth, the remaining protein being degraded. These inter-related processes of protein synthesis, degradation and growth are collectively termed protein metabolism (Simon, 1989). The cost of synthesising protein dominates the overall cost of growth in animals. Jobling (1985) suggested that the published energetic costs of protein synthesis typically lie within the range of 70–100 mmol ATP per gram protein (Section 7), lipids 15–25 mmol ATP g^{-1} and glycogen 10–12 mmol ATP g^{-1}. As an example, we will consider the proximate composition of the Antarctic teleost Notothenia coriiceps Richardson, 1844, an organism that contains 17% protein, 1% lipid, 0.27% carbohydrate, 3% ash with the remainder water (Crawford, 1979). If we scale these proximate compositions to a 100-g fish then the animal will consist of 17 g of protein, 1 g of lipid and 0.27 g of carbohydrate. Simply multiplying the estimated costs of synthesising these body components by their masses demonstrates that about 98% of the costs of producing that animal resulted from the synthesis of proteins, 1–1.5% lipids and less than 1% carbohydrate. In reality the relative costs of synthesising protein are even higher, because a significant proportion of synthesised proteins are degraded, considerably reducing the efficiency of the protein growth process, whereas the turnover of lipids and carbohydrates are likely to be comparatively lower (Reeds et al., 1982; Houlihan et al., 1995a).

An animal's protein pool (total protein mass in the organism) is in a constant state of flux, with proteins entering the pool via protein synthesis and being removed via protein degradation (Fraser et al., 2002a). Therefore, to understand the underlying processes controlling growth, it is important to understand protein synthesis (k_s), degradation (k_d) and growth (k_g). Whether an animal shows net protein growth or de-growth is dependent on the relative balance between protein synthesis and protein degradation (Houlihan et al., 1995a). In growing animals, rates of protein synthesis will be greater than protein degradation and hence protein growth will occur (Figure 1A). In an animal showing negative protein growth, that is loss of protein mass, protein degradation will be greater than protein synthesis (Figure 1B) and in animals maintaining protein

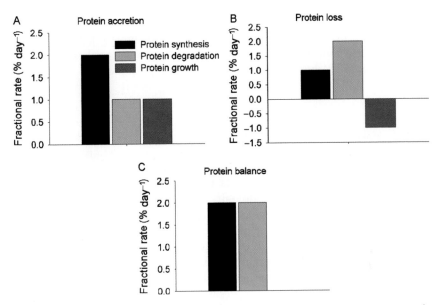

Figure 1 Diagrammatic representation of the relationships between protein synthesis, degradation and growth under the conditions of protein accretion (A), loss (B) and balance (C).

balance, synthesis must equal degradation (Figure 1C). Another term commonly used in protein metabolism studies is protein turnover (k_t) (Waterlow *et al.*, 1978). Protein turnover is the amount of protein that is replaced or 'turned over' from the animals' existing total protein pool. Under conditions of protein accretion, protein turnover is equal to protein degradation, whereas during net protein loss, turnover is equal to protein synthesis. When an animal is in protein balance, that is protein synthesis is equal to protein degradation, then protein turnover will be equal to protein synthesis or degradation. Protein turnover is thought to be important for a number of reasons, including allowing biological systems to adapt to change, altering the concentrations of regulatory proteins, controlling intracellular concentrations of amino acids in concert with protein synthesis, releasing amino acids under conditions of nutritional deprivation and the disposal of defective or redundant proteins (Schoenheimer, 1942; Hershko and Ciechanover, 1982; Ciechanover *et al.*, 1984; Scornik, 1984; Wheatley, 1984; Hawkins, 1991). It has even been suggested that the fundamental basis of heterosis (hybrid-vigour) stems from the fact that more heterozygous individuals have lower rates of protein turnover, resulting in lower energetic requirements for protein metabolism, and a reduced metabolic sensitivity to environmental change (Bayne and Hawkins, 1997; Hawkins and Day, 1999). Selection of individuals on the basis of high efficiencies of protein deposition has been suggested as a method of selecting the fastest growing stock (Hawkins and Day, 1996). The efficiency of protein deposition is typically expressed as the protein synthesis retention efficiency [PSRE (%)]:

$$PSRE = \frac{k_g}{k_s} \times 100$$

where k_g is the fractional protein growth rate (% day^{-1}) and k_s is the fractional protein synthesis rate (% day^{-1}).

Considerable advances have been made in understanding protein metabolism in marine ectotherms over the last 30 years, largely as a result of the development of practical methodologies that have made it possible to make comparatively non-invasive measurements of protein synthesis (Garlick et al., 1980; Hawkins, 1985; Houlihan et al., 1986; Carter et al., 1994; Meyer-Burgdorff and Rosenow, 1995a,b,c).

The aim of this chapter is to examine the current status of our understanding of protein metabolism in marine ectotherms and discuss how the growing field of transcriptomics and proteomics may provide new insights in the field. As the authors have a strong interest in the biology of polar marine organisms, the chapter will also consider whether patterns of protein metabolism in polar species are similar or dissimilar to temperate and tropical species. In writing this chapter, we have had the considerable benefit of several excellent previous reviews of protein metabolism in fish (Haschemeyer, 1978; Fauconneau, 1985; Houlihan et al., 1995c,d; Carter and Houlihan, 2001), crustaceans (Whiteley et al., 2001), ectotherms (Hawkins, 1991; Houlihan, 1991) and animals generally (Houlihan et al., 1995a).

This chapter will concentrate on those species that inhabit the marine environment throughout their lifecycle, but will also include data available on anadromous species during their sea-water phase. Details of protein metabolism in freshwater organisms and endotherms will not generally be discussed, unless the data are of particular relevance. However, details of protein metabolism in many freshwater animal groups can be found in the aforementioned reviews.

1.2. The processes of protein synthesis, degradation and growth

Protein synthesis is the process of translation by which messenger RNA (mRNA) is read by the ribosome and the information converted to a peptide composed of a series of specific amino acids. The decoding of the information held within the mRNA proceeds through a series of three distinct energy-consuming steps: initiation when the start codon (AUG, methionine) is identified and read, elongation during which the internal mRNA codons are translated and the peptide synthesised, and termination when a termination codon on the mRNA is reached, at which point the ribosome separates into its component subunits and the newly synthesised peptide is released (Moldave, 1985; Reeds and Davis, 1992). These three processes are tightly choreographed by a suite of over 140 soluble initiation, translation and termination factors; proteins that catalyse specific elements of the process (London et al., 1987; Hershey, 1989; Rhoads, 1993; Carter and Houlihan, 2001). The rate of protein synthesis is governed by a number of factors, including the rate of transcription, cellular ribosomal concentration (and the formation of poly-ribosomes), ribosomal translational efficiency, hormonal concentrations and amino acid availability (Millward et al., 1973; Hershey, 1989; Millward, 1989; Hershey, 1991; Sugden and Fuller, 1991; Chen and London, 1995). After release from the ribosome, many proteins undergo post-translational modification prior to achieving a functional form (Han and Martinage, 1992; Meek, 1994). Although a highly complex process, many aspects of protein synthesis have been well characterised. Measurement of protein synthesis is considerably simplified by the fact that proteins are only synthesised via a single process.

In contrast, protein degradation is considerably less well understood, in part because a multitude of different protein degradation pathways exist, including lysosomal cathepsins, ubiquitin/ATP-dependent proteases, Ca-dependent calpains, multicatalytic/ATP-dependent proteases, metalloproteinases and chymotrypsin- and trypsin-like proteinases (Mayer and Doherty, 1986; Dice, 1987; Reeds and Davis, 1992; Maurizi, 1993; Hershko and Ciechanover, 1998; Attaix et al., 1999; Mommsen, 2004). Many of these pathways are utilised under specific circumstances such as nutritional stress (lysosomal cathepsins) or the degradation of short-lived or abnormal proteins (poly-ubiquitination and ATP-dependent proteases). The relative importance of the various protein degradation pathways appears to vary within different cellular compartments and tissues, and specific proteins appear to have widely differing half-lives (reviewed in Hershko and Ciechanover, 1982; Reeds and Davis, 1992). Even the means by which specific proteins are identified for degradation is currently far from totally clear, although multiple mechanisms appear to exist (Dice, 1987). There also appears to be differences in the primary pathways used for protein degradation in mammals and fish under some circumstances, for example starvation (Mommsen, 2004). Therefore, the current practical difficulties of estimating overall protein degradation rates in animals are considerable.

Degradation of proteins is thought to be important for many reasons such as allowing environmental adaptation, degradation of exogenous proteins brought into the cell via endocytosis, metabolic regulation, mobilisation of amino acids and elimination of non-functional or denatured proteins (Hershko and Ciechanover, 1982; Millward et al., 1983; Hawkins, 1991).

Degradation can occur at various times during the 'life span' of a protein: during the actual process of protein synthesis, immediately post-synthesis, after secretion or after it has reached its cytomorphological site (Mayer and Doherty, 1986).

Protein growth is simply the change in mass of the protein pool and can be either positive or negative depending on the relative balance between protein synthesis and degradation.

Typically, protein synthesis (k_s), degradation (k_d) and growth (k_g) are expressed as fractional rates, that is the proportion of the protein pool synthesised, degraded or the net change in the protein pool size (growth) per unit time (Waterlow et al., 1978). Fractional rates are therefore usually expressed as % day^{-1} or % hour^{-1}. If the protein mass of the whole animal or tissue under study is known, then the absolute protein synthesis rate (A_s) can also be calculated, and this is usually expressed as mg protein day^{-1}.

2. PROTEIN METABOLISM METHODOLOGIES

2.1. Overview

There is much controversy within the protein metabolism literature regarding methodology (Jahoor et al., 1992; Garlick et al., 1994; Rennie et al., 1994; Davis et al., 1999; Davis and Reeds, 2001). Broadly speaking, three differing methodologies exist for measuring protein synthesis in vivo: (1) constant infusion, (2) flooding-dose and (3) stochastic endpoint. Only the latter two methods have been extensively used to measure protein synthesis rates in marine organisms in the recent past.

2.2. Constant infusion

Fish are the only marine ectotherms in which the constant infusion technique has been used to measure protein synthesis. The technique involves the cannulation of a large blood vessel, usually under anaesthesia, and the slow, constant infusion of a radiolabelled tracer amino acid (typically 5–10% of the tissue amino acid free-pool concentration) into the animal over time periods ranging between 4 and 16 hours (Haschemeyer and Smith, 1979; Haschemeyer and Mathews, 1980; Smith and Haschemeyer, 1980; Smith et al., 1980; Smith, 1981; Rennie et al., 1994). The infusion is continued until a plateau value of specific activity is reached in the body fluids (Haschemeyer, 1978). Cannulation of fish, especially when small, is a technically challenging procedure and evidence from mammals suggests that the stress experienced by the animal may result in depressed rates of protein synthesis (Rennie and MacLennan, 1985; Watt et al., 1988). At some time point after the isotope has been infused, the animal is killed and the required tissues are removed and processed (Waterlow et al., 1978). Fractional protein synthesis rates are then calculated from the relationship between the amounts of radiolabelled amino acid bound in protein and that remaining within the pool of amino acids used to synthesise proteins. In most studies, the precursor pool for protein synthesis is considered to be the amino acids within the intracellular free-pool (Section 2.3). Constant infusion has not been more widely used to measure protein synthesis in marine animals due to the difficulties in ensuring a stable intracellular free-pool of the radiolabelled amino acid. The technique is also comparatively invasive and problematic with animals that cannot easily be restrained, such as fish (Garlick et al., 1994). A second problem is that the tracer dose of radiolabelled amino acid can be significantly diluted by the release of amino acids from protein degradation, as long time periods are required to infuse the animal. This can result in large differences in the specific radioactivities of the blood plasma, extracellular and intracellular free amino acid pools making the selection of an amino acid precursor pool for measurement difficult (Waterlow et al., 1978). The constant infusion method can be used to measure both whole-animal and individual tissue protein synthesis rates (Lobley et al., 1980; Smith, 1981).

2.3. Flooding-dose

The vast majority of protein synthesis measurements reported in marine organisms were made using the flooding-dose methodology. This technique was originally applied to mammals and subsequently to humans, in the latter using stable rather than radioisotopes (Garlick et al., 1980; Martinez, 1987; McNurlan et al., 1991). Since the early 1980s, other workers have applied the methodology to fish (Pocrnjic et al., 1983; Loughna and Goldspink, 1985; Houlihan et al., 1986). The flooding-dose technique involves injecting the animal with a very large bolus (much larger than the endogenous free-pool) of both radiolabelled and unlabelled amino acid (for a review of methodology see Houlihan et al., 1995a). Injections can be delivered via intra-peritoneal (Morgan et al., 2000) or intra-venous injections (Martin et al., 1993), and in larval stages, bathing the animal in a solution of the radiolabelled amino acid has resulted in an acceptable flooding-dose (Houlihan et al., 1995b). Protein synthesis rates are estimated by measuring the specific radioactivity in both the bound protein pool and the intracellular free-pool.

A flooding-dose is used to try and ensure the specific radioactivities of the plasma, extracellular, intracellular and aminoacyl transfer RNA (tRNA) compartments are closely related (Garlick *et al.*, 1994). The true amino acid precursor pool for protein synthesis is the aminoacyl tRNA pool; however, this pool is highly labile, making measurements of radiolabelling problematic (Haschemeyer, 1973). In most measurements of protein synthesis, especially in non-mammalian systems, the specific radioactivity of the intracellular amino acid free-pool has been used as a proxy measurement of the aminoacyl tRNA pool, as the specific radioactivity of the former is much easier to measure. Recent work has suggested that the intracellular free-pool is indeed equilibrated with the aminoacyl tRNA pool, therefore measurements of the specific radioactivity of the former pool after a flooding-dose, should act as a reasonable estimate of the latter (Davis *et al.*, 1999). Typically, a time course of both intracellular free-pool specific radioactivity and protein-bound specific radioactivity are measured. During short-term time courses in ectotherms the intracellular free-pool specific radioactivity usually remains stable, although a slow linearly decreasing specific radioactivity is acceptable if a slightly modified model is used (Houlihan *et al.*, 1995a). Rigorous validation of the flooding-dose methodology is essential to ensure that valid results are generated. The key criteria to ensure during flooding-dose measurements (Davis and Reeds, 2001) are that:

1. The intracellular free-pool specific radioactivities are elevated and stable during the protein synthesis measurement. If the intracellular free-pool specific radioactivities decrease linearly with time, calculation of protein synthesis rates is still possible but a modified model must be used.
2. Incorporation of radiolabelled amino acid into the bound protein pool should be linear and significant over the time course of protein synthesis. If the intercept of a fitted regression line does not pass through the origin (i.e. is significantly different), the time lag before incorporation starts should be subtracted from the incorporation time.
3. The injected amino acid should flood the plasma, intracellular, extracellular and aminoacyl tRNA pools. Typically, successful flooding-doses will elevate intracellular concentrations of the amino acid by about four- to tenfold.
4. Injecting the amino acid should not result in an elevation of the rate of protein synthesis.

The most convenient method to ensure stability of the intracellular free-pool specific radioactivity (criterion 1) and linear radiolabelling of proteins over time (criterion 2) is to sacrifice animals at set time points after injection of the flooding-dose, and plot the measured bound and intracellular free-pool specific radioactivities against time (Figure 2). Simple statistical methods can then be used to confirm criteria 1 and 2 have been met, as shown in Figure 2. If analysis suggests these criteria are not met (Figure 3), valid

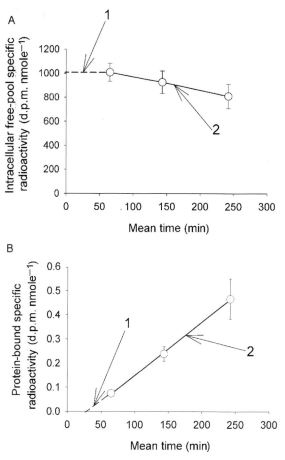

Figure 2 (A) Intracellular free-pool specific radioactivities in the Antarctic limpet *Nacella concinna*. Measurements were made after flooding-dose injections at time zero and are expressed as disintegrations per minute (d.p.m.) nmol^{-1} phenylalanine. All data points are means ± S.E.M., $n = 6$ (first data point) or 7 (last two data points). Note that the intracellular free-pools are rapidly labelled after injection (1), and specific radioactivities remain elevated with no significant decrease during the measurement (2). (B) Incorporation of radiolabelled phenylalanine into body protein. The intercept of the regression line passes through the origin at a point not significantly different to zero (1), and the incorporation of radiolabel is significant and linear with time (2). Fitted regression line $y = 0.0022x - 0.0747$, $r^2 = 0.59$, $p < 0.001$. All other details as in (A).

protein synthesis rates cannot be calculated. In Figure 3, the decrease in intracellular free-pool specific radioactivities with time (Figure 3A) and non-linear labelling of bound protein (Figure 3B) suggests that the time course was too long (for a published example of data with this problem see

Haschemeyer, 1983). Utilisation by protein synthesis of the radiolabelled amino acid results in an observed decrease in the specific radioactivity of the intracellular free-pool (Figure 3A). The effective plateau of protein labelling (Figure 3B) suggests that by 240 min removal of labelled proteins from the

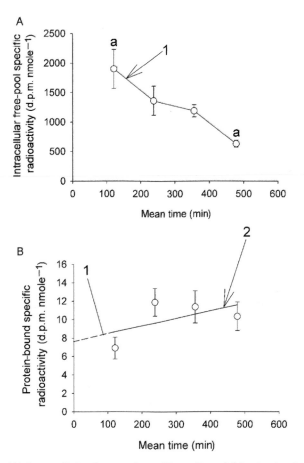

Figure 3 (A) Intracellular free-pool specific radioactivities in the Antarctic fish *Harpagifer antarcticus* (A. Bowgen and K. Fraser, unpublished data). Measurements were made after a flooding-dose at time zero and are expressed as d.p.m. nmole^{-1} phenylalanine. The specific radioactivities are elevated but not stable and decrease significantly during the measurement (1). Data points with the same letter are significantly different to each other (ANOVA $F = 4.44$, $p < 0.05$). (B) Incorporation of radiolabelled phenylalanine into body protein. The intercept of the regression is significantly different to zero (1) and there is no significant incorporation of the radiolabel during the measurement ((2), $p = 0.258$). All data points are means $\pm$ S.E.M., $n = 4$.

protein pool via protein degradation approximately equalled incorporation of newly synthesised radiolabelled proteins. Both of these time course problems were in this case resolved by re-running the experiment with a shorter time course.

To confirm that the flooding-dose has in fact elevated intracellular free-pool concentrations of the amino acid (criterion 3), it is important that baseline amino acid levels are also measured in un-injected animals. Using these data it is straightforward to calculate the degree by which the flooding-dose injection has elevated tissue concentrations of the amino acid (Fraser et al., 2002a).

Criterion 4 is difficult to prove, but highly important if valid protein synthesis rates are to be measured. It appears that when injected at high concentrations, some amino acids, in particular valine and leucine may, in some species, alter protein synthesis rates (Jahoor et al., 1992; Rennie et al., 1994; Smith et al., 1994; Beebe Smith and Sun, 1995). In fact, it has been suggested that the branched chain amino acid leucine may have a role in regulating protein synthesis (Schaefer and Scott, 1993). However, flooding-doses of other amino acids, in particular phenylalanine, are not thought to elevate protein synthesis (Garlick et al., 1980; Loughna and Goldspink, 1985; Rennie et al., 1994; Nyachoti et al., 1998). The choice of amino acid used is therefore critical.

An important advantage of the flooding-dose methodology over constant infusion is that the concentration of the injected amino acid is so high, typically four to ten times the baseline intracellular free-pool amino acid concentration, that dilution by amino acids released via protein degradation back into the free-pool is unlikely to affect the measured protein synthesis rate in the short term (Garlick et al., 1980). Also, as the time course is typically shorter than that used during a constant infusion, recycling of radiolabelled amino acids is reduced (Garlick et al., 1980). However, it is possible that longer term measurements of protein synthesis, such as those made using constant infusion or the stochastic endpoint method (Sections 2.2 and 2.4), may provide a better picture of 'mean' protein synthesis rates, as the measured rates will be less affected by short-term fluctuations. It is clear that short-term diurnal and post-prandial changes in protein synthesis do occur (Garlick et al., 1973; McMillan and Houlihan, 1988; McMillan and Houlihan, 1992).

Some workers have utilised stable rather than radiolabelled amino acids to make flooding-dose protein synthesis measurements (Krawielitzki and Schadereit, 1992; Owen et al., 1999). The use of stable isotopes removes the risks of handling radioisotopes, but the complexity and cost of analysis are considerably increased. The flooding-dose methodology can be used to measure whole-animal, individual tissue, and with modification, cellular protein synthesis rates (McMillan and Houlihan, 1992; Smith and Houlihan, 1995).

2.4. Stochastic endpoint models

Stochastic endpoint models (also known as end-product models) are used to estimate nitrogen flux rates and, in turn, protein synthesis rates (Waterlow *et al.*, 1978; Wolfe, 1992). In these methods, the mechanistic characteristics of the protein synthesis system are ignored; the organism is effectively treated as a 'black box'. The organism is fed a meal containing a known amount of stable isotope-labelled amino acid or protein. The excretion of the oxidised labelled amino acids are then tracked in the relevant excretory pool, which in most marine organisms will be ammonia (Kreeger *et al.*, 1995; Fraser *et al.*, 1998). Previous work in fish has shown that the oxidised labelled amino acids are not excreted via the urea pathway, although this may not be true of all fish species (Fraser *et al.*, 1998). Details of the calculations required to estimate protein synthesis rates using these methods can be found in Waterlow *et al.* (1978), Wolfe (1992), Houlihan *et al.* (1995a) and Fraser *et al.* (1998) together with information on laboratory analysis.

Stochastic endpoint methods have been extensively used in human protein metabolism studies, in part because they generally utilise stable isotopes, such as ^{15}N or ^{13}C, rather than radioisotopes (Young *et al.*, 1991). Endpoint models have now also been used to measure whole-animal protein synthesis rates in several marine and freshwater organisms: *Salmo salar* (L.) (Carter *et al.*, 1994), *Cyprinus carpio* (L.) (Bergner *et al.*, 1993; Meyer-Burgdorff and Günther, 1995; Meyer-Burgdorff and Rosenow, 1995a,b,c), *Hippoglossus hippoglossus* (L.) (Fraser *et al.*, 1998), *Mytilus edulis* (L.) (Hawkins, 1985; Hawkins *et al.*, 1986, 1989b), *Pleuronectes flesus* (L.) (Carter *et al.*, 1998) and *Rhombosolea tapirina* Günther, 1862 (Carter and Bransden, 2001). The major advantages of endpoint models over the flooding-dose and constant infusion methods are as follows:

1. Stable rather than radioisotopes are used, therefore, reducing the risk to the experimenter and removing the need to dispose of radiolabelled biological material. Also, measurements of protein synthesis can be made in situations where it would not be possible to use radioisotopes.

2. Protein synthesis rates are measured over a longer period of time, typically 24–72 hours in ectotherms. Making measurements over longer periods will tend to 'smooth out' short-term changes in protein synthesis and provide a better mean value.

3. The technique is considerably less invasive, with the stable isotope-labelled amino acid or protein usually introduced in the animal's food. However, knowledge of the amount of stable isotope label entering the animal is essential. Therefore, food consumption will need to be measured, for example by radiography in fish, which will cause some disturbance to the animal (Talbot and Higgins, 1983; Carter *et al.*, 1995). Importantly, the

animal does not need to be killed at the end of the experiment and it is, therefore, possible to make repeated measurements on an individual.

4. The labelled amino acid or protein is given as a tracer; therefore, administration is unlikely to perturb protein metabolism in the way that a flooding-dose of some amino acids may (Rennie et al., 1994).

Again a number of assumptions must be met if the methodology is to be reliably applied. The assumptions of the model when used in ectotherms have previously been extensively discussed and a brief overview will only be provided here (Houlihan et al., 1995a; Fraser et al., 1998). The major assumptions are as follows:

1. The labelled end product in which tracer excretion is measured is formed from the same precursor pool as that used for protein synthesis.
2. Nitrogen flux via the non-protein pool is insignificant compared to nitrogen flux through protein metabolism and nitrogen excretion/ oxidation.
3. The label is uniformly distributed throughout the nitrogen pool.
4. The tracer behaviour is representative of the total amino acid nitrogen pool.
5. Recycling of the label during the course of protein synthesis measurement is minimal.
6. The animal is in a steady state of nitrogen flux, that is no net loss or gain of nitrogen.

These assumptions must be either met or discounted as not significant if results obtained using the endpoint methodology are to be considered valid (Waterlow et al., 1978; Wolfe, 1992; Houlihan et al., 1995a; Fraser et al., 1998).

When measuring protein synthesis rates using the stochastic endpoint method, it is important to ensure that the labelled excretory end product is collected over a sufficiently long time period that label enrichment of the end product returns to near pre-labelling values. This can be achieved by plotting the cumulative isotope excretion curve and assessing statistically where the curve reaches an asymptote (Figure 4A). By the end of this time period, the label enrichment of the excreted end product should also have returned to baseline values seen in unlabelled animals (Figure 4B). The label enrichment is usually expressed as the atom percentage excess (APE), which is the level of isotopic abundance above a given background reading that is usually considered as zero. Typically, in studies carried out on temperate fish, and the mussel M. edulis, collecting labelled end product at regular time periods up to 24–72 hours after feeding has proved adequate (Hawkins et al., 1986; Carter et al., 1994; Meyer-Burgdorff and Rosenow, 1995a,b,c; Carter et al., 1998; Fraser et al., 1998; Carter and Bransden, 2001). However, as suggested by Houlihan et al. (1995b), it is highly likely that the length of time required

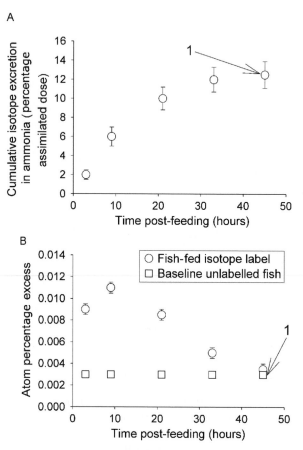

Figure 4 (A) Diagrammatic representation of cumulative excretion of stable isotope-labelled ammonia over time in a marine organism. Isotope excretion is expressed as the percentage of the isotope dose assimilated by the organism. The plateau suggests the excretory ammonia pool is cleared of isotope (1). (B) Diagrammatic representation of the enrichment of excreted ammonia in stable isotope-labelled marine organisms over time, relative to unlabelled control animals. Atom percentage excess (APE) is a measure of isotope enrichment. At 45 hours, the APE has returned to values measured in baseline unlabelled fish, suggesting the excretory ammonia pool is cleared of isotope (1).

for the excretion of the label to decrease to background levels will be temperature dependent.

Providing the stochastic end-point method is correctly validated, and the assumptions of the model are met, then the technique provides an excellent method to measure whole-animal protein synthesis in a comparatively non-invasive way. The fact that repeated protein synthesis measurements can be made on an individual allows researchers to look at the effects of experimental manipulations on protein metabolism without the confounding effect of intra-individual variation. However, stochastic end-point measures can only be used to measure whole-animal protein synthesis.

Although only a limited number of studies in marine organisms have utilised the stochastic endpoint methodology, it appears that the protein synthesis results obtained are broadly comparable with results obtained using the flooding-dose methodology (Houlihan *et al.*, 1995a).

2.5. Protein degradation determinations

Although the development of several comparatively straightforward methods have simplified the accurate measurement of protein synthesis, estimating protein degradation has proved considerably more difficult. The major difficulty in developing a method to measure protein degradation is that many very different pathways exist for the degradation of proteins (Section 1.2). Also when proteins are degraded, a proportion of the released amino acids are recycled for protein synthesis via the cellular amino acid pools. This recycling in effect means it is not possible to simply measure protein degradation by examining the excretion of pre-labelled amino acids in an end product such as ammonia (Botbol and Scornik, 1991). The level of recycling of released amino acids is unclear, but in fasted carp, no significant decrease in radioactivity of pulse-labelled muscle was found 2 weeks after injection of the isotope, suggesting at least during fasting, the majority of amino acids derived from protein degradation were recycled (Fauconneau, 1985).

The method by which whole-animal protein degradation has been estimated in almost all protein metabolism studies in marine organisms is by measuring protein growth over a period of time, measuring protein synthesis at the end of the growth period, and then estimating protein degradation as the difference between protein synthesis and growth (i.e. $k_d = k_s - k_g$) (Millward *et al.*, 1975; Houlihan *et al.*, 1986; Houlihan and Laurent, 1987; Houlihan *et al.*, 1995a; Morgan *et al.*, 2000). The major problem with this solution is that protein growth is measured over weeks to months and protein synthesis is measured over a few hours. Obviously, the assumption then has to be that the measured short-term protein synthesis rate is representative of protein synthesis over the whole period of measured protein growth. Unfortunately, this is currently the only method by which overall protein degradation rates can be measured. To minimise potential errors using this method, protein synthesis should ideally be measured over the longest time period possible, and protein growth measured over the minimum time period during which significant growth can be detected.

Viarengo *et al.* (1992) developed a method to measure protein degradation in the digestive gland of marine mussels in which animals were bathed in sea water containing ^{14}C-leucine for 24 hours before transfer to water containing a high concentration of unlabelled leucine. Protein degradation was subsequently estimated from the decrease in radioactivity of the digestive gland with time. Re-incorporation of ^{14}C-leucine released from protein degradation was minimised, although probably not stopped completely, by the presence of high concentrations of unlabelled leucine in the water. This method appears to offer promise in providing a way of measuring protein degradation in species that can rapidly absorb amino acids from sea water, but does not appear to have been more widely applied.

In cellular *in vitro* systems, it is possible to estimate protein degradation by bathing the cells in a radiolabelled amino acid for a period of time, then washing the cells and transferring them to a second solution containing high concentrations of the same amino acid used for the radiolabelling, but without the radiolabel (Botbol and Scornik, 1991; Land and Hochachka, 1994). The period during which the cells are transferred to the second unlabelled solution of amino acids is called the chase period. The rate of protein degradation is subsequently calculated from the release rate of radiolabelled amino acids from degraded proteins into the solution bathing the cells. Re-incorporation of released radiolabelled amino acids back into proteins during the chase period is greatly reduced by high levels of the unlabelled amino acid. However, it has been suggested that many measurements of protein degradation using this method have been considerable underestimates, due to the fact that large amounts of protein are rapidly degraded after synthesis, and may have been missed due to extended chase periods (Wheatley, 1984). As far as the authors can determine, the above method has not yet been applied in any cell system from a marine organism.

Several methods have been developed in endotherms to measure protein degradation under specific circumstances or in particular tissues. These include, using the excretion of 3-methylhistidine to estimate muscle degradation (Young and Munro, 1978), measuring labelled CO_2 release after a constant infusion, and a method which utilises a combination of organ balance and tracer approaches (Liu and Barrett, 2002). Other workers have used bestatin, an inhibitor of cytosolic peptidases to inhibit this pathway of protein degradation, and used the accumulation of di- and tri-peptide intermediaries as relative measures of protein degradation in rat liver (Botbol and Scornik, 1991). It should be noted the last method only measures a single pathway of protein degradation. None of these methods have yet been applied to marine organisms and in many cases they are unsuitable. The recent development of techniques, utilising new inhibitors of specific protein degradation pathways, will hopefully provide a better future understanding of protein degradation in marine organisms (Lee and Goldberg, 1998; Adams *et al.*, 1999).

2.6. Transcriptomics

Genomic technologies are now being regularly employed to analyse the effects of environmental stress on organisms, including aquatic animals (Gracey *et al.*, 2001; Gracey *et al.*, 2004; Podrabsky and Somero, 2004; Krasnov *et al.*, 2005). Based on expression profiles of genes that code for ribosomal RNAs (rRNAs), structural proteins and specific proteins involved in protein synthesis or degradation, inferences have been made about the impacts of stress on protein metabolism (Gracey *et al.*, 2001, 2004; Podrabsky and Somero, 2004). This approach, however, is controversial since it is well known that the abundance of mRNA transcript is, in many cases, a poor predictor of protein abundance. This is because many proteins can undergo modifications during pre-, co- and post-translation, they can be degraded, or in some cases genes can be alternatively spliced (Watson, 2004; Feder and Walser, 2005). All these factors can result in the weakening of the predicted close-coupling between mRNA transcripts and their coded peptides. Even after synthesis, proteins may not be active, depending on the presence of regulators, specific substrates and products, and the biochemistry of the cell (Feder and Walser, 2005). Even if a protein is active, abundance levels may have little measurable effect on cellular processes, such as protein synthesis, as large numbers of proteins contribute to the overall phenotype.

In addition to the problem of relating mRNA abundance to the synthesis of active proteins, many studies have revealed that substantial numbers of genes show significant differences in expression between individuals within populations (Stammatoyannopoulos, 2004). Oleksiak *et al.* (2005) used microarray technology to examine gene expression in 15 individuals of the estuarine fish, *Fundulus heteroclitus* (L.), from two natural populations, acclimated to the same environmental conditions. This study showed that 18% of the 907 genes assayed showed significant intra-individual differences in mRNA abundance. In these genes, transcript abundance typically varied by a factor of 1.5 but frequently by >2. Only 15 out of 907 genes showed significant differences between fish originating from the two populations, which live in different temperature regimes. These genes showed only small differences in expression levels, suggesting that they were tightly regulated or subject to stabilising selection. This suggested that small but significant changes in transcript abundance might have important biological consequences.

Further work on *Fundulus heteroclitus* analysing subsets of genes involved in metabolism has shown that much higher proportions of genes vary between individuals than between populations (Oleksiak *et al.*, 2002; Whitehead and Crawford, 2005). This is either because groups of functionally related genes show significant intra-individual variations in expression levels or because improving experimental and statistical approaches allow a greater precision in estimating variance in the abundance of transcripts (Whitehead and Crawford, 2005). Analysis of gene expression in different tissues of the same species (liver, brain, heart) from three populations did reveal differential patterns of gene expression consistent with biological function. However, gene expression patterns were not consistent for many genes and over one-third of the genes differentially expressed between tissues were significant in only one of the three populations examined (Whitehead and Crawford, 2005).

A recent analysis of individual variation in cardiac performance, physiology and expression of genes involved in cardiac metabolism in *Fundulus heteroclitus* has revealed further complexity in individual variation. This work showed that gene expression patterns among 119 genes involved in cardiac metabolism clustered into three groups that correlated with substrate-specific metabolism detected among the individuals analysed. A large proportion (94%) of genes showed significant differences in gene expression between individuals, but the variation in mRNA levels was a poor indicator of the biological impact of variation (i.e. genes that showed small but significant variation in mRNA abundance had as much of an influence on cardiac metabolism as those that showed changes of a factor >2).

The prevalence of variation in gene expression between individuals must partially explain variability in protein synthesis rates. This variability is likely to arise from genetic variation and from the irreversible effects of development on gene expression in individuals (Whitehead and Crawford, 2005). Studies on *Drosophila melanogaster* Meigen, 1830, for example, have indicated that sex, and the interaction of sex and genotype, have a strong effect on gene expression profiles of individuals, with the age of animals also having an influence, but on significantly less genes (Jin *et al.*, 2001). However, intra-individual variation in gene expression also represents a major technical barrier to attaching significance to changes in mRNA abundances in terms of major physiological processes. Biological replication of transcriptomic experiments is critical in order to ascribe differences in gene expression to an experimental treatment rather than to individual variation (Whitehead and Crawford, 2005). Furthermore, to generalise results for a species, individuals from more than one population should be analysed (Oleksiak *et al.*, 2002). Problems of relating changes in mRNA abundance to alterations in levels of specific proteins and rate changes in major physiological processes are particularly acute when analysing single genes or when attempting to detect novel proteins involved in specific processes (Feder and Walser, 2005). However, significant changes in the expression of large number of genes coding for proteins involved in specific synthetic or metabolic pathways must be a good indicator of alterations in physiological processes. In terms of protein synthesis, coordinated changes in expression of genes involved in transcriptional regulation, RNA splicing and translation must indicate an increase or decrease in the capacity of cells to synthesise proteins (Gracey *et al.*, 2001, 2004; Podrabsky and Somero, 2004). Coordinated alterations of the expression of genes involved in specific protein degradation pathways, such as those involved in ubiquitination or proteasomal structural proteins, are likely to indicate a change in rates of protein degradation by this pathway (Gracey *et al.*, 2004).

It is clear from the previous discussion that although our knowledge of protein synthesis is fairly good, because of the availability of numerous practical methodologies, our knowledge of protein degradation is severely hindered by the complexity of the processes involved, and the lack of robust methods to measure *in vivo* degradation in marine organisms. Quantitative measurements of protein degradation are especially important as it has been suggested that protein catabolism plays a greater role in determining growth rate than protein anabolism (Sveier *et al.*, 2000). New genomic technologies can provide important information on changes in the capacity for protein metabolism and potentially highlight by which pathways protein degradation is occurring. As such, transcriptomic approaches can point to areas of specific interest for further physiological and biochemical studies.

3. WHOLE-ANIMAL PROTEIN METABOLISM

3.1. Overview

Although protein metabolism can be measured in an organism at a variety of scales from cellular to the whole animal, it is the latter that is of most interest in relating protein metabolism to overall growth. Whole-animal protein metabolism is effectively protein synthesis, degradation and growth measured in an intact organism. It should be remembered that the half-lives of proteins vary from minutes to years and overall measurements of protein metabolism provide an 'averaged' picture of all the proteins within the body as specific proteins may be turned over at significantly different rates (Bachmair *et al.*, 1986; Dice, 1987).

It is apparent from Tables 1 and 2 that our knowledge of whole-animal protein metabolism in marine fish and invertebrates is restricted to a very small number of species. For example, there are in excess of 24,000 known species of fish of which 58% are marine, but at present we have limited whole-animal protein metabolism data available for only 8 marine species (Bone and Marshall, 1992; Table 1). Previous studies have concentrated on commercially important species from the class Osteichthyes and we have no data from any member of the classes Cyclostomata or Chondrichthyes. There is an even greater paucity of data from marine invertebrates, with

Table 1   Whole body protein and RNA metabolism in marine fish held under as near natural conditions as possible

Species	Fresh mass (g)	Temperature (°C)	Feeding regime	k_s (% day^{-1})	k_g (% day^{-1})	k_d (% day^{-1})	RNA to protein ratio (μg mg^{-1})	RNA activity, k_{RNA} (mg mg^{-1} day^{-1})	References
Fish									
Gadus morhua	300	10 ± 0.5	4% BM day^{-1}	3.8	2.0	1.8	8.8	4.3	Houlihan *et al.*, 1988a, 1989
Salmo salar	180	13	1.25–2.5% BM day^{-1}	3.8	1.7	2.1	5.0	7.6	Carter *et al.*, 1993a
Anarhichas lupus	65	9–11	Satiation	1.8	1.0	0.8	10.6	1.7	McCarthy *et al.*, 1999
Hippoglossus hippoglossus	109	8–13	Satiation	2.0	1.3	0.7			Fraser *et al.*, 1998
Rhombosolea tapirina	37.2	16	2% BM day^{-1}	7.0	0.6	6.4			Carter and Bransden, 2001
Pleuronectes flesus	60	7	Unknown	4.1	1.4	2.7			Carter *et al.*, 1998
Limanda limanda	265	10.6 ± 0.3	Satiation three times a week	0.3	0.2	0.1	7.0	0.6	Houlihan *et al.*, 1994
Dicentrarchus labrax	3.5	18 ± 1.0	Daily over 8 hours	5.1	2.4	2.7	6.9	0.7	Langar *et al.*, 1993
	2.5	18 ± 1.0	3.5% BM day^{-1} fed in three meals	6.1 ± 0.6	2.1 ± 0.2	4.1 ± 0.2	10.3 ± 1.0	0.6 ± 0.1	Langar and Guillaume, 1994

BM, body mass.

Table 2 Whole body protein and RNA metabolism in marine invertebrates held under as near natural conditions as possible

Species	Fresh mass (g)	Temperature (°C)	Feeding regime	k_s (% day^{-1})	k_g (% day^{-1})	k_d (% day^{-1})	RNA to protein ratio (μg mg^{-1})	RNA activity, k_{RNA} (mg mg^{-1} day^{-1})	References
Molluscs									
Nacella concinna	2.1 ± 0.7	−1.1 ± 0.6	Feeding	0.6	0.1	0.5	46.9 ± 2.4	0.1 ± 0.0	Fraser, unpublished data
Octopus vulgaris	147.8	22	Feeding	3.8	3.0	0.8	25	1.5	Houlihan *et al.*, 1990a
Mytilus edulis	N/A	N/A	Feeding	0.4	0.2	0.1			Hawkins, 1985
Crustaceans									
Litopenaeus vannamei	2.1	27 ± 1.0	Fasted 1 day	9.8 ± 1.5	9.2 ± 1.0	0.5 ± 0.2			Mente *et al.*, 2002
Homarus gammarus	0.2	19	5% daily	13.3	1.3	12.4	17	3	Mente *et al.*, 2001
Saduria entomon	3.6 ± 0.2	4	Fed	1.5 ± 0.3			10.8 ± 0.5	2.3 ± 0.6	Robertson *et al.*, 2001a
Glyptonotus antarcticus	38.8	0	Satiation	0.2			16.2 ± 3.0	0.2 ± 0.1	Whiteley *et al.*, 1996
Idotea resecata	0.3	14	Not fed?	1.2			30.8 ± 5.5	1.1 ± 0.5	Whiteley *et al.*, 1996

data only available for a few species from two phyla. This section will examine whole-animal protein metabolism in marine organisms together with influencing abiotic and biotic factors.

3.2. Abiotic factors influencing protein metabolism

3.2.1. Temperature

Temperature acts as a 'controlling factor' of growth and is, therefore, the most significant abiotic factor affecting the physiology and growth of fish and other ectothermic marine organisms (Brett, 1979). Hawkins (1996) has suggested that observed temperature-dependent shifts in ectotherm metabolism primarily reflect costs associated with changes in whole-body protein synthesis. Before considering the effects of temperature on protein metabolism, it is important to remember that there are significant conceptual differences in exposing a single species to a range of temperatures and measuring protein metabolism, and comparing the effect of an animal's normal 'living temperature' on protein metabolism across a range of species living at different temperatures. This has been highlighted for metabolic rates but is equally applicable to other physiological measurements (Clarke and Fraser, 2004).

A term often used, and misused, in studies examining physiological processes at differing temperatures, is acclimation. Hazel and Prosser (1974) stated that under conditions where more than one biotic or abiotic factor is changing simultaneously, for example temperature and food consumption, then the term used should be acclimatisation. While acclimation should only be used to describe a situation where a single biotic or abiotic factor is being manipulated, for example temperature when ration is held constant (Hazel and Prosser, 1974). This is the terminology that has been used in this chapter. However, it should be noted that even if only water temperature is altered, many other factors will also change, including dissolved O_2 concentrations and the standard metabolic rate of ectotherms. In reality, it is practically impossible to alter a single biotic or abiotic parameter without altering another.

Growth rate tends to increase with temperature in ectotherms up to the point at which the animal approaches its upper thermal limit, thereafter the growth rate will rapidly decrease (Brett, 1979; McCarthy and Houlihan, 1996; McCarthy et al., 1998, 1999). It has been suggested that maximal protein synthesis rates occur at the optimum growth temperature (Loughna and Goldspink, 1985; Pannevis and Houlihan, 1992; Carter and Houlihan, 2001). In wolffish, McCarthy et al. (1999) demonstrated that protein synthesis rates increased from 5 to 14 °C, while protein growth rates initially increased with temperature, but then decreased at temperatures in excess of 11 °C. As the wolffish approached the upper limits of its thermal range, protein degradation increased, thereby reducing the PSRE and hence growth (McCarthy et al., 1999). In other studies, short-term increases in temperature have resulted in an increase in the rate of protein synthesis, providing the temperature range the animal was exposed to did not approach its upper thermal limit (Whiteley et al., 1996; Robertson et al., 2001a,b; Whiteley et al., 2001). However, in a study of the intertidal isopod Ligia oceanica (L.) neither acute changes in temperature nor longer term acclimatisation (4 weeks) to a range of temperatures resulted in significant alterations in k_s (Whiteley and Faulkner, 2005).

Changes in gene expression of organisms subjected to stress, including low and high temperature, are known to involve large changes in the expression of ribosomal genes and genes coding for proteins involved in RNA metabolism, protein synthesis and cell growth (e.g. yeast, Saccharomyces cerevisiae; Gasch et al., 2000). In the freshwater fish, Austrofundulus limnaeus Schultz, 1949, analysis of transcription profiles when fish were subjected to chronic elevated temperatures showed increases in the abundance of mRNAs for two translation elongation factors—a tRNA synthase and at least one ribosomal protein (Podrabsky and Somero, 2004). There was also a concurrent slight decrease in the expression of a regulatory unit of the 26S proteosome, part of the ubiquitin-dependent protein degradation pathway. While reliance on the interpretation of the expression of a few genes should be viewed with caution, this is suggestive of an increase in the capacity for protein growth both by increasing the translation machinery of cells and decreasing rates of protein degradation. Upper temperatures in these experiments matched those regularly experienced by A. limnaeus in the ponds where it occurs in Africa and South America (Podrabsky and Somero, 2004).

It is essential to consider the thermal history of the organism when investigating the effects of temperature on protein synthesis. Watt et al. (1988) have shown that in carp acclimated to low or high temperatures, muscle protein synthesis rates differ when measured at a common intermediate temperature. Modification of protein synthesis rates at differing temperatures appears in many cases to be achieved, at least in part, by alterations in tissue RNA concentrations or a change in the k_{RNA} (Watt et al., 1988; Whiteley et al., 1996; Robertson et al., 2001a,b; Section 8). After an acute change in temperature, rates of physiological processes will take time to stabilise, and until this occurs true acclimation will not have been reached (Loughna and Goldspink, 1985; Pannevis and Houlihan, 1992). However, given sufficient time, an ectotherm will 'acclimate' to a change in temperature providing the organism has the genotypic capacity, thereby providing the organism with some independence from changes in environmental temperature (Houlihan et al., 1993b; McCarthy and Houlihan, 1996).

Examination of the effects of temperature on protein metabolism rates from different species is complicated by the fact that the animal's mass and nutritional regime will considerably effect rates of protein metabolism (Goldspink and Kelly, 1984; Hawkins, 1985; Houlihan *et al.*, 1988a, 1989; Lyndon *et al.*, 1992; Langar *et al.*, 1993; Langar and Guillaume, 1994; Kreeger *et al.*, 1995; Whiteley *et al.*, 1996; McCarthy *et al.*, 1999; Sveier *et al.*, 2000; Carter and Bransden, 2001; Mente *et al.*, 2001; Robertson *et al.*, 2001a,b; Whiteley *et al.*, 2001; Fraser *et al.*, 2002a; Mente *et al.*, 2002). Scaling the data to correct for body mass differences is straightforward, but care needs to be exercised in comparing studies in which animals were held under widely differing nutritional regimes.

Where possible the protein synthesis data from feeding animals summarised in Tables 1 and 2 have been mass standardised using an ectotherm scaling coefficient of -0.30 (Houlihan *et al.*, 1995a) and plotted in Figure 5. Plotted individual data points can be identified from Table 3. The data currently available are very limited and only provide information on a small range of fish, mollusc and crustacean species. However, it is clear that protein synthesis rates rapidly decrease with temperature below a clear threshold. Interestingly, the decrease in protein synthesis rates does not appear linear across the plotted temperature range (-1.1 to $27\,°C$), and the data are better represented by a cubic rather than a linear regression model. Initially as temperatures decrease, there seems little effect on protein synthesis rates, but at some temperature around $5\,°C$, protein synthesis rates decrease precipitously. It should be noted that data points 8, 12 and 13 on Figure 5 are all invertebrate polar species, while the majority of the other data points plotted are fish, so there may be fundamental phylogenetic differences influencing the interpretation. Notwithstanding, this trend is of potential interest and further comparable data are required to clarify the effect of temperature on protein synthesis. The data so far suggest that protein synthesis rates are considerably lower in polar than temperate species. In turn this suggests a fundamental thermal constraint on the synthesis of proteins at very low temperatures, and therefore growth. If protein synthesis rates are indeed depressed in polar species, this would suggest that the generally observed low growth rates (reviewed in Peck, 2002) are not only the result of resource limitation by a highly seasonal food supply (Clarke, 1988; Clarke and Peck, 1991), but also a reduced ability to synthesise proteins. Although some compensation of protein synthesis rates at low water temperatures occurs via an increase in tissue RNA concentrations, and possibly k_{RNA} (Section 8), it appears that protein synthesis is far from fully compensated at polar water temperatures. There are currently insufficient data to allow a meaningful examination of the effects of temperature on whole-animal protein growth rates in marine organisms.

There are considerable differences in whole-animal fractional protein synthesis rates in ectotherms and endotherms. Similar-sized ectotherms generally have much lower protein synthesis rates than endotherms. For

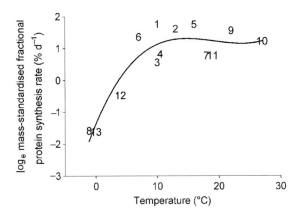

Figure 5 Mass-standardised whole-animal fractional protein synthesis rates for a range of marine ectotherms. Each data point number represents a single species, with its literature source identified from Table 3. Fitted regression line (third-power, cubic) is represented by $y = -1.37 + 0.46x - 0.03x^2 + 0.0004x^3$, $F = 15.88$, $r^2 = 84\%$, $p < 0.001$.

Table 3 Sources of whole-animal protein synthesis data used to plot Figure 5

Data point label	Species	Group	Habitat	References
1	*Gadus morhua*	Teleost	T	Houlihan *et al.*, 1988a, 1989
2	*Salmo salar*	Teleost	T	Carter *et al.*, 1993a
3	*Anarhichas lupus*	Teleost	T	McCarthy *et al.*, 1999
4	*Hippoglossus hippoglossus*	Teleost	T	Fraser *et al.*, 1998
5	*Rhombosolea tapirina*	Teleost	T	Carter and Bransden, 2001
6	*Pleuronectes flesus*	Teleost	T	Carter *et al.*, 1998
7	*Dicentrarchus labrax*	Teleost	T	Langar and Guillaume, 1994; Langar *et al.*, 1993
8	*Nacella concinna*	Mollusc	P	Fraser, unpublished data
9	*Octopus vulgaris*	Mollusc	T	Houlihan *et al.*, 1990a
10	*Litopenaeus vannamei*	Crustacean	Tr.	Mente *et al.*, 2002
11	*Homarus gammarus*	Crustacean	T	Mente *et al.*, 2001
12	*Saduria entomon*	Crustacean	P	Robertson *et al.*, 2001a
13	*Glyptonotus antarcticus*	Crustacean	P	Whiteley *et al.*, 1996

T, temperate; Tr., tropical; P, polar.

example the mean mass of the non-polar ectotherms plotted in Figure 5 is approximately 90 g, the mean protein synthesis rate approximately 2.7% and the mean water temperature approximately 15 °C. In contrast, in 80 g white leghorn chicks (*Gallus domesticus*), whole-body fractional protein synthesis rates were 34% day^{-1} (Muramatsu and Okumura, 1983) and in 50 g rats (*Rattus norvegicus*) 31% day^{-1} (Goldspink and Kelly, 1984). The approximate body temperatures of these organisms are 40.8 and 38.1 °C, respectively. It seems highly unlikely that the 13- and 26-fold difference in protein synthesis rates reported in these two species and the non-polar ectotherms plotted in Figure 5 are only the result of a threefold difference in body temperature. It is not presently clear why fractional protein synthesis rates in ectotherms and endotherms differ to such a great degree, but it seems unlikely temperature is the only factor.

Analysis of changes in the carp (*Cyprinus carpio*) transcriptome with decreasing temperatures suggested an increase in the expression of genes involved in protein synthesis and degradation at lower temperatures (10 °C) (Gracey *et al.*, 2004). The largest functional group of genes induced by low temperature included RNA-polymerase II activators, ribonucleoproteins involved in mRNA processing and splicing, several translation initiation factors and a gene that is involved in the monitoring of the fidelity of the initiation complex (Gracey *et al.*, 2004). Significant up-regulation of the translation machinery was also accompanied by up-regulation of genes coding for 21 proteosomal subunits and several ubiquitin-conjugating enzymes (Gracey *et al.*, 2004). Likewise, in *Austrofundulus limnaeus*, there is a strong up-regulation of the regulatory subunit of the 26S proteosome as temperatures decrease, although in this species mRNA abundance of genes involved in the protein synthetic machinery remain unchanged, or slightly decrease. Temperature cycling in *A. limnaeus* also invokes a very strong response in the regulatory subunit for the 26S proteosome, indicating an important role for protein degradation during short-term (daily) temperature fluctuations. In both of these studies, the minimum water temperatures were still well above the minimum experienced by marine organisms (approximately −1.9 °C), and well above the temperature at which protein synthesis rates decreased rapidly in the data plotted in Figure 5. It may well be that in temperate species, over the temperature ranges that they normally inhabit, compensation of the protein synthesis and degradation processes is possible via modulation of gene expression and other physiological processes. However, in more extreme polar environments full compensation of protein synthesis may not be possible, although for what reason, or reasons is currently unclear.

It is not presently clear whether water temperature affects PSREs (Section 1.1). The mean PSREs for the fish and invertebrate data presented in Tables 1 and 2 are 44% and 53%, respectively. However, there are three species that appear to differ considerably from these mean values: *Nacella concinna* (17%), *Homarus gammarus* (9%) and *Rhombosolea tapirina* (9%) (Tables 1 and 2). It is not clear why protein degradation rates are so high in *Rhombosolea tapirina* in comparison to the other available fish data, but the high rates reported in *Homarus gammarus* may be an artefact, caused by periodic moulting of the animal during the growth measurement and subsequent loss of mass. This would effectively result in an underestimation of growth, and as degradation was calculated as the difference between synthesis and growth, a resultant overestimation of degradation. The low PSRE in the polar limpet *Nacella concinna* is very interesting, as elevated levels of protein chaperoning and an increase in the tissue concentrations of ubiquitin conjugates have been reported in polar fishes (Place *et al.*, 2004; Place and Hofmann, 2005). In Antarctic fish species it appears that HSPs are continually expressed, in contrast to temperate species in which HSPs are only expressed under exposure to specific stressors (Place *et al.*, 2004). Place *et al.*, 2004 have suggested that this evolutionary adaptation has occurred to counteract the denaturing effect of low temperatures on proteins. As polar fishes live in an environment characterised by highly stable water temperatures, the classical heat-shock response has been effectively rendered redundant (Place *et al.*, 2004). Ubiquitin conjugation of proteins usually occurs as a precursor to degradation and has previously been used as a proxy measurement of protein degradation levels (Ciechanover *et al.*, 1984; Place *et al.*, 2004). It therefore appears that at polar water temperatures ectotherms may experience elevated rates of protein chaperoning and protein degradation as a result of a reduction in protein stability. If this is the case, it would mean that growth in polar ectotherms is not only constrained by low protein synthesis rates, but also by low PSREs. Further studies are required to clarify whether protein degradation rates are indeed relatively higher in polar than temperate ectotherms.

It is not currently clear whether there are any differences in the protein synthesis retention efficiencies measured in fish and invertebrates. Houlihan *et al.* (1995a) stated that there was some evidence that invertebrate species may have higher protein synthesis retention efficiencies than fish and endotherms, but that further studies were required to clarify the situation. The mean PSRE for the fish and invertebrate data presented in Tables 1 and 2 is 44% and 61%, respectively (excluding the polar species). Although it appears invertebrates may indeed be more efficient at retaining synthesised proteins for growth, too little data are still available to be able to draw any robust conclusions.

3.2.2. Pollution

Only three studies have examined the effects of pollution on whole-animal protein metabolism in marine organisms. Houlihan *et al.* (1994) examined the effect of sewage sludge exposure on protein metabolism in the dab *Limanda limanda* (L.). Exposure to sewage sludge did not reduce overall growth or protein synthesis rates in dabs, but protein growth rates were reduced suggesting a reduction in the PSRE and an increase in protein turnover. Exposure of *Mytilus galloprovincialis* Lamarck, 1819 to phenanthrene, an organic xenobiotic compound known to adversely affect the stability of lysosomal membranes in molluscan digestive cells, resulted in an initial increase in the rate of protein degradation over the first 2 days followed by a decrease below control values (Viarengo *et al.*, 1992). Unfortunately, protein synthesis was not measured in this study; therefore, conclusions about the overall effect on protein metabolism cannot be drawn. Hawkins *et al.* (1989a) examined the effects of terminal concentrations of copper on *Mytilus edulis*. Individuals with higher rates of protein turnover appeared more susceptible to copper poisoning. Protein turnover was not measured simultaneously with copper exposure, so it is not possible to establish whether protein turnover in high turnover individuals increased further during copper exposure (Hawkins *et al.*, 1989a). Shepard and Bradley (2000) also report an increase in lysosomal damage with increasing concentrations of copper but again no information on protein synthesis was reported.

While both genomic and proteomic technologies are being developed for analysis of the effects of pollutants on the transcriptome and proteome of marine organisms (Shepard and Bradley, 2000; Shepard *et al.*, 2000; Venier *et al.*, 2003; Azumi *et al.*, 2004; Schalburg *et al.*, 2005), no specific results relevant to processes associated with protein synthesis and turnover are currently available. Analysis of gene expression in the gills of rainbow trout (*Oncorhynchus mykiss* Walbaum, 1792) exposed to zinc has detected up-regulation of genes coding for ribosomal proteins and those involved in the initiation of translation (Hogstrand *et al.*, 2002). In addition, at least one ubiquitination-mediating protein was also up-regulated. This suggests that rates of protein metabolism may change in response to zinc exposure in this species.

3.2.3. Seasonality

In many areas of the world outside of the tropics, animals experience significant seasonal changes in environmental conditions, in particular, temperature, food availability and light (Cushing, 1975; Clarke, 1988). Protein synthesis rates in the Antarctic limpet *Nacella concinna* Strebel, 1908 decreased by a maximum of 52% from summer (December $0.56 \pm 0.04\%$ day^{-1}, February $0.40 \pm 0.06\%$ day^{-1}) to winter (July $0.27 \pm 0.03\%$ day^{-1}, October $0.35 \pm 0.03\%$ day^{-1}). The reader should note that Antarctica is in the Southern Hemisphere and hence the seasons are reversed in comparison to the Northern Hemisphere. The decrease in protein synthesis was probably a result of the tenfold decrease in food consumption, measured via faecal egestion, rather than the 2 °C decrease in water temperature (Fraser *et al.*, 2002a,b). *Mytilus edulis* also showed a seasonal pattern of protein synthesis, with summer rates up to 3.5-fold higher than winter rates (Hawkins, 1985; Kreeger *et al.*, 1995). PSREs in *M. edulis* varied considerably throughout the season. In March, rates of protein degradation were greater than synthesis, regardless of whether the animals were fed or starved. In June and October, fed animals had much reduced protein degradation, elevated protein synthesis and thus exceptionally high PSREs of 87.5% and 97.5%, respectively. However, in starved animals at the same times of year, protein was again degraded faster than it was synthesised (Hawkins, 1985). In the intertidal isopod *Ligia oceanica*, protein synthesis rates in summer-sampled animals were approximately twofold higher than in animals collected in winter (Whiteley and Faulkner, 2005). A strong seasonal pattern of protein metabolism, therefore, appears to exist in the marine species examined so far. In contrast, in endotherms the situation is less clear. For example, in winter dormant bears [*Ursus americanus* Pallas, 1780 and *U. arctos* (L.)], no significant seasonal change in protein synthesis has been reported, although protein degradation did increase during non-feeding periods (Barboza *et al.*, 1997).

At present the direct effect on protein metabolism of seasonal changes in light period are unknown. However, it appears unlikely that the effects will be as large as those resulting from changes in water temperature and food availability (Sections 3.2.1 and 3.3.1). Evidence from Japanese quail (*Coturnix coturnix japonica*) suggests that the primary mechanism by which light period affects protein synthesis is by determining the length of time the birds can feed (Boon *et al.*, 2001). The effect of light period on marine organisms that only feed during daylight hours will probably be similar.

3.3. Biotic factors influencing protein metabolism

3.3.1. Food consumption and nutrition

Two of the most significant biotic factors affecting protein metabolism are the rate at which an animal consumes food and the specific nutritional value of the food source. Because of the increasing interest in marine aquaculture, a large proportion of the current protein metabolism literature has concentrated on diet development and maximising animal growth by optimising feeding strategy.

Accurate information on food intake is essential to understanding how various biotic and abiotic factors can affect protein metabolism (Carter and Houlihan, 2001). Protein metabolism can be related to food consumption at either the local level, that is all the animals in a tank, or at the level of an individual animal. The latter is of greater use in protein metabolism studies. With some species it is possible to maintain individuals separately, making measurements of food consumption straightforward, but many species will only feed when held in groups (Houlihan *et al.*, 1988a; Lyndon *et al.*, 1992;

Mente *et al.,* 2001). The measurement of individual food consumption in groups of fish has been made possible by the development of X-radiographic methods (Talbot and Higgins, 1983; Thorpe *et al.,* 1990; McCarthy *et al.,* 1993; Carter *et al.,* 1995).

Protein synthesis rates in starved *Saduria entomon* (L.), an Arctic isopod, increased after feeding by 2.7-fold at 4 °C and 1.9-fold at 13 °C parallel with a two- to threefold increase in O_2 consumption (Robertson *et al.,* 2001a). This suggests that a significant proportion of the post-prandial specific dynamic action (SDA) was the result of an increase in protein synthesis (Lyndon *et al.,* 1992; Robertson *et al.,* 2001a). Likewise, in other species that have been examined, whole-animal protein synthesis rates increase after feeding by similar magnitudes (Lyndon *et al.,* 1992; Robertson *et al.,* 2001b). In cod starved for 6 days, re-feeding a single meal resulted in a rapid increase in protein synthesis, doubling at 3 hours and peaking at 18 hours, at a level four times higher than in starved animals. From 24 hours after the meal, protein synthesis rates gradually decreased (Lyndon *et al.,* 1992). In the studies discussed previously, animals were starved for extended periods before re-feeding, so it is possible that in animals that are fed continually a large post-prandial peak in protein synthesis may not occur, or at least may be much reduced (Houlihan, 1991).

The frequency of feeding affects both overall growth and protein metabolism in animals. In the European lobster, *Homarus gammarus* (L.), animals fed continually on smaller rations, or fed less frequently, showed elevated protein synthesis but also elevated degradation resulting in high turnover and reduced growth (Mente *et al.,* 2001). Maximal growth rates were achieved by feeding a fairly large ration (10% BW day^{-1}) on a daily basis, fast growth being achieved by a reduction in protein degradation, rather than an elevation of protein synthesis. A similar result has also been reported in shrimps (Mente *et al.,* 2001). Feeding greenback flounder (*Rhombosolea tapirina,* 1% BW day^{-1}) resulted in reduced protein synthesis and degradation rates compared to animals fed 2% BW day^{-1} in a single meal, but animals fed the 2% BW day^{-1} in two separate meals showed synthesis and degradation not significantly different to those animals fed 1% BW day^{-1} (Carter and Bransden, 2001). Net protein growth was not significantly different between any of the treatments. In sea bass (*Dicentrarchus labrax*), fed three equal meals a day instead of the same amount of food continuously, protein synthesis rates doubled but there was also a fourfold increase in protein degradation. The net result being the same protein growth rates for both feeding methods, although presumably the overall cost of growth in the meal-fed fish was higher because of the reduced protein synthesis retention efficiency (Langar and Guillaume, 1994). Although the authors know of no comparable work in marine species, it appears that in freshwater species the time of day fish are fed also has a significant effect on protein synthesis rates (Bolliet *et al.,* 2000). In this study, rainbow trout fed at 07:00 hours had significantly higher post-prandial rates of protein synthesis than those fed at 19:00 hours.

As food consumption increases above the maintenance ration, there is a linear increase in both whole-animal protein synthesis and growth rates (Houlihan *et al.*, 1988a, 1989, 1995d). However, it is still not totally clear whether protein degradation rates also increase. In some studies, protein degradation rates do not increase with an increase in feeding, and it has been suggested that degradation rates are genetically fixed (Tomas *et al.*, 1991; Carter *et al.*, 1992; McCarthy *et al.*, 1994, Houlihan *et al.*, 1995d; Carter *et al.*, 1998). While in other studies protein degradation rates increase with ration, but often at a lower rate than synthesis or growth, the net result is an increase in the PSRE as food consumption increases (Houlihan *et al.*, 1988a, 1989; Carter and Bransden, 2001). Therefore, the amount of food eaten by an animal has a fundamental role in determining both protein synthesis and protein growth and, in turn, these processes will determine both the overall animal growth rate and, in some cases, efficiency. Carter *et al.* (1998) have previously demonstrated in a study using repeated, non-invasive measurements of protein metabolism that 92% of the variation in individual growth rates of juvenile flounder (*Pleuronectes flesus*) can be explained by food consumption (Carter *et al.*, 1998). Interestingly, the highest growth efficiencies of individual flounder were achieved by having both very low rates of both protein synthesis and degradation, rather than just low degradation, the net result being low net turnover and low energetic costs of protein metabolism (Carter *et al.*, 1998).

Because of the high cost of using fish meal as the protein source in aquaculture diets, several studies have examined the effects on protein metabolism of reducing the protein content, or replacing a proportion of the fish meal with an alternative protein source (Langar *et al.*, 1993; Langar and Guillaume, 1994; Sveier *et al.*, 2000; Mente *et al.*, 2002). Replacing up to 37% of fish meal with 'greaves' meal (defatted collagen meal) did not reduce protein growth rates in sea bass, although both protein synthesis and degradation rates were elevated (Langar and Guillaume, 1994). However, in a previous longer-term study by the group, using the same 'greaves'-based diet and other protein alternatives to fish meal, growth was reduced when high concentrations of the alternative protein sources were used in the diet (Langar *et al.*, 1993). The difference between the two studies was possibly because of the former study being of too short a duration to show a reduction in growth rate (Langar *et al.*, 1993; Langar and Guillaume, 1994).

In fish larvae, when the dietary amino acid profile fed is balanced, amino acid retention is increased and oxidation of amino acids is reduced (Aragão *et al.*, 2004). In shrimps, the use of casein as a complete replacement for fish/squid/shrimp meal resulted in a decrease in specific growth rate, no change in protein synthesis rates, but an increase in protein degradation resulting in a reduction in the PSRE (Mente et al., 2002). Similarly, in pigs (Fuller et al., 1987) and chickens (Tesseraud et al., 1992), protein degradation rates are increased in animals fed lysine-deficient diets. It appears likely that small amounts of natural protein sources such as fishmeal can be replaced in aquaculture diets with alternative protein sources, but if high concentrations of the latter are utilised, growth is typically reduced and protein metabolism less efficient.

Under conditions of nutritional stress, Hawkins (1991) has demonstrated that Mytilus edulis can degrade proteins, and recycle amino acids required for specific biosynthetic processes with considerable efficiency, while catabolising less essential amino acids for energy. Mussels also appear to be able to respond to the requirement for endogenous proteins by altering their utilisation of exogenous proteins (Kreeger et al., 1995). The term 'the protein sparing effect' has been used by Hawkins (1985) to describe the highly efficient protein metabolism seen in M. edulis, with very low rates of protein degradation and a high proportion (<89%) of degradation products being recycled (Hawkins and Bayne, 1991).

3.3.2. Genetic diversity

There is much interest in understanding the underlying genetic factors controlling intra-individual variation in growth, particularly where such knowledge could be used to increase the efficiency of animal production (for review see Oddy, 1999). There is evidence that animals and plants with a higher degree of heterozygosity show enhanced growth rates (reviewed in Hawkins et al., 1986). However, the underlying mechanisms behind this observation are only recently becoming clear. In Mytilus edulis, it has now been demonstrated that higher growth rates in heterozygous individuals are achieved by a reduction in both protein synthesis and degradation rates, resulting in high PSREs, this is in contrast to mammals where faster growing individuals have fast protein synthesis and degradation rates (Millward et al., 1981; Hawkins et al., 1986; Hawkins, 1996; Hawkins and Day, 1996; Bayne and Hawkins, 1997; Hawkins and Day, 1999). Faster growth is achieved in mussels by a reduction in the energetic costs of protein metabolism, resulting from reduced turnover, which in turn leads to a reduction in basal metabolic rate, so leaving a larger proportion of the energy budget available for growth (Hawkins et al., 1986).

How reduced protein turnover affects the ability of more heterozygous individuals to adapt to environmental change is unclear, as protein turnover is thought to be important for adaptation to changing environments (Hawkins, 1991). Hawkins et al. (1986) suggest more heterozygous animals may have an advantage over more homozygous ones when exposed to environmental change, as they have lower basal metabolic rates and hence a greater capacity to cope with change. It is not currently clear whether the adoption of a low protein turnover strategy carries with it any negative factors, but if turnover of proteins within an animal's body fulfils a required role this seems likely.

Higher heterozygosity is also traditionally considered to be correlated with increased fitness through masking of deleterious recessive alleles in hetero-zygotes and the summation of the effects of dominant alleles (Gibson and Weir, 2005). However, higher heterozygosity may also confer greater pheno-typic plasticity, enabling individuals and populations to maintain normal function across a wider range of environmental conditions. This may explain why genetic diversity, in some cases, appears to be correlated with environ-mental variability or stress (Nevo, 2001; Gabriel, 2005). Heterozygosity may, therefore, increase fitness during environmental stress by reducing the proportion of basal metabolism required by protein metabolism and by conferring increased phenotypic plasticity.

There is evidence that levels of heterozygosity are also related to develop-mental stability (reviewed in Møller, 1997). Growth rates and metabolism are also often positively correlated with phenotypic symmetry, a trait com-monly associated with developmental stability (Møller, 1997). Developmen-tal instability can be caused by a variety of environmental as well as genetic factors, including stressful environmental conditions, nutritional deficiency and the effects of pollutants and parasitism (Møller, 1997). This means that the developmental history of an organism will influence developmen-tal stability and, in turn, is likely to influence biochemical processes includ-ing protein synthesis. Variation in gene expression between individuals of *Fundulus heteroclitus* has been attributed to a mix of genotypic factors and irreversible environmental effects on development (Whitehead and Crawford, 2005). Parental diet, for example, is well known to influence growth rates and survival in marine invertebrate larvae (Utting and Millican, 1997). Such environmental effects on development and parental physiology may partially explain inter-individual variations in measures of protein synthesis.

Direct experimental evidence for biochemical mechanisms involved in inbreeding depression or heterosis is limited. Inbreeding has been demon-strated to influence expression of the chaperone protein Hsp70 in *Drosophila melanogaster*. Inbred larvae have higher expression levels of Hsp70 expres-sion than outbred controls and a lower resistance to heat (Pedersen *et al.*, 2005). This suggests that Hsp70 is induced in inbred lines to counteract the deleterious effects of inbreeding, possibly dealing with increased levels of abnormal proteins or as a more general mechanism to restore protein homeostasis (Pedersen *et al.*, 2005).

It should also be considered that the original work on heterozygosity–growth relationships was carried out by observing the frequencies of alleles of enzyme loci that code for isozymes with potentially different functional characteristics. We now know that, as well as variation in the structure and function of proteins, there is also heritable genetic variation in levels of gene expression (Knight, 2004; Stammatoyannopoulos, 2004; Gibson and Weir, 2005; Vasemägi and Primmer, 2005). This is of functional significance and must play a role in evolution and adaptation.

4. TISSUE AND ORGAN PROTEIN METABOLISM

4.1. Overview

Although whole-animal protein metabolism measurements are of the greatest interest when relating protein metabolism to animal growth, knowledge of the underlying tissue protein synthesis rates are also important. Whole-animal protein metabolism rates are an amalgam of the specific rates for each tissue. The influence of each tissue on whole-animal protein metabolism will depend on the relative mass of that tissue, the protein content and the rate of protein synthesis, degradation or growth. Measurement of tissue protein metabolism rates can allow the detection of significant biotic- or abiotic-induced changes which it is not possible to determine at the whole-animal level (Houlihan et al., 1990b; Hewitt, 1992; Martin et al., 1993; El Haj et al., 1996). Although tissue protein synthesis rates have been measured in considerably more marine species than whole-animal rates, our knowledge is still fragmentary, and little data is available for protein growth or degradation (Tables 4 and 5).

4.2. Abiotic factors influencing protein metabolism

4.2.1. Temperature

Temperature affects protein metabolism at the tissue and organ level in a similar manner to which it does at the whole-animal level (Section 3.2.1). Evidence from trout suggests that elongation rates in a range of tissues decrease with temperature (Simon, 1987). Elongation rates in tissues from fish that have previously been acclimated to cold water temperatures are generally higher than those acclimated to warm water, when rates are measured at the same temperature. This suggests some form of partial cold acclimation of the elongation step of protein synthesis, although the cold-induced increase in elongation rate may also be partially the result of an increase in the cellular concentration of RNA and associated enzymes (Simon, 1987). Evidence does

Table 4 Tissue protein synthesis rates in marine fish held under a variety of experimental conditions

Species	Temperature (°C)	Feeding regime	White muscle (k_s, % day^{-1})	Liver (k_s, % day^{-1})	Intestine (k_s, % day^{-1})	Stomach (k_s, % day^{-1})	Heart (k_s, % day^{-1})	Brain (k_s, % day^{-1})	Gill (k_s, % day^{-1})	References
General tissues										
Pleuronectes flesus	12	*ad libitum*							6.2	Lyndon and Brechin, 1999
Mugil cephalus	28	Fed	0.5	20					23	Haschemeyer and Smith, 1979
Lactophrys bicaudalis	28	N/A	0.4							Smith *et al.*, 1980
Haemulon plumieri	28	N/A	0.6 ± 0.2						22 ± 5	Smith *et al.*, 1980
Ocyurus chrysurus	28	N/A	0.6						25	Smith *et al.*, 1980
Halichoeres radiatus	28	N/A	0.9							Smith *et al.*, 1980
Scarus ghobban	28	N/A	0.3 ± 0.1						13 ± 3	Smith *et al.*, 1980
Vomer declivifrons	28	N/A	0.6 ± 0.2						16 ± 3	Smith *et al.*, 1980
Sufflamen verres	28	N/A	0.7 ± 0.3						15 ± 5	Smith *et al.*, 1980
Paranthias furcifer	28	N/A	0.8 ± 02						23 ± 5	Smith *et al.*, 1980
Orthopristis cantharinus	28	N/A	0.8 ± 0.3						14 ± 3	Smith *et al.*, 1980
Lutjanus viridis	28	N/A	0.8 ± 0.3						16 ± 4	Smith *et al.*, 1980
Anisotremus interruptus	28	N/A	1.1 ± 0.4						23 ± 4	Smith *et al.*, 1980
Caranx caballus	28	N/A							14 ± 2	Smith *et al.*, 1980
Nutrition										

(Continued)

Table 4 (Continued)

Species	Temperature (°C)	Feeding regime	White muscle (k_s, % day⁻¹)	Liver (k_s, % day⁻¹)	Intestine (k_s, % day⁻¹)	Stomach (k_s, % day⁻¹)	Heart (k_s, % day⁻¹)	Brain (k_s, % day⁻¹)	Gill (k_s, % day⁻¹)	References
Salmo salar	12.9 ± 0.6 12.9 ± 0.6 7.6 ± 0.2	Starved 0.6% BW day⁻¹ Excess three times daily	1.0 ± 0.2 1.4 ± 0.3 5.0 ± 1.5 (30%)[a] 7.1 ± 1.5 (35%) 3.9 ± 1.3 (45%) 8.4 ± 1.2 (m) 4.3 ± 1.2 (c)							Carter et al., 1993b Carter et al., 1993b Sveier et al., 2000
Gadus morhua	8–12	Starved then fed *ad libitum*		1.2 ± 0.2 (0)[b] 8.5 ± 1.4 (3) 18.3 ± 7.0 (6) 7.0 ± 1.1 (12) 2.9 ± 0.6 (18) 1.7 ± 0.3 (24) 2.4 ± 0.4 (48)		1.2 ± 0.3 (0) 7.8 (6) 3.8 ± 0.7 (12) 1.4 ± 0.2 (18) 1.2 ± 0.3 (24) 3.1 ± 0.4 (48)				Lyndon et al., 1992
Notothenia coriiceps	2	Fed/Starved	0.4 ± 0.0 (f)[c] 0.2 ± 0.0 (s)	10.4 ± 1.5 (f) 11.5 ± 1.3 (s)			1.4 ± 0.1 (f) 1.1 ± 0.1 (s)	0.4 ± 0.1 (f) 0.5 ± 0.1 (s)	1.6 ± 0.2 (f) 2.9 ± 0.4 (s)	Haschemeyer, 1983
Chaenocephalus aceratus	2	Fed/Starved	1.1 (f)[c] 0.9 (s)	0.5 (f) 0.2 (s)			1.2 (f) 1.1 (s)	1.0 (f) 1.0 (s)	1.1 (f) 1.0 (s)	Haschemeyer, 1983
Trematomus bernacchii	−1.5	Fed/Starved	0.2 ± 0.0 (f) 0.1 ± 0.0 (s)	5.3 ± 3.4 (f) 6.9 ± 3.3 (s)					5.3 ± 1.0 (f) 3.9 ± 1.6 (s)	Smith and Haschemeyer, 1980
Trematomus hansoni	−1.5	Fed/Starved	0.2 ± 0.1 (f) 0.04 ± 0.0 (s)	4.0 ± 1.0 (f) 8.4 ± 3.2 (s)	12.1 ± 4.5 (f) 12.3 ± 4.2 (s)				3.2 ± 1.6 (f) 2.2 ± 1.7 (s)	Smith and Haschemeyer, 1980

	Temperature								Reference
Trematomus newnesi	−1.5	Fed/Starved	0.1 ± 0.0 (f)	6.0 ± 0.8 (f)	18.3 ± 1.0 (f)			1.5 ± 1.1 (f)	Smith and Haschemeyer, 1980
Gymnodraco acuticeps	−1.5	Fed/Starved	0.1 ± 0.0 (f)	9.2 ± 3.0 (f)	12.5 ± 1.5 (f)			1.3 ± 1.0 (f)	Smith and Haschemeyer, 1980
Temperature									
Anarhichas lupus	5	Satiation	0.6						McCarthy et al., 1999
	8	Satiation	0.8						McCarthy et al., 1999
	11	Satiation	1.0						McCarthy et al., 1999
	14	Satiation	1.2						McCarthy et al., 1999
Gadus morhua	5	3.0 ± 0.1% BW day^{-1}			5.3	2.0	2.8	8.0	Foster et al., 1992
	15	3.0 ± 0.0% BW day^{-1}			6.3	1.6	3.4	8.7	Foster et al., 1992
Opsanus tau (summer)	11	Fed	0.1 ± 0.0	2.4 ± 0.8					Pocrnjic et al., 1983
	22	Fed	0.7 ± 0.3	17.0 ± 7.0					Pocrnjic et al., 1983
Opsanus tau (spring)	10	Starved	0.1 ± 0.0	2.2 ± 0.4			0.7 ± 0.2	2.0 ± 0.3	Pocrnjic et al., 1983
	20	Starved	0.2 ± 0.0	13.7 ± 4.4			2.3 ± 0.6	7.7 ± 3.5	Pocrnjic et al., 1983
Sufflamen verres	20	N/A	0.4 ± 0.2	13 ± 2				5.5 ± 1.2	Haschemeyer et al., 1979
	26	N/A	0.7 ± 0.3	21 ± 5				15 ± 5	Haschemeyer et al., 1979
	30	N/A	1.1 ± 0.1	36 ± 4				14 ± 5	Haschemeyer et al., 1979

(Continued)

Table 4 (Continued)

Species	Temperature (°C)	Feeding regime	White muscle (k_s, % day^{-1})	Liver (k_s, % day^{-1})	Intestine (k_s, % day^{-1})	Stomach (k_s, % day^{-1})	Heart (k_s, % day^{-1})	Brain (k_s, % day^{-1})	Gill (k_s, % day^{-1})	References
Sexual maturation (sea water fish returning to freshwater)										
Salmo salar (July)	15 ± 2	Starved 5 days	2.0	8.2		1.4	2.0		5.4	Martin *et al.*, 1993
Salmo salar (October)	(15 ± 2)-(8 ± 2)	Starved 95 days	2.0	13.2		8.6	2.5		5.7	Martin *et al.*, 1993
Pollution										
Limanda limanda (control)	10.6 ± 0.3	Satiation three times a week	0.2 ± 0.0	2.5						Houlihan *et al.*, 1994
Limanda limanda (sewage treated)	10.6 ± 0.3	Satiation three times a week	0.1 ± 0.2	2.6						Houlihan *et al.*, 1994

[a]Figure in parentheses represents protein percentage in diet; m or c signifies a micro or course ground diet.
[b]Figure in parentheses represents number of hours after feeding a meal protein synthesis was measured.
[c]Letter in parentheses signifies whether the fish were fed (f) or starved (s).

Table 5 Tissue protein synthesis rates in marine invertebrates under a variety of experimental conditions

Species	Group	Temperature (°C)	Feeding regime	Muscle (k_s, % day^{-1})	Hepato-pancreas (k_s, % day^{-1})	Heart (k_s, % day^{-1})	Digestive tract (k_s, % day^{-1})	Gill (k_s, % day^{-1})	Notes	References
Seasonal										
Heterocucumis steineni	Echinoderm	−1.2 ± 0.3	Feeding	0.2 ± 0.0 (February)					Body wall	Fraser et al., 2004
			Starved	0.2 ± 0.0 (July)					Body wall	Fraser et al., 2004
			Feeding	0.4 ± 0.0 (December)					Body wall	Fraser et al., 2004
Nutrition										
Carcinus maenas	Crustacean	15	Starved	0.2 (c)[a] 0.4 (l)	5.4	0.9				Mente et al., 2003
			2 hours after meal	0.5 (c) 0.7 (l)	7.5	1.9				Mente et al., 2003
			5 hours after meal	0.7 (c) 0.7 (l)	11.2	3.3				Mente et al., 2003
			24 hours after meal	0.6 (c) 0.6 (l)	13.7	2.8				Mente et al., 2003
			48 hours after meal	0.7 (c) 0.7 (l)	4.4	2.3				Mente et al., 2003
		15	Starved	0.2 (c) 0.2 (l)	2.7	1.7	2.0	2.4		Houlihan et al., 1990b
			3 hours after meal	0.6 (c) 0.6 (l)	9.4	5.2	2.6	4.9		Houlihan et al., 1990b
			9 hours after meal	0.4 (c) 0.4 (l)	3.4	1.7	2.5	2.3		Houlihan et al., 1990b
Carcinus maenas	Crustacean	15	Fed then starved 2 days	0.6(c) 0.2 (l)	4.6	1.3	1.5	2.4		Houlihan et al., 1990b
Octopus vulgaris	Mollusc	22	Starved 48 hours	2.9 ± 0.5 (arm)		3.7 ± 0.6 (ventricle)	3.0 ± 0.5 (stomach)	3.0 ± 0.9		Houlihan et al., 1990a

(Continued)

Table 5 (Continued)

Species	Group	Temperature (°C)	Feeding regime	Muscle (k_s, % day⁻¹)	Hepato-pancreas (k_s, % day⁻¹)	Heart (k_s, % day⁻¹)	Digestive tract (k_s, % day⁻¹)	Gill (k_s, % day⁻¹)	Notes	References
Penaeus esculentus	Crustacean	30 ± 0.2	30% protein twice daily	1.3	58					Hewitt, 1992
			40% protein twice daily	1.0	53					Hewitt, 1992
			50% protein twice daily	1.4	49					Hewitt, 1992
Moulting										
Homarus americanus	Crustacean	15–18	N/A	0.4 (l) 0.2 (a)					Post-moult	El Haj *et al.*, 1996
				0.4 (c)						
				0.3 (l) 0.4 (a)					Inter-moult	El Haj *et al.*, 1996
				0.3 (c)						
				0.5 (l) 0.7 (a)					Pre-moult	El Haj *et al.*, 1996
				0.9 (c)						
Homarus americanus	Crustacean	15–18	N/A	0.4 (l) 0.2 (a)					Control	El Haj *et al.*, 1996
				0.2 (c)						
				0.6 (l) 0.5 (a)					Injected moult hormone	El Haj *et al.*, 1996
				0.7 (c)						
Carcinus maenas	Crustacean	15	Daily	0.1					ᵇD1–2	El Haj and Haulihan, 1987
				1.7					D3–4	El Haj and Haulihan, 1987
				1.0					A1–2	El Haj and Haulihan, 1987
				0.9					B1–2	El Haj and Haulihan, 1987
				0.4					C1–2	El Haj and Haulihan, 1987
				0.3					C3–4	El Haj and Haulihan, 1987
Hypoxia/hyperoxia										
Carcinus maenas	Crustacean	15	Starved	0.2 (c) 0.4 (l)	5.6	0.9			O₂ 4 kPa	Mente *et al.*, 2003
			2 hours after meal	0.4 (c) 0.5 (l)	2.2	1.4			O₂ 4 kPa	Mente *et al.*, 2003
			5 hours after meal	0.4 (c) 0.5 (l)	3.3	2.0			O₂ 4 kPa	Mente *et al.*, 2003

		10 hours after meal	0.7 (c) 0.6 (l)	5.6	3.0	O_2 4 kPa	Mente et al., 2003
		24 hours after meal	0.5 (c) 0.6 (l)	7.0	1.0	O_2 4 kPa	Mente et al., 2003
		48 hours after meal	0.5 (c) 0.8 (l)	4.8	1.6	O_2 4 kPa	Mente et al., 2003
Carcinus maenas	Crustacean 15	Starved	0.2 (c) 0.4 (l)	5.6	0.9	O_2 3 kPa	Mente et al., 2003
		2 hours after meal	0.3 (c) 0.7 (l)	3.3	1.5	O_2 3 kPa	Mente et al., 2003
		5 hours after meal	0.6 (c) 0.6 (l)	3.0	0.9	O_2 3 kPa	Mente et al., 2003
		24 hours after meal	0.3 (c) 0.4 (l)	1.5	0.9	O_2 3 kPa	Mente et al., 2003
		48 hours after meal	0.4 (c)	1.1	1.4	O_2 3 kPa	Mente et al., 2003
		24 hours after meal	0.5 (c) 0.6 (l)	13.2	2.6	O_2 60 kPa	Mente et al., 2003

[a]Letter in parentheses signifies: c, claw muscle; l, leg muscle; a, abdomen muscle.
[b]Moult stages.

suggest that there is an increase in the concentration of elongation factor 1 (EF1) at low temperatures (Haschemeyer, 1978).

In cod *Gadus morhua* (L.) acclimated to 5 or 15 °C for 40 days and fed similar rations, overall growth rates and tissue protein synthesis rates were not significantly different when measured at the respective acclimation temperature (Foster *et al.*, 1992; Table 4). Protein synthesis rates at the lower temperature were maintained by an increase in tissue RNA concentrations to counteract a decrease in the k_{RNA}. In contrast, in *Anarhichas lupus* (L.), the wolffish, white muscle protein synthesis rates in fish fed to satiation increased linearly with water temperature over a range of 5–14 °C (McCarthy *et al.*, 1999; Table 3). Protein growth rates, although initially increasing with temperature, peaked at 11 °C before decreasing at 14 °C, suggesting an increase in protein degradation rates at temperatures towards the upper thermal limit of the species (McCarthy *et al.*, 1999). It has been suggested that maximal protein synthesis rates occur at the optimal temperature for that species, although in wolffish at least, maximal protein growth appears to occur at a lower temperature (McCarthy and Houlihan, 1996; McCarthy *et al.*, 1999). Protein synthesis rates increased with temperature in McCarthy *et al.* (1999), rather than remaining constant as in Foster *et al.* (1992), because animals in the former study were fed to satiation at each temperature, rather than being fed a fixed ration irrespective of temperature. In another study on cod that were fed to satiation, protein synthesis rates did indeed increase with temperature (Treberg *et al.*, 2005). Similarly, increases in protein synthesis with temperature have been reported in freshwater fish fed to satiation (Loughna and Goldspink, 1985; Watt *et al.*, 1988).

If investigating the effect of temperature on protein metabolism, the previous thermal history of the animal should be considered. When red and white muscle protein synthesis rates were measured over a range of water temperatures in groups of carp acclimatised to low or high temperatures, rates increased with temperature in both groups, but the synthesis rates in the low-temperature group were always higher than the high temperature group (Watt *et al.*, 1988). The carp acclimated to low temperatures did not show an increase in muscle RNA concentrations and Watt *et al.* (1988) suggested elevated protein synthesis rates were possibly a result of increased elongation factor activity or initiation by initiation factor I. Loughna and Goldspink (1985) found that carp acclimatised for 1 month to a given temperature showed elevations in white muscle protein synthesis at low temperatures and reductions in protein synthesis at higher temperatures in comparison to animals acclimatised to the same temperatures for only 24 hours. In contrast, rainbow trout white muscle protein synthesis rates were not significantly different whether the animals were acclimatised for 1 month or 24 hours. The reasons for the contrasting results between the two species are unclear.

Many studies examining the effects of temperature on tissue protein synthesis rates have examined alterations in protein synthesis after acute changes in temperature (Haschemeyer, 1968; Mathews and Haschemeyer, 1978; Haschemeyer et al., 1979). Typically, acute increases or decreases in temperature will result in a parallel increase or decrease in protein synthesis, providing the temperature is within the thermal tolerance range of the species (Jackim and La Roche, 1973; Mathews and Haschemeyer, 1978; Haschemeyer et al., 1979; Haschemeyer and Mathews, 1982). At the limits of the species thermal tolerance, both upper and lower rates of protein synthesis rapidly decrease (Mathews and Haschemeyer, 1978; Haschemeyer et al., 1979; Haschemeyer and Mathews, 1982). In tropical marine fish, it has been shown that radiolabel incorporation rates decrease rapidly at $32\,^{\circ}C$, even though this temperature is approximately $5\,^{\circ}C$ lower than the upper lethal limits of the experimental species (Haschemeyer et al., 1979).

There appears to be a seasonal effect on the thermal range over which fish show a linear increase in protein synthesis rates with increasing temperature, irrespective of any temperature the fish have been acclimatised to (Jackim and La Roche, 1973). Treberg et al. (2005) examined protein synthesis in cod, either acclimated to a control temperature ($8–11\,^{\circ}C$) or an artificial seasonal cycle of changing water temperatures ranging between -0.3 and $11\,^{\circ}C$. Protein synthesis rates were measured three times over a year at ambient temperatures of -0.3, 4.5 and $11\,^{\circ}C$ as well as at the control temperature. Protein synthesis rates were also measured in control animals acutely exposed to experimental temperatures, and experimental animals transferred to the control temperature. Generally, at lower temperatures cod had lower white muscle and liver protein synthesis rates, but in animals which were measured at the same temperature, cold acclimated animals generally had elevated rates of muscle but not liver protein synthesis. This suggests some thermal compensation of muscle but not liver protein synthesis. In toadfish, Opsanus tau (L.), acclimated to 10 or $20–23\,^{\circ}C$, liver protein synthesis rates in the cold-adapted animals were 75% higher than the warm-acclimatised animals, when protein synthesis rates were measured at the higher temperature (Haschemeyer, 1968). Overall, data so far suggest animals have an ability to at least partially compensate tissue protein synthesis rates at low temperatures.

4.2.2. Pollution

Exposure of the dab, Limanda limanda, to sewage sludge did not affect white muscle protein synthesis rates, but protein degradation rates increased significantly and body protein content was reduced (Houlihan et al., 1994). Likewise, in the liver, kidney and spleen, protein synthesis rates were not significantly reduced by sewage exposure, but protein degradation rates were elevated leading to significantly lower PSREs (Houlihan *et al.*, 1994). The authors concluded that sewage sludge exposure induces increased protein turnover.

When the crab, *Oziotelphusa senex senex* Fabricius, 1791, was exposed to Fenvalerate, an insecticide, hepatopancreas and muscle protein content decreased significantly, while free amino acids concentrations increased (Reddy and Bhagyalakshmi, 1994). Protease activity increased in both the hepatopancreas and muscle, suggesting an elevation in protein degradation.

In the few studies carried out on marine organisms so far, exposure to pollutants appears to result in no significant alteration in protein synthesis rates, but an increase in protein degradation, resulting in a reduction in the PSRE and hence reduced growth.

4.2.3. Diurnal and seasonal variations

Little is known regarding temporal variation in tissue protein metabolism of marine organisms. ^{14}C-glycine incorporation in the gulf killifish, *F. grandis* Baird and Girard, 1853, held under a light:dark cycle of 12:12 hours, shows tissue-dependent diurnal variation (Negatu and Meier, 1993). In males, the highest rates of incorporation in scales occurred during the dark, during early light-phase in muscle and liver and during late dark and early light-phase in the intestine. Similar results were also found in the scales and muscle of female fish, but in the liver peak incorporation occurred in the dark-phase, while in the intestine incorporation increased during the light-phase and peaked at the start of the dark-phase. Care should be taken in interpreting these results, as in all incorporation studies, as no assessment of the stability of the precursor pools for protein synthesis was made, and tissue-specific variations in incorporation could simply be the result of tissue-specific differences in the availability of ^{14}C-glycine.

A seasonal study of the Antarctic holothurian, *Heterocucumis steineni* Ludwig, 1898, has shown that the species ceases feeding completely for 4 months in winter, while body wall protein synthesis rates decrease from $0.35 \pm 0.03\%$ day^{-1} in summer to $0.23 \pm 0.03\%$ day^{-1} in winter (Fraser *et al.*, 2004). Body wall RNA to protein ratios also decreased significantly from 32.8 ± 1.2 to 27.9 ± 1.3 μg RNA mg^{-1} protein. The decrease in winter protein synthesis rates and RNA to protein ratios was probably the result of a decrease in food consumption, rather than the small seasonal water temperature variations seen at these latitudes (Clarke, 1988; Fraser *et al.*, 2004). Few studies have examined seasonal changes in tissue protein metabolism in endotherms. In the big brown bat, *Eptesicus fuscus* Beauvois, 1796, protein synthesis is completely arrested in the liver and pectoralis muscle during hibernation, in contrast to the continuation of protein synthesis in the tissues of non-feeding ectotherms (Yacoe, 1983).

4.2.4. Hyperoxia and hypoxia

A single study has specifically addressed the effect of water O_2 concentration on protein synthesis in marine organisms (Mente et al., 2003). At a reduced water pO_2 of 4 kPa, post-prandial increases in protein synthesis in the leg and claw muscle of Carcinus maenas (L.) were delayed, but maximum protein synthesis rates were similar to those measured under normoxic conditions (21 kPa). In contrast, in the hepatopancreas during hypoxia, protein synthesis rates initially decreased after feeding and then showed little or no increase, which was in strong contrast to normoxic animals in which protein synthesis rates approximately doubled after feeding (Mente et al., 2003). In the hypoxic heart, protein synthesis rates were slow to increase after feeding and did not reach levels seen in normoxic animals. At a reduced water pO_2 of 3 kPa, protein synthesis rates did not increase after feeding in any tissues, with the exception of a small transient increase in claw muscle protein synthesis rates. Exposing the crabs to hyperoxic water increased the rate of protein synthesis in the claw muscle, heart and hepatopancreas but not the leg muscle.

Gene expression profiling has been used to examine the effects of hypoxia on different tissues in the estuarine fish Gillichthys mirabilis Cooper, 1864 (Gracey et al., 2001). Marked differences in gene expression were observed in brain, liver, skeletal and cardiac muscle tissue in response to hypoxia when the pO_2 was lowered from 100% to 10% over 90 min and maintained at reduced levels for up to 6 days.

In muscle, there was a strong down-regulation of gene-encoding components of the protein translation machinery (ribosomal proteins and EF2), suggesting a significant decrease in capacity for protein synthesis. Genes coding for several contractile proteins were also repressed, possibly resulting in reduced locomotory activity in the fish as part of an energy saving strategy. In the liver, a number of genes involved in suppression of cell growth and proliferation were induced. Genes encoding for enzymes involved in amino acid catabolism were also induced in the liver, along with glutamine synthetase, an enzyme which catalyses the synthesis of glutamine from glutamate (ammonia detoxification) and glucose-6-phosphatase (dephosphorylation of glucose-6-phosphate to glucose). Together, these changes in expression suggest that amino acid catabolism and gluconeogenesis in the liver may represent a mechanism to maintain blood glucose levels during hypoxia. The decrease in rates of protein synthesis detected in muscle tissue may also release amino acids for gluconeogenesis (Gracey et al., 2001). Cessation of protein synthesis during anoxia has also been observed in various tissues of turtles (Fraser *et al.*, 2001). The possibility of a link between rates of protein synthesis and energy metabolism in response to stress is worthy of further exploration.

4.3. Biotic factors influencing protein metabolism

4.3.1. Tissue-specific rates

It is clear that protein synthesis rates in different tissues vary greatly (Simon, 1989; Houlihan, 1991; Carter and Houlihan, 2001; Tables 4 and 5). However, there is an observable general pattern in marine ectotherms that is similar to that seen in non-marine ectotherms and endotherms (Preedy *et al.*, 1985, 1988; Simon, 1989; Beebe Smith and Sun, 1995; Fraser *et al.*, 2001). Mean tissue protein synthesis rates (calculated from Table 4) rank in the following order, from the highest rate of protein synthesis to the lowest: intestine, gill, liver, stomach, heart, white muscle and brain. For marine invertebrates, however, the tissue rank from the highest to lowest rates are: hepatopancreas, gill, digestive tract, heart and muscle (Table 5). Similar rankings, albeit with some sequential differences, have previously been reported in other ectotherms (Houlihan *et al.*, 1988a, 1990a, 1991; Lyndon and Houlihan, 1998; Carter and Houlihan, 2001). Some care should be taken in interpreting these rankings, as animals were held under a wide variety of conditions and were at different life stages, factors that are likely to affect tissue protein synthesis rates (Tables 4 and 5). It is likely that the ranking of these tissues in terms of protein synthesis rates will produce a very similar result to that obtained by ranking the same tissues based on O_2 consumption (Houlihan, 1991).

In general, metabolically active tissues such as the gastrointestinal tract, liver and gills appear to synthesise more protein than muscular tissues and the brain. This is not particularly surprising as the gastrointestinal tract produces considerable quantities of both mucopolysaccharides and enzymes, while the liver in fish and the hepatopancreas in invertebrates have important roles in amino acid metabolism and the synthesis and export of many proteins (Glass, 1953; Barrington, 1957; Barnard, 1973; Houlihan *et al.*, 1990b; Hewitt, 1992; Barnes *et al.*, 1996; Carter and Houlihan, 2001). Likewise, the gill in marine organisms is a highly metabolically active tissue with a crucial, multifunctional role in homeostasis, and high gill protein synthesis rates have previously been highlighted (Lyndon and Houlihan, 1998; Lyndon and Brechin, 1999). Interestingly, rates of protein synthesis even appear to differ significantly between individual fish holobranchs (a gill filament found in jawed fishes), possibly suggesting a difference in physiological function (Lyndon and Brechin, 1999). Limited protein synthesis data for other metabolically active tissues such as the head kidney and kidney in fish also suggest comparatively high protein synthesis rates (Haschemeyer, 1983).

Protein synthesis rates in the brain, heart and white muscle are generally considerably lower than in more metabolically active tissues (Tables 4 and 5). Only a single study has reported brain protein synthesis rates in two marine species, so conclusions are difficult to draw (Haschemeyer, 1983; Table 4). However, in previous endotherm and non-marine ectotherm studies, brain protein synthesis rates also appear low, although often slightly higher than the muscle and heart (Preedy et al., 1985, 1988; Sayegh and Lajtha, 1989; Simon, 1989; Beebe Smith and Sun, 1995; Fraser et al., 2001).

Detailed studies of striated and smooth muscle protein synthesis rates in marine organisms are lacking, but in rainbow trout it appears that muscle synthesis rates generally rank in the following order: white muscle < red skeletal muscle < heart (smooth) (Fauconneau et al., 1995; Carter and Houlihan, 2001). The majority of muscle protein synthesis occurs in the myofibrils. In white/red muscle 60–64% of protein synthesis is myofibrillar while in the heart 50% (Fauconneau et al., 1995). In Carcinus maenas, higher fractional rates of protein synthesis occur in slow tonic fibres than in the fast phasic muscle fibres (El Haj and Houlihan, 1987). While within the arms of Octopus vulgaris Cuvier, 1797, which are primarily composed of white muscle, there can be considerable variations in protein synthesis rates depending on which part of the arm is utilised for the measurement (Houlihan et al., 1990a).

Houlihan (1991) has suggested that white muscle protein synthesis rates provide a good proxy measurement of whole-animal growth rates, as protein degradation rates in this tissue are very low, muscle is the major site of protein accretion and protein synthesis is highly sensitive to dietary changes (Smith, 1981; Fauconneau, 1985; Lied et al., 1985; Houlihan et al., 1986; Houlihan and Laurent, 1987; Houlihan et al., 1988a).

4.3.2. Food consumption and nutrition

There is a strong relationship between food consumption, O_2 consumption and protein metabolism in many tissues and organs (Smith and Haschemeyer, 1980; Haschemeyer, 1983; Lyndon et al., 1992; Carter et al., 1993a). Animals typically show a post-prandial increase in O_2 consumption that rises to a peak before gradually declining to pre-feeding levels over hours or days (Houlihan et al., 1990b; Chapelle et al., 1994; Boyce and Clarke, 1997; Clarke and Prothero-Thomas, 1997; Peck and Veal, 2001; Mente et al., 2003). The duration of the SDA peak has been shown to be proportional to meal size and temperature (Beamish, 1974; Houlihan et al., 1990b; Peck, 1998). Previous workers have suggested a significant proportion of the SDA response is a direct result of an increase in protein synthesis and growth (Jobling, 1985; Houlihan et al., 1990b; Robertson et al., 2001a,b; Whiteley et al., 2001). Strong evidence for the importance of protein synthesis in the SDA response is provided by the fact that an infusion of essential amino acids results in both an increase in O_2 consumption and increased rates of protein synthesis in liver and white muscle, while injection of the protein synthesis inhibitor cycloheximide negates the SDA response (Brown and Cameron, 1986; Brown, 1988).

The relationship between food consumption and tissue protein synthesis rates in marine ectotherms is complex and still not fully understood. However, it is clear that post-prandial changes in tissue protein synthesis rates are to some degree dependent on the previous nutritional history of the organism and the size of the meal (Houlihan, 1991). Different tissues appear to up-regulate protein synthesis at varying times after feeding and to differing degrees in both ectotherms and endotherms (McMillan and Houlihan, 1989; Houlihan et al., 1990b; Lyndon et al., 1992; Danicke et al., 1999; Mente et al., 2003). In starved rainbow trout and cod, tissues with high protein synthesis rates such as the gill, liver and intestine show rapid increases in protein synthesis rates after a single meal, with measured rates comparable to those seen in continually feeding animals (McMillan and Houlihan, 1989; Lyndon et al., 1992). In contrast, tissues with generally lower protein synthesis rates, such as white muscle and the heart, show slower increases in protein synthesis rates that do not reach levels seen in continually fed fish (McMillan and Houlihan, 1989; Lyndon et al., 1992). In both cod and rainbow trout, there appears to be a large transient increase in liver protein synthesis prior to elevations in other tissues (McMillan and Houlihan, 1989; Lyndon et al., 1992). Considerably less is understood about the effects of fasting and feeding on protein degradation, although a study suggests the activity of the proteasome pathway, a major mechanism of protein degradation, is reduced after 2 weeks starvation in rainbow trout (Martin et al., 2002). In contrast, the activities of lysosomal enzymes appear to increase during starvation (Martin et al., 2001). Another study on trout has shown a negative correlation between liver, but not white muscle, 20S proteasome activity and specific growth rate (Dobly et al., 2004).

Detailed studies of post-prandial tissue protein synthesis rates have only been reported in one marine invertebrate, the shore crab, Carcinus maenas (Houlihan et al., 1990b; Mente et al., 2003). Tissue protein synthesis rates and O_2 consumption increased rapidly 3–5 hours after feeding in the shore crab (Houlihan et al., 1990b; Mente et al., 2003). Protein synthesis rates increased in the claw and leg extensor muscles, midgut gland, gill and heart, although only in the first tissue was the increase in protein synthesis sustained (Houlihan et al., 1990b). In another study, C. maenas post-prandial protein synthesis rates were sustained in the claw muscle, leg muscle and heart but not in the hepatopancreas (Mente *et al.*, 2003). Generally speaking, *C. maenas* muscular tissues appear to show smaller but sustained increases in protein synthesis after feeding, while more metabolically active tissues show larger more transient increases. In endotherms, increases in protein synthesis rates after a meal are generally more rapid and shorter lived than in ectotherms, for example in growing rats protein synthesis rates in muscle and liver increase by 30 min after feeding, but start to decrease after only 2 hours (Danicke *et al.*, 1999).

Many studies have demonstrated that protein synthesis rates generally decrease in starved marine organisms to some relatively constant level (Houlihan *et al.*, 1988a, 1990b; Lyndon *et al.*, 1992; Carter *et al.*, 1993a; Mente *et al.*, 2003; Fraser *et al.*, 2004). It is also usually evident that even during total nutritional deprivation, some level of tissue protein synthesis is essential in all animals (Garlick *et al.*, 1973; Smith and Haschemeyer, 1980; Smith, 1981; Haschemeyer, 1983; Houlihan *et al.*, 1988a; McMillan and Houlihan, 1988; Preedy *et al.*, 1988; Watt *et al.*, 1988; McMillan and Houlihan, 1989; Lyndon *et al.*, 1992; Martin *et al.*, 1993; Fraser *et al.*, 2004), although in hibernating bats (*E. fuscus*) protein synthesis does appear to cease (Yacoe, 1983). White muscle in particular appears very sensitive to starvation, with protein synthesis rates decreasing rapidly when feeding ceases (Smith and Haschemeyer, 1980; Haschemeyer, 1983; Pocrnjic *et al.*, 1983; Fauconneau, 1985; Houlihan *et al.*, 1988a). The sensitivity of white muscle protein synthesis is particularly important to overall growth. For example, white muscle in cod has been shown to account for about 30% of total animal protein synthesis and 40% of growth; therefore, any decrease in white muscle protein synthesis will have a very large effect on both growth and metabolism (Houlihan *et al.*, 1988a). It appears likely that at least in fish, white muscle is utilised as a protein store during times of nutritional stress, providing more metabolically active tissues with amino acids (Houlihan *et al.*, 1988a). The observed decrease in white muscle protein synthesis during starvation appears to be highly protein specific, with a 90% decrease in myofibrillar protein synthesis observed during starvation in *Paralabrax nebulifer* Girard, 1854, without a significant change in the synthesis of total sarcoplasmic protein (Lowery and Somero, 1990).

The responses of individual tissues to fasting are complex and poorly understood. In some studies protein synthesis rates have increased during starvation. For example, after 2 weeks of starvation in the Antarctic fish, *Notothenia coriiceps*, protein synthesis rates appeared to increase in the liver, gill, head kidney, kidney, brain and gonad, while in the spleen, pectoral muscle, heart and white muscle rates decreased (Haschemeyer, 1983). Two other species of Antarctic fish also showed an increase in liver protein synthesis rate after 5–15 days of starvation, while protein synthesis rates in the gill, white muscle and spleen decreased and the gut remained unchanged (Smith and Haschemeyer, 1980).

Female Atlantic salmon returning to rivers to spawn do not feed, often for many months, and are therefore acting as an extreme example of naturally occurring fasting (Martin *et al.*, 1993). During the spawning phase white muscle acts as the major energy source once lipids are depleted (Love, 1980; Mommsen *et al.*, 1980; Ando *et al.*, 1986; Mommsen, 2004). Martin *et al.* (1993) reported a decrease in body mass of 10% and 15% in sexually maturing male and female salmon, respectively, as a result of specific changes in individual tissue masses. White muscle, intestinal caecae and the stomach declined in mass, while the liver, gills, gonad and red muscle increased. The heart mass remained unchanged. Tissue protein synthesis and degradation rates were controlled at the tissue level, with protein synthesis rates decreased in the red muscle, elevated in the stomach, liver, ovaries and ventricle, and unchanged in the white muscle and gill. Net protein losses occurred in all tissues except the ovary. The increase in stomach protein synthesis rates when the animals were not eating is perhaps surprising (Martin *et al.*, 1993). Amino acid supply for both energy requirements and anabolic processes in tissues, such as the ovary, appears to occur at the expense of other tissues, in particular white muscle. It is interesting to note that in starving salmon and other fish, protein degradation appears to occur via the lysosomal pathways, rather than the ATP–ubiquitin proteasome pathway primarily used in mammals during starvation (Mommsen, 2004).

An important point is that in almost all of the reported studies examining the responses of tissue protein synthesis to feeding, the animals have been starved for a period of time prior to feeding. In continually feeding animals, observed post-prandial increases in protein synthesis may be reduced or absent. For example, Houlihan *et al.* (1990b) showed that tissue protein synthesis rates in continually fed crabs were higher than in fasted animals, while in crabs re-fed after fasting, rates were as much as sixfold higher than in continually fed animals, albeit not significantly so, due to large inter-individual variability. Generally, it appears that there is an increase in protein synthesis rates after feeding, with the previous nutritional history of the animal, and the specific characteristics of each tissue determining the rapidity and duration of the response (Carter and Houlihan, 2001).

Two other factors that appear to influence tissue protein metabolism are the amino acid composition of the diet and the protein to energy ratio (von der Decken and Lied, 1992; de la Higuera *et al.*, 1998, 1999). The majority of studies on fish in this area have utilised cultivated freshwater species, with little or no research carried out on marine species. Readers with an interest in the effects of dietary amino acid composition and protein to energy ratios on protein metabolism are therefore directed to a review of this subject which includes freshwater species (Carter and Houlihan, 2001).

The overall protein content of the diet also affects protein metabolism. In *Penaeus esculentus* Haswell, 1879, feeding a diet containing 40% protein achieves better muscle growth than diets containing 30% or 50% protein, this is achieved not by increasing rates of protein synthesis, but by greatly reducing protein degradation (Hewitt, 1992). The protein source also appears important, in a study feeding three isonitrogenous and isoenergetic diets to *Litopenaeus vannamei* Boone, 1931, protein synthesis rates were the same for all diets, but protein degradation was significantly higher for the diet containing only casein as the protein source, rather than meal or a combination of meal and soya (Mente *et al.*, 2002). Interestingly, gilthead sea bream fed on a range of diets with between 50% and 100% of the fish meal in the diet replaced with plant protein did not show any increase in the activity of three hepatic enzymes associated with amino acid catabolism (Gómez-Requeni *et al.*, 2004). It is still possible, however, that replacing part, or all, of the fish meal with the plant protein did result in elevated protein degradation rates via other pathways or in other tissues.

4.3.3. Sexual maturation

Only a single study has examined the effects of sexual maturation on protein metabolism in marine organisms. Martin *et al.* (1993) examined protein metabolism during sexual maturation in salmon. However, interpreting the effects of maturation on protein metabolism in salmon is complicated by the fact that the animals were also fasting (Section 4.3.2). During maturation, the liver mass in female salmon increases by 20% relative to body mass due to elevated protein synthesis rates. A large proportion of the synthesised protein is exported from the liver, but it is not currently clear whether the exported protein is primarily vitellogenin being exported to the ovary. In turn, the ovary shows an 82% increase in protein synthesis, resulting in a tenfold increase in the protein mass. In maturing salmon, white muscle loses 54% of its protein content. In male salmon, the liver mass decreases by 35% relative to body mass, while the testes show only small increases in mass (Martin *et al.*, 1993). It appears likely that the primary source of amino acids used to fuel elevated protein synthesis rates in the liver and ovary is white muscle.

4.3.4. Moulting

Growth in crustaceans is intermittent, with frequent shedding of the exoskeleton through ecdysis. Lengthening of the leg muscles, presumably via protein synthesis, is associated with ecdysis in many crustaceans (El Haj *et al.*, 1984; Houlihan and El Haj, 1985). Two studies have examined changes in crustacean protein metabolism during the moult cycle (El Haj and Houlihan, 1987; El Haj *et al.*, 1996). An *in vivo* study of protein synthesis in *Carcinus maenas* showed a tenfold increase in extensor muscle protein synthesis in the D3–4 stage just prior to ecdysis, before protein synthesis rates gradually decreased to near inter-moult levels by C3–4. The increase in protein synthesis prior to ecdysis may be mediated by high circulating concentrations of the hormone 20-hydroxyecdysone, which has been shown to elevate protein synthesis rates when injected (El Haj and Houlihan, 1987; El Haj et al., 1996; Table 5). Elevated protein synthesis rates during the pre-ecdysis and imme-diate post-ecdysis periods probably result in muscle fibre lengthening via the addition of additional sarcomeres at the external cuticular attachment re-gion of the fibres (El Haj and Houlihan, 1987). Increased rates of protein synthesis during moulting appear to be controlled by alterations in k_{RNA} rather than an increase in tissue RNA concentration (El Haj et al., 1996). Similar results have also been reported in *Homarus americanus* Milne Edwards, 1837 where pre-moult protein synthesis rates were higher in the leg, abdominal and claw muscles than during the post- or inter-moult periods (El Haj et al., 1996; Table 4).

5. *IN VITRO* AND CELL-FREE PROTEIN METABOLISM

5.1. Overview

The use of *in vitro* or cell-free systems to examine protein metabolism provides a useful means to reduce the number of variables present when measuring whole-animal *in vivo* protein metabolism. It is also possible to make measurements of, for example, the major cellular-energy-consuming processes using specific inhibitors, which would not be possible in the intact animal (Wieser and Krumschnabel, 2001). However, it is important to realise that *in vitro* rates of protein metabolism are usually much lower than *in vivo* measurements and must be interpreted with care, although in more recent studies, optimised cell-free systems have been developed with synthesis rates which appear similar to intact animals (Haschemeyer and Mathews, 1983; Storch et al., 2003, 2005). Also, because of differing experi-mental conditions, comparisons of *in vitro* measurements made in different studies can be problematic.

5.2. Abiotic factors influencing protein metabolism

5.2.1. Temperature

Haschemeyer and Mathews (1983) used a collagenase perfusion to isolate hepatocytes from two species of Antarctic fish. The extracted hepatocyte suspension showed protein synthesis rates of 0.6% day^{-1}, only 15% of *in vivo* rates. Plated hepatocyctes had higher protein synthesis rates, but still only reached 50% of *in vivo* rates (Smith and Haschemeyer, 1980). Hepatocytes in suspension showed increases in the rate of protein synthesis with acute changes in temperature up to 20 °C, but protein synthesis rates decreased above this temperature and cell lysis had occurred by 30 °C (Haschemeyer and Mathews, 1983). In cell-free protein synthesis systems from Antarctic fish, protein synthesis rates also increased with temperatures between 0 and 37 °C (Haschemeyer and Williams, 1982). In contrast, plated hepatocytes, cultured for extended periods at elevated temperatures, showed a rapid decrease in protein synthesis at temperatures in excess of 7 °C, thereby showing similar stenothermality to that seen in the whole animal (Haschemeyer and Mathews, 1983). In an optimised, gill cell-free system, protein synthesis rates in the Antarctic scallop *Adamussium colbecki* Smith, 1902 measured at 15 °C were over 10 times higher than those measured in the temperate scallop *Aequipecten opercularis* (L.) at the same temperature (Storch and Pörtner, 2003). This was in part a result of the twofold higher RNA concentrations in the lysate of the Antarctic species, but mainly because of ninefold higher RNA activities. *A. colbecki* and *A. opercularis* had previously been acclimated to 0 and 10 °C, respectively.

5.2.2. Anoxia

Many intertidal marine invertebrates experience diurnal or semi-diurnal emersion, often with resultant hypoxia or anoxia. In the periwinkle, *Littorina littorea* (L.), hepatopancreas cell-free protein synthesis rates in anoxia-exposed animals were reduced by 49% after 30 min of anoxia (Larade and Storey, 2002). Anoxic animals showed protein synthesis rates not significantly different to control animals after 1 hour of recovery in normoxia. Soluble protein concentrations did not change in the hepatopancreas of anoxia-exposed animals, suggesting a coordinated decrease in protein degradation. Usually during protein synthesis, each mRNA molecule is simultaneously read by several ribosomes, so forming what is termed a polysome or poly-ribosome. The decrease in *L. littorea* protein synthesis under anoxia was associated with a decrease in the number of polysomes and an increase in monosomes, that is each mRNA molecule only being read by a single ribosome. During recovery from anoxia the proportion of polysomes returned to pre-anoxia values after 3 hours (Larade and Storey, 2002). Phosphorylation of eIF-2α under anoxia, so preventing the binding of tRNA to the 40S ribosome subunit, also appeared to be at least partially responsible for the reduction in the rate of protein synthesis. During recovery from anoxia, phosphorylation of eIF-2α was greatly reduced (Larade and Storey, 2002).

5.2.3. Salinity

Oyster hemocytes exposed to elevated salinities of 32–98 PSU and then allowed to recover for 2–4 hours (20–25 PSU) had amino acid incorporation rates that were up to 56% lower than control (20–25 PSU) cells (Tirard et al., 1997). Likewise, exposure to hyposalinity (3.5–4.0 PSU) reduced incorporation by 82%. Gel electrophoresis showed no change in the suite of proteins synthesised during hypersaline exposure, in comparison to control cells, but under hyposaline conditions the pattern of synthesised proteins altered.

5.2.4. pH

In hepatocytes from two Antarctic fish species, *Lepidonotothen kempi* Norman, 1937 and *Pachycara brachycephalum* Pappenheim, 1912, protein synthesis was shown to account for 20% of aerobic metabolism at a pH of 7.90 (Langenbuch and Pörtner, 2003). On exposure to severe intracellular and extracellular acidosis, pH 6.50, protein synthesis rates were depressed by 80%. The authors in turn suggest that observed global increases in CO_2 and proposed deep-sea dumping of anthropogenic CO_2 may, via altering seawater pH, depress liver biosynthesis in fish (Langenbuch and Pörtner, 2003).

5.3. Biotic factors influencing protein metabolism

5.3.1. Reproductive status

Hepatocyte protein synthesis rates in female, sexually mature *Trematomus hansoni* Boulenger, 1902 were approximately twofold higher than in immature fish (Haschemeyer and Mathews, 1983). Elevated protein synthesis rates were probably the result of an increase in the synthesis rate of the egg yolk precursor vitellogenin.

5.3.2. Moulting

El Haj and Houlihan (1987) used incubated, whole carpopodite extensor muscles to measure muscle protein synthesis during the moult cycle in *Carcinus maenas*. Whole legs were bathed in aerated Ringer's solution containing amino acids and radiolabelled phenylalanine; protein synthesis was effectively measured using the flooding-dose method (Section 2.3). A mean extensor-muscle protein synthesis rate of $0.37 \pm 0.06\%$ day^{-1} was determined, which was only one-third of the mean *in vivo* protein synthesis rate.

In vitro muscle protein synthesis rates increased significantly prior to moulting before decreasing post-moult and gradually returning to inter-moult levels.

6. LARVAL PROTEIN METABOLISM

6.1. Fish

Two reviews have concentrated on fish larval protein metabolism: Houlihan *et al.* (1995c) reviewed methods of measuring protein synthesis rates, while Pedersen (1997) reviewed the energetic costs of larval fish growth. Making measurements of *in vivo* protein metabolism in larval marine organisms is generally problematic because of the difficulties of administering radio-labelled amino acids. *In vitro* polyribosome assays have been developed for fish larvae, although only by using large numbers of pooled larvae to ensure measurable levels of incorporated radiolabel (Hansen *et al.*, 1989). A major disadvantage with using polyribosomes from pooled larvae is that individual protein synthesis rates cannot then be measured. Using the polyribosome assay, higher *in vitro* protein synthesis rates were found in whole *Gadus morhua* larvae than when using the same methodology in adult cod white muscle (Hansen *et al.*, 1989). This is perhaps not surprising, as white muscle has one of the lowest reported protein synthesis rates of any tissue, while the larval cod would contain a significant proportion of other tissues, many of which will have higher protein synthesis rates than white muscle. It is, therefore, open to speculation whether cod larval protein synthesis rates are indeed higher than adult rates, or whether observed differences are simply the result of the differing tissues utilised in the study. It is, however, well established that there is a general ontogenetic reduction in protein synthesis rates in animals (reviewed in Houlihan *et al.*, 1995a).

Early data generated by bathing freshwater fish larvae in radiolabelled amino acids suggested very high protein synthesis rates coupled with high protein turnover and low net growth (Fauconneau, 1984; Fauconneau *et al.*, 1986a,b). However, more recent measurements have suggested that fish larval protein metabolism rates are higher than, but comparative with, juvenile and adult rates, and broadly fit the same scaling relationships, while protein synthesis retention efficiencies are also similar to adults (Houlihan *et al.*, 1992, 1993a, 1995c).

Larval protein metabolism has only been measured in a few larval marine species under a very limited range of biotic and abiotic conditions (Table 6). Larval herring fed live copepods at three different ration levels had mean protein synthesis rates of 7.80% day^{-1} (1.31–18.1% day^{-1}) and mean protein

Table 6 Whole-body protein and RNA metabolism in marine larvae and blastulae held under as near natural conditions as possible

Species	Fresh mass (g)	Temperature (°C)	Feeding regime	k_s (% day⁻¹)	k_g (% day⁻¹)	k_d (% day⁻¹)	RNA to protein ratio (µg mg⁻¹)	RNA activity, k_{RNA} (mg mg⁻¹ day⁻¹)	Method used to measure k_s	References
Fish larvae										
Clupea harengus	8		Continuous	7.8	5.5	2.3	41.8	0.7–6.1	FD Bath	Houlihan *et al.* 1995b
Scophthalmus maximus	N/A	18.0 ± 0.5	Twice a day	49.0	35.1	14.0	45.9	12.3	FD Bath	Conceição *et al.*, 1997
	6.5 × 10⁻⁴	15–18	Continuous	35.6 ± 4.5	18.2 ± 4.6	17.4 ± 0.1			¹⁵N FD[a]	Conceição *et al.*, 2001
Mollusc larvae										
Haliotis rufescens	N/A	15 ± 0.1	Non-feeding	14–40					Modified FD Bath	Vavra and Manahan, 1999
Echinoderm larvae										
Sterechinus neumayeri	Blastula	−1.5	Non-feeding	52.8	0	52.8			FD bath	Marsh *et al.*, 2001
	Gastrulation	−1.5	Non-feeding	240	0	240			FD bath	
	Pluteus	−1.5	Non-feeding	<24	0	<24			FD bath	
Strongylocentrotus purpuratus	Blastula	16	Non-feeding	26.4	0	26.4			FD bath	Goustin and Wilt, 1981
	Gastrula	25	Non-feeding	26	0	26			FD bath	Fry and Gross, 1970
Lytechinus pictus	Blastula	19	Non-feeding	21	0	24			FD bath	Berg and Mertes, 1970
Arbacia punctulata	Early gastrula	19	Non-feeding	23	0	23			FD bath	
	Blastula	25	Non-feeding	45.6	0	45.6			FD bath	Fry and Gross, 1970

[a]The authors have used the flooding-dose model without evidence that they have successfully flooded the animal's intracellular free-pools, therefore the data should be treated with caution. FD, flooding-dose.

growth rates of 5.53% day^{-1} (3.34–8.22% day^{-1}) (Houlihan et al., 1995b). Feeding the herring larvae after a period of starvation increased protein synthesis rates by 5- to 11-fold within a few hours. Protein synthesis retention efficiencies were around 50%, similar to those measured in adult species and in strong contrast to the very low efficiencies reported by Fauconneau et al. (1986a,b). Cycloheximide inhibition of protein synthesis showed that protein synthesis accounted for 79% of the larvae's total O_2 consumption.

Conceição et al. (1997) examined the relationship between protein metabolism and diet type in larval turbot Scopthalmus maximus (L.). However, the results need to be interpreted with caution as it appears the intracellular free-pools were not stable during the protein synthesis measurements (see authors' Figure 2). Turbot larvae fed on zooplankton showed higher rates of protein synthesis, growth and degradation than larvae fed artemia, although the latter had a higher PSRE. Protein synthesis, degradation and growth increased from day 11 to day 17 in zooplankton-fed larvae with a resultant decrease in the PSRE.

The effect on protein metabolism of an immunostimulating alginate, which is rich in mannuronic acid, has also been examined in turbot larvae (Conceição et al., 2001). The control larvae were fed a rotifer diet that had been grown on stable isotope-labelled algae. A second group of larvae were fed rotifers that had previously been fed on both stable isotope-labelled algae and mannuronic acid-rich alginate capsules. Protein synthesis rates were estimated using the flooding-dose equations of Garlick et al. (1980), but using stable rather than radioisotopes. However, it appears from the data presented that stable-isotope enrichment of the intracellular free-pool, expressed as the atom percentage (at %, the absolute number of atoms of the stable isotope in 100 atoms of the element), actually decreased during the time course. The time course also only consisted of two time points making statistical analysis impossible. The decrease in enrichment of the intracellular free-pool with time suggests the intracellular free-pools were not stable and the measured protein synthesis rates may be unreliable (Section 2.3). Stable isotope enrichment of the bound-protein pool appears to increase over the time course, but as before, this cannot be tested statistically because only two time points were utilised. The following discussion should be considered bearing in mind the concerns expressed above. In larvae fed on rotifers that had previously been fed the alginate-containing diet, protein synthesis and degradation were elevated, but protein growth was reduced because of increased protein turnover, resulting in a 72% reduction in the PSRE compared to the control group. Food consumption in the alginate-fed larvae was higher than control larvae, which may have counteracted the reduction in growth efficiency. Conceição et al. (2001) suggest elevated rates of protein turnover may be advantageous in allowing the alginate-fed larvae

an increased ability to respond to environmental stress and disease (Skjermo
et al., 1999, cited in Conceição et al., 2001).

6.2. Invertebrates

A large proportion of the non-fish larval protein metabolism measurements
have been made using sea urchin larvae (Berg and Mertes, 1970; Fry and
Gross, 1970; Goustin and Wilt, 1981; Marsh et al., 2001). Protein synthesis
rates in blastula appear high, but rates gradually decrease during development. In *Strongylocentrotus purpuratus* Stimpson, 1857 embryos, there
appears to be considerable variation in the rates of protein synthesis through
development. There are two peaks in the rate of protein synthesis, the first
probably associated with the mobilisation of maternal mRNAs, while the
second, possibly, due to the input of newly synthesised mRNAs (Goustin
and Wilt, 1981). In larvae from the Antarctic sea urchin *Sterechinus neumayeri* Meissner, 1900, maximum rates of protein synthesis appear to occur
at gastrulation (Marsh et al., 2001). Interestingly, reported protein synthesis
rates in larval *Sterechinus neumayeri* appear at least as high, if not higher,
than protein synthesis rates reported in larvae from temperate urchin species, even though the Antarctic species was maintained at $-1.5\,°C$ while the
temperate species were maintained at temperatures ranging between 16 and
$25\,°C$ (Berg and Mertes, 1970; Fry and Gross, 1970; Goustin and Wilt, 1981;
Marsh et al., 2001; Table 6). These results are in contrast to the relationship
seen in data from adults of other polar species where protein synthesis rates
are considerably lower than those seen in temperate species (Figure 5).
If protein synthesis rates in larvae from *Sterechinus neumayeri* are similar
to those seen in temperate species, this would suggest full temperature
compensation of protein synthesis rates has occurred, which does not appear
to be the case in the adult Antarctic species so far examined. Further studies
are required to explain these conflicting results.

In the single molluscan species examined, fractional protein synthesis rates
in the lecithotrophic larvae of *Haliotis rufescens* Swainson, 1822, the abalone,
decreased from 40% day^{-1} at 1-day old to 14% day^{-1} at 7-day old (Vavra and
Manahan, 1999). As the animals were not feeding during this period and
losing mass, the protein synthesis rates quoted equate to protein turnover
(Vavra and Manahan, 1999; Section 1.1) During the course of development
the larvae lost body protein at a linear rate, thus by day 8 they had lost 34% of
their original egg protein mass.

With the paucity of data available for marine larval protein metabolism,
little in the way of conclusions can be drawn. However, it does appear that
improvements in methods for measuring larval protein metabolism have

demonstrated that larval protein synthesis rates are higher, but comparable with juvenile and adult rates, and that PSREs also follow a similar pattern.

7. ENERGETIC COSTS OF PROTEIN SYNTHESIS

It is well established that the energetic costs of protein synthesis account for a fairly significant proportion of an animals' energy budget. Typically, values of between 11% and 42% of total O_2 consumption are quoted (Houlihan et al., 1995a). Mammalian studies have also shown that there is an energetic cost to intracellular protein degradation, but because of the existence of multiple degradation pathways, estimating the overall costs remains difficult (Hershko and Ciechanover, 1982; Ciechanover et al., 1984; Gronostajski et al., 1985; Section 2.5). A few cellular studies in mammals have utilised inhibitors to block some degradation pathways. Estimates indicate that protein degradation costs account for between 8% and 19% of O_2 consumption (Siems et al., 1984; Müller et al., 1986). At present no studies have examined protein degradation costs in marine organisms.

Estimating the energetic costs of protein synthesis has been achieved by using three different methods which differ in exactly what they measure:

1. Use of minimal theoretical costs of protein synthesis and measured absolute rates of protein synthesis to allocate an energetic cost to the mass of protein synthesised. Typically, the minimal costs of protein synthesis are taken as 4ATP (2ATP + 2GTP) equivalents per peptide bond plus another ATP equivalent for transport processes. This results in an estimated cost of 8.3 mmol O_2 per gram of protein synthesised. These calculations presume a mean peptide molecular weight of 110 and 6 mmol of ATP being synthesised per mmol of O_2 (Reeds et al., 1985; Houlihan et al., 1995d). There are many assumptions underlying the calculations and therefore many potential concomitant sources of error (Reeds et al., 1985). This method only accounts for the direct synthesis and transport costs of manufacturing a peptide.

2. The energetic cost of protein synthesis can also be estimated from the regression slope relating O_2 consumption and absolute protein synthesis rate. This method should also include other processes associated with an increase in protein synthesis such as protein degradation, RNA synthesis and so on (Reeds et al., 1985; Houlihan, 1991; Houlihan et al., 1995a,d). It has been pointed out that the correlative method may not provide a reliable cost estimate of protein synthesis because there is no logical reason to suppose that a change in metabolism is solely determined by a change in protein metabolism (Waterlow, 1995).

3. A protein synthesis inhibitor, typically cycloheximide, can be used to block protein synthesis while simultaneously measuring respiration. If the

O_2 consumption of the organism is measured before and after the application of the inhibitor, then the proportion of metabolism utilised for protein synthesis can be estimated. In order to estimate the suppression of protein synthesis by the inhibitor, protein synthesis rates are measured in a control group of animals, or cells, and in the cycloheximide-treated animals. Protein synthesis rates cannot be measured in the same animals pre- and post-cycloheximide administration, as the flooding-dose technique is by necessity terminal. The most important assumption made when using an inhibitor is that only the specific process being inhibited is affected. Although this may be true in the short term, blocking a major cellular process such as protein synthesis will inevitably lead to effects on other cellular processes, in turn, possibly affecting O_2 consumption and hence estimates of protein synthesis costs. Consideration should be given when using inhibitors to ensure concentrations used are minimised and exposure times are as short as is practicable. There appears to be growing evidence that the degree to which metabolism is depressed after the administration of cycloheximide is dose dependent, which in turn suggests non-specific effects (Fuery et al., 1998; Wieser and Krumschnabel, 2001). There are also other potential concerns with the use of inhibitors to measure the energy costs of specific cellular processes (Wieser and Krumschnabel, 2001; reviewed in Rolfe and Brown, 1997). It should be noted that cycloheximide only blocks cytosolic protein synthesis, while mitochondrial protein synthesis is not affected (Alberts et al., 1989).

Costs of protein synthesis have been measured using the above methods in a small number of marine organisms. Minimum theoretical costs of peptide bond formation have been used to estimate the proportion of O_2 consumption used for protein synthesis in only one species, the Antarctic limpet *Nacella concinna* (Fraser et al., 2002a). The proportion of O_2 consumption allocated to protein synthesis appeared to vary little with season (February 34%, July 36%, October 35%, December 40%), even though food consumption was estimated to decrease by tenfold during winter (Fraser et al., 2002a,b). Lyndon et al. (1992) used peptide bond costs derived in chickens to estimate the proportion of metabolism allocated to protein synthesis in *Gadus morhua* (Aoyagi et al., 1988). In post-prandial *G. morhua*, the liver and stomach alone account for 11% of O_2 consumption, while whole-body protein synthesis accounted for 44% of O_2 consumption (Lyndon et al., 1992). Robertson et al. (2001b) used energetic costs of protein synthesis measured in the Antarctic isopod *Glyptonotus antarcticus* Eights, 1853 by another author (147 mmol O_2 per gram of protein) to estimate that 68% of the post-prandial increase in O_2 consumption resulted from protein synthesis (Whiteley et al., 1996).

The relationship between O_2 consumption and the absolute rate of protein synthesis (method 2) has been used to measure the costs of protein synthesis in the shore crab *Carcinus maenas* (Houlihan *et al.*, 1990b), larvae from the Antarctic sea urchin *Sterechinus neumayeri*, the white sea urchin *Lytechinus pictus* Verrill, 1879, the Antarctic starfish *Odontaster validus* Koehler, 1906 (Marsh *et al.*, 2001; Pace and Manahan, 2001; Pace *et al.*, 2004) and adult mussels (*M. edulis*) (Hawkins *et al.*, 1989b). It should be noted that calculations of the costs of protein synthesis using this method should include the cost of related processes and hence be higher than estimates calculated from minimal theoretical costs (8.3 mmol O_2 g^{-1} protein) or the use of inhibitors. In *C. maenas*, the energy cost of protein synthesis was estimated at 39 mmol O_2 g^{-1} (Houlihan *et al.*, 1990b). While the costs of protein synthesis in larvae of the temperate urchin, *L. pictus*, were 12.2 mmol O_2 g^{-1} in fed animals and 13.10 mmol O_2 g^{-1} in starved larvae at a variety of developmental stages (Pace and Manahan, 2001, 2006). The cost of synthesising a unit of protein was also the same in *L. pictus* larvae over a range of protein synthesis rates that varied by two orders of magnitude (Pace and Manahan, 2006). In contrast, other authors have suggested that the costs of protein synthesis per unit of protein are dependent on the rate of synthesis, with a decreasing cost of protein synthesis as rates increase (Pannevis and Houlihan, 1992; Smith and Houlihan, 1995; Pedersen, 1997). In rainbow trout (*Oncorhynchus mykiss*) hepatocytes, Pannevis and Houlihan (1992) hypothesised that the costs of protein synthesis are composed of fixed and variable costs. At low rates of protein synthesis the costs are dominated by fixed costs, while at higher rates the variable costs dominate, resulting in a net decrease in the cost of synthesising a unit of protein. However, it is difficult to see how this could be concluded from the data, as the authors used the inhibitor cyclo-heximide to estimate the costs of protein synthesis, and this method should only measure the costs of the specific peptide bonds, and no other associated costs should be included (Pannevis and Houlihan, 1992).

Using the correlative method, protein synthesis was shown to account for at least 20% of energy expenditure in mussels fed a maintenance ration, with the cost of protein synthesis estimated at 23.5 mmol O_2 g^{-1} and protein turnover at 36.5 mmol O_2 g^{-1} (Hawkins *et al.*, 1989b). In contrast, Marsh *et al.* (2001) investigated the costs of protein turnover in larval *Sterechinus neumayeri* and suggested that the energetic costs of synthesising proteins at polar water temperatures were only 0.92 mmol O_2 g^{-1} protein, which is about 1/25th of the costs reported in other animals. The same authors also reported protein synthesis costs of 4.0 mmol O_2 g^{-1} in larvae from the Antarctic starfish *Odontaster validus*, again much lower than reported for temperate species (Pace *et al.*, 2004). However, a study of protein synthesis costs in an Antarctic and a temperate scallop species have suggested no significant differences in the costs of protein synthesis at temperate or polar water temperatures (Storch and Pörtner, 2003). Reported costs of protein synthesis in endotherms measured using the correlative method appear similar to those reported in ectotherms. In pigs, the cost of protein synthesis has been estimated at 7.4–31.6 mmol O_2 g^{-1} protein and in rats 24–72 mmol O_2 g^{-1} protein (Coyer et al., 1987; Fuller et al., 1987).

The protein synthesis inhibitor cycloheximide has been used to measure protein synthesis costs in several marine species, while a study has utilised pancreatic ribonuclease A to inhibit protein synthesis in a cell-free system derived from pectinids (Storch and Pörtner, 2003). The use of cycloheximide has produced cost estimates of protein synthesis that vary widely. For example, rates of 147.5 mmol O_2 g^{-1} protein in the Antarctic giant isopod *Glyptonotus antarcticus*, 39.5 mmol O_2 g^{-1} protein in the isopod *Idotea resecata* Stimpson, 1857, 98 mmol O_2 g^{-1} protein in herring larvae, 20–24 mmol O_2 g^{-1} protein in the urchin *Lytechinus pictus* and 14 mmol O_2 g^{-1} protein in flounder gills have been reported (Houlihan et al., 1995b; Whiteley et al., 1996; Lyndon and Houlihan, 1998). While *in vitro* cell-free costs of protein synthesis were 4.3 ± 0.7 (±SE) and 5.6 ± 0.6 ATP equivalents in *Adamussium colbecki* (Antarctic scallop) and *Aequipecten opercularis* (temperate scallop), respectively, this equates to 7–9 mmol O_2 g^{-1} protein, very similar to theoretical costs (Storch and Pörtner, 2003). It should be noted that in the latter study, this cost estimate only accounts for the cost of peptide bond synthesis, with any additional amino acid transport costs not included (see method 1). In non-marine ectotherms, costs of protein synthesis of 7.3 mmol O_2 g^{-1} protein have been estimated in liver slices from the cane toad *Bufo marinus* (L.) and 3.9 mmol O_2 g^{-1} protein in the turtle *Chrysemys picta bellii* Gray, 1831 (Land and Hochachka, 1994; Fuery et al., 1998). While in chickens (*Gallus domesticus*) and human cells (HepG2), protein synthesis costs have been estimated at 11 mmol O_2 g^{-1} protein and 8.3 mmol O_2 g^{-1} protein, respectively (Aoyagi et al., 1988; Wieser and Krumschnabel, 2001). It appears that within the constraints of the available data, the cost of synthesising proteins in amphibians, reptiles and mammals appears in the same range as those reported in marine ectotherms. It should be noted, however, that there appears to be an alarmingly large variation in reported costs of protein synthesis.

It appears highly unlikely that the energetic costs of synthesising protein varies between marine organisms (or indeed all animals), to the degree that reported measurements indicate, that is, from 0.92 to 147.5 mmol O_2 g^{-1} protein, a range of greater than two orders of magnitude (Whiteley et al., 1996; Marsh et al., 2001). One would expect costs estimated from the use of inhibitors to be lower than those calculated from correlations of absolute protein synthesis rates and O_2 consumption. In fact, reported costs from studies utilising inhibitors range between 14 and 147.5 mmol O_2 g^{-1} protein, while costs measured using correlation range between 0.92 and

39 mmol O_2 g^{-1}. There are clearly considerable inconsistencies in reported costs of protein synthesis, and carefully validated studies are required in the future to resolve these differences. This is important because a clear understanding of the costs of protein synthesis has wider implications in key areas such as pure and applied animal energetics and, in particular, the cost of production.

8. PROTEIN SYNTHESIS AND RNA

There is generally a linear relationship between the concentration of RNA within an organism, or tissue, and the observed rate of protein synthesis and/or growth in both ectotherms (McMillan and Houlihan, 1988; Houlihan et al., 1989, 1990a,b; Houlihan, 1991; Foster et al., 1992; Lyndon et al., 1992; Mathers et al., 1992b; Carter et al., 1993a; Houlihan et al., 1993b, 1994, 1998; Carter and Houlihan, 2001; Vrede et al., 2004) and endotherms (Millward et al., 1973), although in some studies this relationship may be absent, particularly in starved animals (Lyndon et al., 1992; Sveier et al., 2000). Increasing the concentration and/or activity of the RNA within an organism appears to provide a mechanism to alter the rate of protein synthesis (McMillan and Houlihan, 1989; Carter and Houlihan, 2001). Typically, in studies of protein metabolism, whole-animal or tissue RNA concentrations are expressed as a ratio relative to protein concentration (μg RNA mg^{-1} protein), although in some cases RNA concentrations are expressed relative to the fresh weight of tissue (μg RNA mg^{-1} fresh weight) or DNA (μg RNA μg^{-1} DNA). The benefits and drawbacks of each of these metrics have been discussed by Houlihan et al. (1993b). RNA activity (k_{RNA}, mg protein mg^{-1} RNA day^{-1}), also sometimes termed the RNA translational efficiency, is a measure of the amount of protein synthesised per unit mass of RNA and is calculated using the following equation (Goldspink and Kelly, 1984; Preedy et al., 1988):

$$k_{RNA} = 10 \times \frac{k_s}{\text{RNA to protein ratio}}$$

where k_s is the fractional protein synthesis rate (% day^{-1}) and the RNA to protein ratio is expressed as μg RNA mg^{-1} protein. RNA to protein ratios and k_{RNA} appear to vary between species, between different tissues within an organism and even within specific regions of an organ or tissue, resulting in tissue-specific protein synthesis rates (Houlihan et al., 1988b; McMillan and Houlihan, 1988; Houlihan et al., 1990a,b; Houlihan, 1991; Foster et al., 1992; Hewitt, 1992; Lyndon et al., 1992; Foster et al., 1993b; Martin et al., 1993; Houlihan et al., 1994, 1998; Lyndon and Brechin, 1999; Mente et al., 2003;

Storch *et al.*, 2005; Table 1). There is some evidence that tissues such as white muscle have low k_{RNA} values, while more metabolically active tissues such as the liver, gut and gill have higher values. It has been suggested that tissue-specific differences in protein synthesis are mediated solely via changes in the RNA content, while alterations in the k_{RNA} are simply related to temperature or short-term alterations in nutrition (see below, McCarthy and Houlihan, 1996).

That RNA to protein ratios are closely related to rates of protein synthesis, and in most cases growth, is perhaps not surprising. The total RNA concentration within an organism, at least within mammals, is dominated by rRNA (87% of the total RNA concentration), while mRNA and tRNA only comprise approximately 3% and 10% of the total RNA concentration, respectively (Sugden and Fuller, 1991). It also appears that rRNA accounts for 80–85% of total RNA in diploid fish and sea urchins (Goustin and Wilt, 1981; Leipoldt *et al.*, 1984). The close relationship between protein synthesis/growth and various metrics of RNA concentration has resulted in several authors suggesting that these metrics are prime candidates for proxy indicators of field growth rates, for example in wild caught animals (Houlihan, 1991; Mathers *et al.*, 1992a; Houlihan *et al.*, 1993b; Melzner *et al.*, 2005). However, interpretation of RNA to protein ratios and their relationship to growth in wild animals is complex (Mathers *et al.*, 1992a). It is impossible to know whether the RNA to protein ratio measured at capture is representative of the previous growth history of the organism. In order to try and counteract this problem, Mathers *et al.* (1992a) have suggested the use of multiple rather than single growth predictors to estimate previous growth history, for example RNA to protein ratios, RNA to DNA ratios, citrate synthase or cytochrome oxidase concentrations.

RNA to protein ratios are sensitive to several biotic and abiotic factors, in particular temperature, nutritional status and body size. At reduced temperatures, it appears that tissue RNA to protein ratios are elevated in an effort to counteract a thermally induced reduction in RNA activity (Goolish *et al.*, 1984; Houlihan, 1991; Foster *et al.*, 1992, 1993a,b; McCarthy *et al.*, 1999; Robertson *et al.*, 2001b; Fraser *et al.*, 2002a; Storch *et al.*, 2003; Treberg *et al.*, 2005), although in a few studies no alteration in the RNA to protein ratio with temperature was observed (Watt *et al.*, 1988; Storch *et al.*, 2005; Whiteley and Faulkner, 2005). There is also evidence that at polar water temperatures k_{RNA} may be compensated in some species (Storch *et al.*, 2003). In a large study examining the effects of temperature on protein metabolism in *Anarhichas lupus*, the wolffish, both whole body and white muscle RNA to protein ratios decreased linearly with increasing temperature (McCarthy *et al.*, 1999). The k_{RNA} increased with increasing temperature but the relationship was exponential rather than linear. The observed increase in RNA to protein ratios at reduced temperatures appears to occur at both the intra-specific and inter-specific scales. In a meta-analysis examining the effects of temperature on RNA to protein ratios and k_{RNA} in a range of phyla, there was a significant decrease in whole-animal RNA to protein ratios as temperatures increased (Fraser *et al.*, 2002a; Figure 6; for sources of RNA to protein ratio data used to plot Figure 6 see Table 7). This relationship was still significant ($p < 0.05$) even if data from the rat, *Rattus norvegicus*, were excluded (Figure 6). Conversely, k_{RNA} significantly increased with temperature (McCarthy and Houlihan, 1996; Fraser *et al.*, 2002a; Figure 7; for sources of k_{RNA} data used to plot Figure 7 see Table 8). In the current analysis, the increase in k_{RNA} with temperature was linear in contrast to the non-linear relationship seen in *A. lupus* (McCarthy *et al.*, 1999), the reason for this difference is currently unclear. We therefore speculate that, at both the intra-specific and inter-specific scales, as temperatures decrease, RNA to protein ratios are increased to offset a thermally induced reduction in the k_{RNA} (Fraser *et al.*, 2002a). Whole-animal RNA to protein ratios also appear to follow the same relationship with temperature in endotherms (Figure 6).

Some studies on Antarctic organisms have suggested that k_{RNA} may be cold-compensated and that protein synthesis rates are maintained at low temperatures primarily via this mechanism, rather than elevated tissue

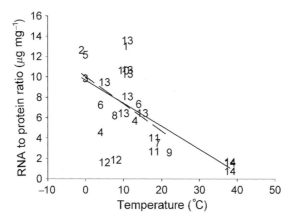

Figure 6 Mass-standardised whole-body RNA to protein ratios plotted against temperature. All values were standardised to a mean body mass of 129 g, the mean body mass of all the animals in the data set, using a scaling coefficient of -0.24. The scaling coefficient was calculated by fitting a least-squares regression model to the natural log-transformed body mass and RNA to protein ratio data for all species. The solid regression line relating temperature and RNA to protein ratios was fitted to all data in the plot using least-squares regression analysis ($y = 9.69 - 0.23x$, $r^2 = 0.42$, $p < 0.001$). The broken regression line was fitted to the full data set but excluding data from *Rattus norvegicus* ($y = 10.1 - 0.27x$, $r^2 = 0.20$, $p < 0.05$). For sources of plotted data see Table 7. Data redrawn from Fraser *et al.* (2002a). Reproduced with permission of the Company of Biologists.

Table 7   Sources of RNA to protein ratio data used to plot Figure 6. Reproduced with the permission of the Company of Biologists

Data point label	Species	Group	Habitat	References
1	*Limanda limanda*	Teleost	T	Houlihan *et al.*, 1994
2	*Nacella concinna*	Gastropod	P	Fraser *et al.*, 2002a
3	*Glyptonotus antarcticus*	Isopod	P	Robertson *et al.*, 2001b
4	*Saduria entomon*	Isopod	P	Robertson *et al.*, 2001a
5	*Glyptonotus antarcticus*	Isopod	P	Whiteley *et al.*, 1996
6	*Idotea resecata*	Isopod	T	Whiteley *et al.*, 1996
7	*Homarus gammarus*	Decapod	T	Mente *et al.*, 2001
8	*Oncorhynchus mykiss*	Teleost	T	McCarthy *et al.*, 1994
9	*Ctenopharyngodon idella*	Teleost	T	Carter *et al.*, 1993b
10	*Gadus morhua*	Teleost	T	Lyndon *et al.*, 1992
11	*Dicentrarchus labrax*	Teleost	T	Langar and Guillaume, 1994
12	*Gadus morhua*	Teleost	T	Raae *et al.*, 1988
13	*Oncorhynchus mykiss*	Teleost	T	Mathers *et al.*, 1993
14	*Rattus norvegicus*	Mammal	L	Goldspink and Kelly, 1984

T, temperate; P, polar; L, terrestrial.

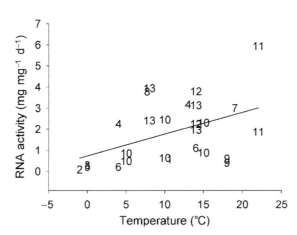

Figure 7   Whole-body RNA activities (k_{RNA}) plotted against temperature. The plotted regression line was fitted by least-squares regression analysis ($y = 0.74 + 0.10x$, $r^2 = 0.20$, $p < 0.05$) to all data within the data set. For literature sources of plotted data see Table 8. Data redrawn from Fraser *et al.* (2002a). Reproduced with the permission of the Company of Biologists.

Table 8   Sources of RNA activity data used to plot Figure 7. Reproduced with the permission of the Company of Biologists

Data point label	Species	Group	Habitat	References
1	*Limanda limanda*	Teleost	T	Houlihan *et al.*, 1994
2	*Nacella concinna*	Gastropod	P	Fraser *et al.*, 2002a
3	*Glyptonotus antarcticus*	Isopod	P	Robertson *et al.*, 2001b
4	*Saduria entomon*	Isopod	P	Robertson *et al.*, 2001a
5	*Glyptonotus antarcticus*	Isopod	P	Whiteley *et al.*, 1996
6	*Idotea resecata*	Isopod	T	Whiteley *et al.*, 1996
7	*Homarus gammarus*	Decapod	T	Mente *et al.*, 2001
8	*Oncorhynchus mykiss*	Teleost	T	McCarthy *et al.*, 1994
9	*Dicentrarchus labrax*	Teleost	T	Langar and Guillaume, 1994
10	*Oncorhynchus mykiss*	Teleost	T	Mathers *et al.*, 1993
11	*Ctenopharyngodon idella*	Teleost	T	Carter *et al.*, 1993b
12	*Salmo salar*	Teleost	T	Houlihan *et al.*, 1995c
13	*Oncorhynchus mykiss*	Teleost	T	Houlihan *et al.*, 1995c

T, temperate; P, polar.

RNA concentrations (Storch *et al.*, 2003, 2005). These workers have concluded that the Antarctic fish species, *Pachycara brachycephalum*, has evolved cold-adapted RNA translational machinery with reduced Arrhenius activation energies, allowing efficient function at low temperatures (Storch *et al.*, 2003, 2005). When the closely related temperate zoarcid species, *Zoarces viviparus* (L.) and *P. brachycephalum*, were both acclimated to 5 °C, both species had similar protein synthesis rates but the former species had considerably higher RNA to protein ratios than the Antarctic species. Therefore, *P. brachycephalum* did not show the same low-temperature elevation of RNA to protein ratios previously seen in temperate species and Antarctic invertebrates (reviewed in Houlihan, 1991; Fraser *et al.*, 2002a, 2004; Storch *et al.*, 2005). Further work is required to resolve whether in other Antarctic fish, which have evolved to live at low temperatures for millions of years, protein synthesis rates are compensated via an increase in RNA concentration or k_{RNA}, or some combination of both. Fraser *et al.* (2002a) suggested that the apparent requirement to maintain elevated tissue concentrations of RNA in Antarctic limpets may further elevate the total costs of manufacturing proteins. However, other authors have suggested no net increase in costs may be incurred if turnover rates are reduced (Storch *et al.*, 2005). At present, it is unknown whether turnover of RNA is reduced at low temperatures.

When temperate marine organisms are exposed to acute change in temperature, any immediate change in the rate of protein synthesis appears to be the result of an alteration in the k_{RNA} rather than the RNA to protein ratio (Pannevis and Houlihan, 1992; Treberg et al., 2005). However, if the animals are exposed to a longer term change in temperature the RNA to protein ratio will alter (Pannevis and Houlihan, 1992). Surprisingly, RNA concentrations appear unaffected by temperature in feeding fish larvae (Mathers et al., 1993). Mathers et al. (1993) suggested RNA to protein ratios in larvae are already very high, considerably higher than in adults, and it may be impossible to further elevate concentrations, even at low temperatures. In turn, this indicates protein synthesis, and hence growth in larvae, may be highly temperature dependent.

What is not currently clear is whether observed elevations in RNA to protein ratios at low temperatures are the result of an increase in all RNA species equally (i.e. rRNA, mRNA and tRNA), or whether concentrations of specific RNA species are preferentially increased. Evidence from Antarctic echinoderm larvae has shown a tenfold elevation in poly(A$^+$) mRNA, in comparison to larvae from temperate species, while total RNA concentrations increased only by threefold (Marsh et al., 2001). Intriguingly, this suggests that at low temperatures, compensation of protein synthesis may not only result via an increase in the 'protein synthesis machinery', rRNA, but also by a disproportionate increase in the availability of the 'protein blueprint', mRNA. The same study also demonstrated thermal compensation of total RNA synthesis rates in Antarctic larvae, resulting in synthesis rates similar to those seen in temperate larvae, while poly(A$^+$) mRNA synthesis rates were much higher than in temperate species. In the Antarctic species, 42% of total RNA synthesis was allocated to mRNA synthesis compared to only 9% in temperate larvae. This indicated that at low temperatures increasing the concentrations of mRNA, or counteracting higher rates of mRNA turnover, may be critical. If the concentrations of mRNA are increased relative to other RNA species at low temperatures, the overall cost of RNA synthesis are also likely to increase, as the turnover rates of mRNA are thought to be considerably higher than tRNA or rRNA (Waterlow and Millward, 1989). A study with carp has showed total tRNA concentrations were elevated in fish at summer water temperatures, but the proportion of aminoacyl-tRNA was reduced (Zuvic et al., 1980). The interpretation was that elevated rates of protein synthesis occurred in carp during summer (Oñate et al., 1987). The same authors also showed carp hepatocyte transcription rates in cells held at the same temperature were higher in summer samples than winter ones (Saez et al., 1982). Further studies are required examining the changes in relative proportions of rRNA, mRNA and tRNA under a variety of physiological conditions, in order to help understand how alterations in the concentration of specific RNA species are involved with modifying rates of protein synthesis.

Increases in RNA concentration metrics with elevated feeding, and decreases with starvation, suggest that the RNA concentration provides an indication of previous feeding history as well as a measure of the capacity of the tissues to synthesise proteins, and thereby grow (Millward *et al.*, 1973; McMillan and Houlihan, 1988; Watt *et al.*, 1988; Houlihan *et al.*, 1990b; Lowery and Somero, 1990; Houlihan, 1991; Arndt *et al.*, 1996; Houlihan *et al.*, 1998; Mente *et al.*, 2002; Melzner *et al.*, 2005). In fact, white muscle RNA concentrations have been shown to correlate with gut fullness in salmon, although after a change in feeding regime at least 4 days were required before a change in the RNA concentration could be measured (Arndt *et al.*, 1996). Broadly speaking, an increase in food consumption appears to drive an increase in RNA concentrations and subsequently an increase in protein synthesis in both ectotherms and endotherms (Millward *et al.*, 1973; Mathers *et al.*, 1992b; Carter *et al.*, 1993a; Houlihan *et al.*, 1993b). During re-feeding after a period of starvation, or after an increase in food availability, initial increases in protein synthesis are mediated via an elevation in the k_{RNA} (Millward *et al.*, 1973). Subsequently, if the ration level remains elevated over a longer period of time, the tissue RNA concentration will gradually increase (McMillan and Houlihan, 1988; Houlihan *et al.*, 1989; McMillan and Houlihan, 1989; Houlihan, 1991; Houlihan *et al.*, 1993b; Mente *et al.*, 2001; Robertson *et al.*, 2001a,b). There is some evidence that for a given tissue, as RNA to protein ratios become elevated after increased feeding, the k_{RNA} also increases, suggesting k_{RNA} is not fixed (McMillan and Houlihan, 1988). However, it seems likely that in non-feeding animals, or those on reduced rations, RNA may not be synthesising protein at the maximum possible rate, and hence measurements of k_{RNA} will not be the maximal rates for the available RNA. As protein synthesis rates increase after feeding, initial increases in protein synthesis will result from an increase in k_{RNA}, as available RNA becomes fully utilised. Subsequently, if feeding continues at an increased rate, RNA concentrations may increase. In cod (*Gadus morhua*), alteration of tissue RNA concentrations to a new level after a change in nutrition has been shown to take 2 weeks (Foster, 1990). However, this time period appears to be species specific and is likely to be dependent on the environmental temperature. For example, in fasted *Carcinus maenas*, initial increases in tissue protein synthesis rates after a meal were the result of elevated k_{RNA}, but by 9 hours, in the majority of tissues, RNA to protein ratios were significantly higher than in fasted animals (Houlihan *et al.*, 1990b). In cod (*G. morhua*), withholding food for 35 days resulted in RNA to protein ratios significantly lower than those of fish fed daily or every other day (Foster *et al.*, 1993b). After the starved fish were re-fed, RNA to protein ratios increased, with the fastest rates of increase seen in the intestine and stomach (Foster *et al.*, 1993b). In contrast, in the same species Lyndon *et al.* (1992) found no significant decrease in RNA to protein ratios after 14 days of fasting, and only the liver showed an elevation of RNA to protein ratios up to 48 hours after re-feeding. It appears likely, therefore, that in the latter study a starvation period of 14 days was too short to cause a reduction in tissue RNA concentrations in cod at the experimental temperature. In contrast, juvenile shrimps, starved for 6 days and then fed, had elevated RNA to protein ratios 2 hours after feeding while by 4 hours k_{RNA} had also increased (Mente *et al.*, 2002). Starvation for 24 hours or 6 days decreased RNA to protein ratios significantly, but the length of starvation had no significant effect.

The effect of starvation on RNA concentrations also appears to be tissue specific. In *Paralabrax nebulifer* Girard, 1854, RNA to DNA ratios decreased by 56% in white muscle but only 21% in red muscle after 23 days of starvation (Lowery and Somero, 1990). In non-feeding larvae, it has been suggested that when newly hatched they contain all the RNA necessary for early protein synthesis, and subsequent alterations in food consumption, as long as they are not as extreme as starvation, do not affect RNA concentrations in the short term (Mathers *et al.*, 1993; Houlihan *et al.*, 1995b).

The protein source used in commercial fish diets also appears to affect the RNA to protein ratio. Fish that are fed temporally discrete meals, rather than fed continuously, appear to have both higher RNA to protein ratios and k_{RNA}s (Langar and Guillaume, 1994). Fish fed a diet with half of the fish-meal protein replaced with 'greaves' meal had elevated RNA to protein ratios, but k_{RNA}s were unchanged (Langar and Guillaume, 1994). In another study by the same authors, RNA to protein ratios were elevated in fish fed a diet containing hydrolysed feather meal in comparison to fish fed on diets containing fish meal, fish protein hydrolysate or 'greaves' meal, while k_{RNA}s were similar for all diets (Langar *et al.*, 1993). The overall proportion of the diet fed as protein can also affect RNA concentrations. Increasing proportions of dietary energy provided as protein to cod resulted in a parallel increase in white muscle RNA concentration (von der Decken and Lied, 1992). However, in the prawn *Penaeus esculentus,* dietary protein contents varying between 30% and 50% had no effect on RNA to protein ratios (Hewitt, 1992). In general, it appears that a reduction in the quality of protein in the diet results in an increase in RNA concentrations, without significant alterations in the k_{RNA}, while an increase in dietary total protein content can in some cases result in an increase in RNA concentrations.

The frequency of feeding and meal size affects RNA to protein ratios but not k_{RNA} in juvenile lobsters (Mente *et al.*, 2001). For a range of feeding regimes (20% daily, 10% daily, 5% daily, 20% every 2 days or 20% every 4 days) based on a percentage of the body masses of the lobsters, RNA to protein ratios were not significantly different, except for the 10% daily feeding regime, which resulted in an elevated RNA to protein ratio. With the small amount of data available, it appears that feeding frequency and meal size can under some circumstances affect RNA to protein ratios but not k_{RNA}.

In non-feeding Atlantic salmon during maturation the picture is complex. Generally in tissues where the protein content was reducing, RNA content behaved in a similar manner (Martin *et al.*, 1993). However, in the liver of female fish, RNA content increased twofold while the protein content remained constant. In the ovary, protein content increased while the RNA content remained constant, but in the stomach and white muscle, protein content decreased by 50% with RNA concentrations decreasing even further. It is possible that the liver was synthesising large amounts of protein, possibly vitellogenin, hence liver RNA to protein ratios were high. However, if the protein was subsequently exported from the liver to the ovary, no net increase in liver protein content would have been measured. The increase in ovary protein content may, therefore, have been the result of protein import rather than *in situ* synthesis. In other tissues, protein and RNA were probably broken down to provide amino acids, nucleotides and energy to fuel maturation (Martin *et al.*, 1993).

Seasonal alterations in RNA to protein ratios have been observed in several species, with RNA concentrations reducing during winter (Mathers *et al.*, 1992a; Fraser *et al.*, 2002a, 2004). It has not been established conclusively whether the observed winter reductions in RNA to protein ratios are the result of a reduction in water temperature or food consumption, the two most likely driving factors. It has been suggested in an Antarctic holothurian, which ceases feeding completely in winter, that the observed 23% decrease in RNA to protein ratios was most likely because of the cessation of feeding rather than the small seasonal change in water temperature ($\sim$3 °C) (Fraser *et al.*, 2004). A winter reduction in RNA synthesis rates has also been reported for carp (Saez *et al.*, 1982). Aquarium acclimatisation to summer or winter water temperatures and light:dark cycles in cod fed *ad libitum* resulted in an increase in white muscle RNA to protein ratios in winter acclimatised fish, presumably to counteract the thermally induced reduction in k_{RNA} (Foster *et al.*, 1993a). In contrast, seasonal field collections of the saithe *Pollachius virens* (L.) have shown winter decreases in white muscle RNA to protein ratios, RNA concentrations and RNA to DNA ratios, suggesting a winter decrease in protein synthesis and growth (Mathers *et al.*, 1992a,b). In the latter study, it was not possible to say whether field caught fish were feeding year round. It is likely that in the study on cod by Foster *et al.* (1993a), *ad libitum* feeding resulted in the maintenance of high RNA to protein ratios even at 'winter' water temperatures, while the saithe were probably either not feeding or had reduced feeding during winter.

There is a significant decrease in whole-animal and tissue RNA to protein ratios as animals grow, with larvae having considerably higher RNA concentrations than adult animals (Mathers *et al.*, 1992a,b, 1993; Houlihan *et al.*, 1993b, 1998). The decrease in RNA concentrations parallels the observed decrease in weight-specific growth and protein synthesis rates

(Mathers *et al.*, 1993). There is also an intra-specific gradient of RNA to protein ratios with larger animals having lower RNA to protein ratios than smaller animals (Fraser *et al.*, 2002a). It is, therefore, important that in studies comparing either intra-specific or inter-specific RNA to protein ratios, if a significant body mass effect exists, data should be mass standardised prior to analysis.

A few studies have examined the effects of various biotic and abiotic factors on RNA metabolism. In dab (*Limanda limanda*), it was found that exposure to sewage sludge resulted in a reduction in the white muscle, but not whole animal, RNA to protein ratios, while k_{RNA}s were unaffected (Houlihan *et al.*, 1994). Hypoxia had no effect on RNA to protein ratios in *Carcinus maenas*, the reported changes in protein synthesis rates were controlled by changes in k_{RNA} alone (Mente *et al.*, 2003). While moulting in the lobster, *Homarus americanus*, does not appear to result in any significant alteration in tissue RNA to protein ratios. Changes in protein synthesis being mediated via alterations in k_{RNA} (El Haj *et al.*, 1996). In a study on the land crab *Gecarcinus lateralis* Freminville, 1835, rRNA concentrations in the muscle, midgut gland and epidermis were elevated during the moult cycle, and for several weeks post-moult closely mimicking rates of protein synthesis (Skinner, 1968).

9. SUMMARY AND FUTURE RESEARCH DIRECTIONS

9.1. Protein metabolism of marine organisms

This summary will concentrate on advances in our understanding of protein metabolism since the previous major reviews of protein metabolism in fish (Carter and Houlihan, 2001) and invertebrates (Houlihan *et al.*, 1995a). Protein metabolism in polar species and comparisons with temperate data will be discussed separately (Section 9.2). It is clear that there have been larger increases in our knowledge of protein metabolism in invertebrates than fish since the last reviews (Tables 1–4). In particular, several studies have recently expanded our knowledge of crustacean and echinoderm protein metabolism.

Transcriptomics and proteomics are now helping us to understand the mechanisms underlying changes in protein metabolism induced by alterations in biotic or abiotic factors. For example, proteomic techniques have been used to show that the elevation of protein degradation during starvation is not simply the result of a general upgrading of all degradation pathways (Martin *et al.*, 2001, 2002; Mommsen, 2004). It appears that degradation in starving ectotherms occurs primarily via the lysosomal pathway, rather than the ATP–ubiquitin proteasome pathway which dominates in feeding animals. This result contrasts with mammals in which the ATP–ubiquitin proteasome pathway is the major pathway for degrading proteins during starvation. Transcriptomics and proteomics will undoubtedly play a major role in helping us understand the changes in gene and protein expression underlying the control of protein metabolism.

The majority of protein metabolism studies have examined species held under laboratory conditions, with little consideration of their ecology. Also, many studies have focussed on a narrow range of species, mostly fish of interest to the aquaculture industry. In recent years, several studies have examined physiological processes such as protein synthesis, RNA concentrations, ammonia and urea excretion, food and O_2 consumption in more diverse organisms, and related these to the environmental conditions in which the animal lives. In particular, these studies have demonstrated large effects of seasonal changes in environmental conditions on physiological parameters including protein metabolism (Fraser et al., 2002a,b, 2004; Whiteley and Faulkner, 2005). This more integrative, eco-physiological approach is vital if we are to gain a better understanding of how animals actually grow within their natural environments.

It is now clear that protein synthesis rates increase in marine invertebrates at both the whole-animal and tissue levels after a meal, in the same manner as seen in fish (Houlihan et al., 1990b, Robertson et al., 2001a,b; Mente et al., 2003). In turn, this suggests a significant proportion of the increase in metabolic rate seen after feeding is associated with the cost of protein synthesis, as previously reported in fish. It also appears that the frequency with which fish and invertebrates are fed can affect both protein synthesis and degradation (Carter and Bransden, 2001; Mente et al., 2001).

The reported costs of synthesising proteins in marine organisms appear within the range of those reported in both non-marine ectotherms and endotherms. However, it is also clear that the range of costs reported in animals vary by an unfeasibly large degree. It is difficult to understand how the costs of synthesising a unit of protein can vary by two orders of magnitudes between species. Data suggest that high concentrations of certain protein synthesis inhibitors used to estimate the cost of protein synthesis could affect the cost estimate. Carefully constructed, rigorous studies are required to ascertain whether protein synthesis costs do indeed vary between species to such a high degree.

The effect of temperature on protein synthesis rates has been well studied. However, the analysis presented in Figure 5 demonstrates that in similarly fed, mass standardised temperate and tropical ectotherms, the variations in whole-body protein synthesis rates between different species are not great. In turn, this suggests that species have the ability to compensate for the effects of water temperature over a considerable temperature range. In contrast, in polar ectotherms, protein synthesis rates are greatly reduced (Section 9.2). It is also clear that alteration of tissue and whole-animal RNA concentrations is a major factor in counteracting the reduction in k_{RNA} at low temperatures, and thereby compensating protein synthesis rates in most species (Figures 6 and 7).

9.2. Protein metabolism of polar marine organisms

This chapter has highlighted some interesting differences between protein metabolism in polar and temperate organisms. It appears that at very low water temperatures ($<5\,^{\circ}C$) protein synthesis rates decrease precipitously (Section 3.2.1, Figure 5). A temperate organism living at $10\,^{\circ}C$ would typically have a whole-body fractional protein synthesis rate of approximately 2.7% day^{-1} (data calculated from Figure 5). In contrast, a polar organism living at $0\,^{\circ}C$ would have a whole-body fractional protein synthesis rate in the order of 0.2% day^{-1}, a 13-fold reduction that occurs over a temperature range of only $10\,^{\circ}C$. It appears that although marine ectotherms are able to compensate for the effects of temperature on protein synthesis at temperate and tropical temperatures, full compensation at polar water temperatures is not possible (Figure 5). However, it should be noted that in Antarctic echinoderm larvae, protein synthesis rates appear similar, or in some cases higher than temperate species, suggesting that in larvae some compensation of protein synthesis may be possible (Section 6.2). The reason for these apparently contradictory results in adult and larval Antarctic ectotherms is currently unclear.

Protein synthesis retention efficiencies are very low in the single Antarctic species examined so far (Table 2). High levels of ubiquitinated proteins have also been reported in Antarctic fishes, together with the permanent expression of HSPs, both findings that suggest elevated rates of protein degradation (Section 3.2.1). It appears that high levels of protein degradation may be a prerequisite of living at polar water temperatures. Low protein synthesis rates coupled with high rates of protein degradation must in turn result in low growth rates. It seems likely that inefficient protein metabolism is a significant factor in at least partially determining the low growth rates observed in polar organisms (Peck, 2002).

The cost of synthesising proteins at polar water temperatures is currently unclear. One study has suggested the costs are extremely low (0.92 and 4.0 mmol O_2 g^{-1} protein in *Sterechinus neumayeri* and *Odontaster validus*, respectively) (Marsh et al., 2001; Pace et al., 2004), another very high (*Glyptonotus antarcticus*, 147 mmol O_2 g^{-1} protein) (Whiteley et al., 1996) and the only other study suggests costs are similar in both temperate and polar scallops [7 and 9 mmol O_2 g^{-1} protein in *Adamussium colbecki* (Antarctic) and *Aequipecten opercularis* (temperate), respectively] (Storch *et al.*, 2003). It is difficult to understand why the cost of synthesising proteins at polar water temperatures should be either higher or lower than in temperate and tropical species, or indeed vary over two orders of magnitude. Further research is required to resolve this issue, as the cost of synthesising proteins will have a major influence on our understanding of energy partitioning in Antarctic organisms.

Thermal compensation of protein synthesis rates in temperate organisms appears to be primarily achieved by alterations in tissue concentrations of RNA to counter a thermally induced reduction in k_{RNA} (Figures 6 and 7). This mechanism is also utilised by some polar species, although as mentioned above, full compensation of protein synthesis rates is not achieved. There is also evidence that compensation of k_{RNA} may have evolved in an Antarctic fish and scallop (Storch *et al.*, 2003, 2005). This is in contrast to both temperate ectotherms and other Antarctic species in which altering RNA concentrations appears to be the primary mechanism for thermal compensation of protein synthesis.

In summary, it appears that Antarctic organisms, with the possible exception of larvae, have very low protein synthesis rates, much lower than would be predicted from simple Q_{10} relationships. There is also growing evidence that degradation rates are elevated at polar water temperatures. The net result of this will be highly constrained, slow growth rates. It is well established that many Antarctic ectotherms are particularly vulnerable to even small increases in water temperature (Pörtner *et al.*, 1999; Peck, 2002; Peck *et al.*, 2002, 2004), and there is now growing evidence that near-shore water temperatures are increasing (Meridith and King, 2005). Antarctic species are characterised by slow growth, increased longevity and deferred maturity and as a result have long generation times, typically of 10–20 years (Peck *et al.*, 2004). The ability of these species to evolve adaptations to predicted sea water temperature increases is therefore poor, making Antarctic marine species some of the most vulnerable to environmental change. It appears that inefficient protein metabolism may play a major part in limiting growth rates in Antarctic species, and in doing so increase their vulnerability to climate change.

9.3. The future

The accurate measurement of protein degradation is still the most significant methodological problem in protein metabolism studies. The most commonly utilised method (the estimation of protein degradation as the difference between protein synthesis and protein growth) can potentially lead to large errors and is far from ideal. However, the presence of multiple protein degradation pathways means that the development of a simple practical technique to accurately measure total protein degradation is unlikely (Sections 1.2 and 2.5). Protein degradation inhibitors that inhibit specific degradation pathways, together with transcriptomics and proteomics, may provide important insights into the relative importance of the various degradation pathways (Lee and Goldberg, 1998; Adams *et al.*, 1999; Martin *et al.*, 2002).

The developing field of transcriptomics now offers the opportunity for workers to investigate the transcriptional and translational processes underpinning overall protein metabolism in an organism. Prior to transcriptomics, studies investigating the expression of specific genes associated with protein metabolism were confined to examining one, or a few genes at most. The development of microarrays has allowed alterations in the expression of hundreds to thousands of genes to be measured simultaneously (Oleksiak and Crawford, 2006). In turn, we are beginning to realise that even apparently small changes in the expression of a single gene can result in significant changes in physiological function. We are also gaining a growing understanding that to fully utilise genomic technology, powerful statistical and computing resources are required to help us understand complex networks of gene expression associated with physiological processes. Careful experimental design is also critical to determine what represents significant changes in gene expression, resulting from experimental manipulations, in the face of individual variation. Understanding the relationships between mRNA transcripts and functional proteins is a continuing challenge, and transcriptomics data do need to be related to protein metabolism data with caution. However, broad scale changes in the expression of related groups of genes may be more reliable in showing underlying patterns in protein synthesis and degradation than changes in the expression of one or two genes.

Proteomics offers the opportunity to examine the expression of specific proteins associated with protein metabolism and should provide results that are more straightforward to interpret than genomic data. Proteomic methods that allow the examination of changes in expression of many proteins simultaneously, such as difference gel electrophoresis (DiGE; Lilley and Friedman, 2004), have not been applied to studies of protein synthesis in marine animals to date. Over the next few years, transcriptomic and proteomic methods, in combination with established methods of measuring protein synthesis, should provide a large step forward in our understanding of the processes controlling protein metabolism, and the influencing abiotic and biotic factors.

ACKNOWLEDGEMENTS

This chapter was written as a contribution from the British Antarctic Surveys, MOA (Marine Organismal Adaptations) and BIOREACH (Biological

Responses to Extreme Antarctic Conditions and Hyper-extremes) projects which were within the core science programmes LATEST (Life at the Edge: Stresses and Thresholds) and BIOFLAME (Biodiversity, Function, Limits and Adaptation from Molecules to Ecosystems), respectively. Funding for these programmes was provided by the Natural Environment Research Council.

REFERENCES

Adams, J., Palombella, V. J., Sausville, E. A., Johnson, J., Destree, A., Lazarus, D. D., Maas, J., Pien, C. S., Prakash, S. and Elliot, P. J. (1999). Protease inhibitors: A novel class of potent and effective antitumour agents. *Cancer Research* **59**, 2615–2622.

Alberts, B., Bray, D., Lewis, J., Raff, M., Roberts, K. and Watson, J. D. (1989). "Molecular Biology of the Cell". Garland Publishing Inc., London.

Ali, K. S., Dorgai, L., Abraham, M. and Hermesz, E. (2003). Tissue- and stressor-specific differential expression of two *hsc70* genes in carp. *Biochemical and Biophysical Research Communications* **307**, 503–509.

Ando, S., Yamazaki, F., Hatano, M. and Zama, K. (1986). Deterioration of chum salmon (*Oncorhynchus keta*) muscle during spawning migration. III. Changes in protein composition and protease activity of juvenile chum salmon muscle upon treatment with sex steroids. *Comparative Biochemistry and Physiology* **83B**, 325–330.

Aoyagi, Y., Tasaki, I., Okumura, J.-I. and Muramatsu, T. (1988). Energy cost of whole-body protein synthesis measured *in vivo* in chicks. *Comparative Biochemistry and Physiology* **91A**, 765–768.

Aragão, C., Conceição, L. E. C., Martins, D., Rønnestad, I., Gomes, E. and Dinis, M.-T. (2004). A balanced dietary amino acid profile improves amino acid retention in post-larval Senegalese sole (*Solea senegalensis*). *Aquaculture* **233**, 293–304.

Arndt, S. K. A., Benfey, T. J. and Cunjak, R. A. (1996). Effect of temporary reductions in feeding on protein synthesis and energy storage of juvenile Atlantic salmon. *Journal of Fish Biology* **49**, 257–276.

Attaix, D., Combaret, L. and Taillandier, D. (1999). Mechanisms and regulation in protein degradation. *In* "Protein Metabolism and Nutrition" Proceedings of the VIIIth International Symposium on Protein Metabolism and Nutrition, EAAP Publication No. 96, Wageningen Pers, Wageningen.

Azumi, K., Fujie, M., Usami, T., Miki, Y. and Satoh, N. (2004). A cDNA microarray technique applied for analysis of global gene expression profiles in tributyltin-exposed ascidians. *Marine Environmental Research* **58**, 543–546.

Bachmair, A., Finley, D. and Varshavsky, A. (1986). *In vivo* half-life of a protein is a function of its amino-terminal residue. *Science* **234**, 179–186.

Barboza, P. S., Farley, S. D. and Robbins, C. T. (1997). Whole-body urea cycling and protein turnover during hyperphagia and dormancy in growing bears (*Ursus americanus* and *U. arctos*). *Canadian Journal of Zoology* **75**, 2129–2136.

Barnard, E. A. (1973). Comparative biochemistry and physiology of digestion. *In* "Comparative Animal Physiology" (C. L. Prosser, ed.), pp. 136–164. W. B. Saunders Company, London.

Barnes, R. S. K., Calow, P., Olive, P. J. W. and Golding, D. W. (1996). "The Invertebrates: A New Synthesis". Blackwall Science Ltd., Oxford.

Barrington, E. J. W. (1957). The alimentary canal and digestion. In "The Physiology of Fishes, Vol. I Metabolism" (M. E. Brown, ed.), pp. 109–162. Academic Press Inc., New York.

Bayne, B. L. and Hawkins, A. J. S. (1997). Protein metabolism, the costs of growth, and genomic heterozygosity: Experiments with the mussel Mytilus galloprovincialis Lmk. Physiological Zoology 70, 391–402.

Beamish, F. W. H. (1974). Apparent specific dynamic action of largemouth bass, Micropterus salmoides. Journal of the Fisheries Research Board of Canada 31, 1763–1769.

Beebe Smith, C. and Sun, Y. (1995). Influence of valine flooding on channelling of valine into tissue pools and on protein synthesis. American Journal of Physiology 268, E735–E744.

Berg, W. E. and Mertes, D. H. (1970). Rates of synthesis and degradation of protein in the sea urchin embryo. Experimental Cell Research 60, 218–224.

Bergner, H., Götz, K.-P., Simon, M. and Rennert, B. (1993). ^{15}N-Markierung von fischen über ^{15}N-isotope im aquarienwasser und der einfluss einer unterschiedlichen eiweissernährung auf die ^{15}N-eliminierung nach der markierungsperiode. Archives of Animal Nutrition 45, 139–154.

Bolliet, V., Cheewasedtham, C., Houlihan, D. F., Gélineau, A. and Boujard, T. (2000). Effect of feeding time on digestibility, growth performance and protein metabolism in the rainbow trout Oncorhynchus mykiss: Interactions with dietary fat levels. Aquatic Living Resources 13, 107–113.

Bone, Q. and Marshall, N. B. (1992). "Biology of Fishes". Chapman & Hall, London.

Boon, P., Watt, P. W., Smith, K. and Visser, G. H. (2001). Day length has a major effect on the response of protein synthesis rates to feeding in growing quail. Journal of Nutrition 131, 268–275.

Botbol, V. and Scornik, O. A. (1991). Measurement of instant rates of protein degradation in the livers of intact mice by the accumulation of bestatin-induced peptides. The Journal of Biological Chemistry 266, 2151–2157.

Boyce, S. J. and Clarke, A. (1997). Effect of body size and ration on specific dynamic action in the Antarctic plunderfish, Harpagifer antarcticus Nybelin 1947. Physiological Zoology 70, 679–690.

Brett, J. R. (1979). Environmental factors and growth. In "Fish Physiology and Growth" (W. S. Hoar, D. J. Randall and J. R. Brett, eds), Vol. 8, pp. 599–667. Academic Press, New York.

Brown, C. R. (1988). "The physiological basis of specific dynamic action: The relationship between protein synthesis and oxygen consumption in Ictalurus punctatus. Ph.D. Thesis, University of Texas at Austin, Texas.

Brown, J. R. and Cameron, J. N. (1986). The relationship between specific dynamic action (SDA) and protein synthesis and oxygen consumption in the channel catfish. American Zoologist 26, 124.

Carter, C. G. and Bransden, M. P. (2001). Relationships between protein-nitrogen flux and feeding regime in greenback flounder, Rhombosolea tapirina (Günther). Comparative Biochemistry and Physiology A 130, 799–807.

Carter, C. G. and Houlihan, D. F. (2001). Protein synthesis. In "Fish Physiology: Nitrogen Excretion, Vol. 20" (P. A. Wright and P. M. Andersen, eds), pp. 31–75. Academic Press, London.

Carter, C. G., Houlihan, D. F. and Owen, S. F. (1998). Protein synthesis, nitrogen excretion and long-term growth of juvenile Pleuronectes flesus. Journal of Fish Biology 53, 272–284.

Carter, C. G., Houlihan, D. F., Brechin, J. and McCarthy, I. D. (1993b). The relationships between protein intake and protein accretion, synthesis, and retention efficiency for individual grass carp, *Ctenopharyngodon idella* (Valenciennes). *Canadian Journal of Zoology* **71**, 392–400.

Carter, C. G., Houlihan, D. F., Buchanan, B. and Mitchell, A. I. (1993a). Protein-nitrogen flux and protein growth efficiency of individual Atlantic salmon (*Salmo salar* L.). *Fish Physiology and Biochemistry* **12**, 305–315.

Carter, C. G., Houlihan, D. F., McCarthy, I. D. and Braefield, A. E. (1992). Variation in the food intake of grass carp, *Ctenopharyngodon idella* (Val.), fed singly or in groups. *Aquatic Living Resources* **5**, 225–228.

Carter, C. G., McCarthy, I. D., Houlihan, D. F., Fonesca, M., Perera, W. M. K. and Sillah, A. B. S. (1995). The application of radiography to the study of fish nutrition. *Journal of Applied Ichthyology* **11**, 231–239.

Carter, C. G., Owen, S. F., He, Z.-Y., Watt, P. W., Scrimgeour, C., Houlihan, D. F. and Rennie, M. J. (1994). Determination of protein synthesis in rainbow trout, *Oncorhynchus mykiss*, using a stable isotope. *Journal of Experimental Biology* **189**, 279–284.

Chapelle, G., Peck, L. S. and Clarke, A. (1994). Effects of feeding and starvation on the metabolic rate of the necrophagous Antarctic amphipod *Waldeckia obesa* (Chevreux, 1905). *Journal of Experimental Marine Biology and Ecology* **183**, 63–76.

Chen, J.-J. and London, I. M. (1995). Regulation of protein synthesis by heme-regulated eIF-2α kinase. *Trends in Biochemical Sciences* **20**, 105–108.

Ciechanover, A., Finley, D. and Varshavsky, A. (1984). The ubiquitin-mediated proteolytic pathway and mechanisms of energy-dependent intracellular protein degradation. *Journal of Cellular Biochemistry* **24**, 27–53.

Clarke, A. (1988). Seasonality in the Antarctic marine environment. *Comparative Biochemistry and Physiology* **90B**, 461–473.

Clarke, A. and Fraser, K. P. P. (2004). Why does metabolism scale with temperature? *Functional Ecology* **18**, 243–251.

Clarke, A. and Peck, L. S. (1991). The physiology of polar marine zoo-plankton. *In* "Proceedings of the Pro-Mare Symposium on Polar Marine Ecology, Trond-heim" (E. Sakshaug, C. C. E. Hopkins and N. A. Øritslan, eds), *Polar Research* **10**, 355–369.

Clarke, A. and Prothero-Thomas, E. (1997). The influence of feeding on oxygen consumption and nitrogen excretion in the Antarctic Nemertean *Parbolasia corrugatus*. *Physiological Zoology* **70**, 639–649.

Conceição, L. E. C., van der Meeren, T., Verreth, J. A. J., Evjen, M. S., Houlihan, D. F. and Fyhn, H. J. (1997). Amino acid metabolism and protein turnover in larval turbot (*Scophthalmus maximus*) fed natural zooplankton or *Artemia*. *Marine Biology* **129**, 255–265.

Conceição, L. E. C., Skjermo, J., Skjak-Braek, G. and Verreth, J. A. J. (2001). Effect of an immunostimulating alginate on protein turnover of turbot (*Scophthalmus maximus* L.) larvae. *Fish Physiology and Biochemistry* **24**, 207–212.

Cowey, C. B. (1992). Nutrition: Estimating requirements of rainbow trout. *Aquaculture* **100**, 177–189.

Coyer, P. A., River, J. P. and Millward, D. J. (1987). The effects of dietary protein and energy restriction on heat production and growth costs in the young rat. *British Journal of Nutrition* **58**, 73–85.

Crawford, R. E. (1979). Effect of starvation and experimental feeding on the proximate composition and calorific content of an Antarctic teleost, *Notothenia coriiceps neglecta*. *Comparative Biochemistry and Physiology* **62A**, 321–326.

Cushing, D. H. (1975). "Marine Ecology and Fisheries". Cambridge University Press, Cambridge.

Danicke, S., Nieto, R., Lobley, G. E., Fuller, M. F., Brown, D. S., Milne, E., Calder, A. G., Chen, S., Grant, I. and Bottcher, W. (1999). Responses to the absorptive phase in muscle and liver protein synthesis rates of growing rats. *Archives of Animal Nutrition–Archiv fur Tierernahrung* **51**, 41–52.

Davis, T. A. and Reeds, P. J. (2001). Of flux and flooding: The advantages and problems of different isotopic methods for quantifying protein turnover *in vivo*: II. Methods based on the incorporation of tracer. *Current Opinion in Clinical Nutrition and Metabolic Care* **4**, 51–56.

Davis, T. A., Fiorotto, M. L., Nguyen, H. V. and Burrin, D. G. (1999). Aminoacyl-tRNA and tissue free amino acid pools are equilibrated after a flooding-dose of phenylalanine. *American Journal of Physiology* **277**, E103–E109.

de la Higuera, M., Garzón, A., Hidalgo, M. C., Peragón, J., Cardenete, G. and Lupiáñez, J. A. (1998). Influence of temperature and dietary-protein supplementation either with free or coated lysine on the fractional protein-turnover rates in the white muscle of carp. *Fish Physiology and Biochemistry* **18**, 85–95.

de la Higuera, M., Akharbah, H., Hidalgo, M. C., Peragón, J., Lupiáñez, J. A. and García-Gallego, M. (1999). Liver and white muscle protein turnover rates in the European eel (*Anguilla anguilla*): Effects of dietary protein quality. *Aquaculture* **179**, 203–216.

Dice, J. F. (1987). Molecular determinants of protein half-lives in eukaryotic cells. *FASEB Journal* **1**, 349–357.

Dobly, A., Martin, S. A. M., Blaney, S. C. and Houlihan, D. F. (2004). Protein growth rate in rainbow trout (*Oncorhynchus mykiss*) is negatively correlated to liver 20S proteasome activity. *Comparative Biochemistry and Physiology A* **137**, 75–85.

El Haj, A. J. and Houlihan, D. F. (1987). *In vitro* and *in vivo* protein synthesis rates in a crustacean muscle during the moult cycle. *Journal of Experimental Biology* **127**, 413–426.

El Haj, A. J., Clarke, S. R., Harrison, P. and Chang, E. S. (1996). *In vivo* muscle protein synthesis rates in the American lobster *Homarus americanus* during the moult cycle and in response to 20-hydroxyecdysone. *The Journal of Experimental Biology* **199**, 579–585.

El Haj, A. J., Govind, C. K. and Houlihan, D. F. (1984). Growth of lobster leg muscle fibres over intermoult and moult. *Journal of Crustacean Biology* **4**, 536–545.

Fauconneau, B. (1984). The measurement of whole body protein synthesis in larval and juvenile carp (*Cyprinus carpio*). *Comparative Biochemistry and Physiology* **78B**, 845–850.

Fauconneau, B. (1985). Protein synthesis and deposition in fish. *In* "Nutrition and Feeding in Fish" (C. B. Cowey, A. M. Mackie and J. G. Bell, eds), pp. 17–46. Academic Press, London.

Fauconneau, B., Aguirre, P. and Bergot, P. (1986a). Protein synthesis in early life of coregonids. Influence of temperature and feeding. *Archiv für Hydrobiologie–Beiheft* **22**, 171–188.

Fauconneau, B., Aguirre, P., Dabrowski, K. and Kaushik, S. J. (1986b). Rearing of sturgeon (*Acipenser baeri* Brandt) larvae 2. Protein metabolism: Influence of fasting and diet quality. *Aquaculture* **51**, 117–131.

Fauconneau, B., Gray, C. and Houlihan, D. F. (1995). Assessment of individual protein turnover in three muscle types of rainbow trout. *Comparative Biochemistry and Physiology* **111B**, 45–51.

Feder, M. E. and Walser, J.-C. (2005). The biological limitations of transcriptomics in elucidating stress and stress responses. *Journal of Evolutionary Biology* **18**, 901–910.

Foster, A. R. (1990). Growth and protein turnover in fish. Ph.D. Thesis, University of Aberdeen, Aberdeen.

Foster, A. R., Hall, S. J. and Houlihan, D. F. (1993a). The effects of seasonal acclimatization on correlates of growth rate in juvenile cod, *Gadus morhua*. *Journal of Fish Biology* **42**, 461–464.

Foster, A. R., Houlihan, D. F. and Hall, S. J. (1993b). Effects of nutritional regime on correlates of growth rate in juvenile Atlantic cod (*Gadus morhua*): Comparison of morphological and biochemical measurements. *Canadian Journal of Fisheries and Aquatic Science* **50**, 502–512.

Foster, A. R., Houlihan, D. F., Hall, S. J. and Burren, L. J. (1992). The effects of temperature acclimation on protein synthesis rates and nucleic acid content of juvenile cod (*Gadus morhua* L.). *Canadian Journal of Zoology* **70**, 2095–2102.

Fraser, K. P. P., Lyndon, A. R. and Houlihan, D. F. (1998). Protein synthesis and growth in juvenile Atlantic halibut, *Hippoglossus hippoglossus* (L.): Application of ^{15}N stable isotope tracer. *Aquaculture Research* **29**, 289–298.

Fraser, K. P. P., Houlihan, D. F., Lutz, P. L., Leone-Kabler, S., Manuel, L. and Brechin, J. G. (2001). Complete suppression of protein synthesis during anoxia with no post-anoxia protein synthesis debt in the red-eared slider turtle *Trachemys scripta elegans*. *Journal of Experimental Biology* **204**, 4353–4360.

Fraser, K. P. P., Clarke, A. and Peck, L. S. (2002a). Low-temperature protein metabolism: Seasonal changes in protein synthesis and RNA dynamics in the Antarctic limpet *Nacella concinna* Strebel 1908. *Journal of Experimental Biology* **205**, 3077–3086.

Fraser, K. P. P., Clarke, A. and Peck, L. S. (2002b). Feast and famine in Antarctica: Seasonal physiology in the limpet *Nacella concinna*. *Marine Ecology Progress Series* **242**, 169–177.

Fraser, K. P. P., Peck, L. S. and Clarke, A. (2004). Protein synthesis, RNA concentrations, nitrogen excretion and metabolism vary seasonally in the Antarctic holothurian *Heterocucumis steineni* (Ludwig 1898). *Physiological and Biochemical Zoology* **77**, 556–569.

Fry, B. J. and Gross, P. R. (1970). Patterns and rates of protein synthesis in sea urchin embryos. II. The calculation of absolute rates. *Developmental Biology* **21**, 125–146.

Fuery, C. J., Withers, P. C. and Guppy, M. (1998). Protein synthesis in the liver of *Bufo marinus*: Cost and contribution to oxygen consumption. *Comparative and Biochemical Physiology* **119A**, 459–476.

Fuller, M. F., Reeds, P. J., Cadenhead, A., Sève, B. and Preston, T. (1987). Effect of the amount and quality of dietary protein on nitrogen metabolism and protein turnover of pigs. *British Journal of Nutrition* **58**, 287–300.

Gabriel, W. (2005). How stress selects for reversible phenotypic plasticity. *Journal of Evolutionary Biology* **18**, 873–883.

Garlick, P. J., McNurlan, M. A. and Preedy, V. R. (1980). A rapid and convenient technique for measuring the rate of protein synthesis in tissues by injection of [^{3}H] phenylalanine. *Biochemical Journal* **192**, 719–723.

Garlick, P. J., McNurlan, M. A., Essen, P. and Wernerman, J. (1994). Measurement of tissue protein synthesis rates *in vivo*: A critical analysis of contrasting methods. *American Journal of Physiology* **266**, E287–E297.

Garlick, P. J., Millward, D. J. and James, W. P. T. (1973). The diurnal response of muscle and liver protein synthesis *in vivo* in meal-fed rats. *Biochemical Journal* **136**, 935–945.

Gasch, A. P., Spellman, P. T., Kao, C. M., Carmel-Harel, O., Eisen, M. B., Storz, G., Botstein, D. and Brown, P. O. (2000). Genomic expression programs in the response of yeast cells to environmental changes. *Molecular Biology of the Cell* **11**, 4241–4257.

Gibson, G. and Weir, B. (2005). The quantitative genetics of transcription. *Trends in Genetics* **21**(11), 616–623.

Gille, S. T. (2002). Warming of the Southern Ocean since the 1950s. *Science* **295**, 1275–1277.

Glass, G. B. J. (1953). Gastric mucin and its constituents: Physico-chemical characteristics, cellular origin, and physiological significance. *Gastroenterology* **23**, 636–658.

Goldspink, D. F. and Kelly, F. J. (1984). Protein turnover and growth in the whole body, liver and kidney of the rat from foetus to senility. *Biochemical Journal* **217**, 507–516.

Gómez-Requeni, P., Mingarro, M., Calduch-Giner, J. A., Médale, F., Martin, S. A. M. and Houlihan, D. F. (2004). Protein growth performance, amino acid utilisation and somatotrophic axis responsiveness to fish meal replacement by plant protein sources in gilthead sea bream (*Sparus aurata*). *Aquaculture* **232**, 493–510.

Goolish, E. M., Barron, M. G. and Adelman, I. R. (1984). Thermoacclimatory response of nucleic acid and protein content of carp muscle tissue: Influence of growth rate and relationship to glycine uptake by scales. *Canadian Journal of Zoology* **62**, 2164–2170.

Goustin, A. S. and Wilt, F. H. (1981). Protein synthesis, polyribosomes, and peptide elongation in early development of *Strongylocentrotus purpuratus*. *Developmental Biology* **82**, 32–40.

Gracey, A. Y., Fraser, E. J., Weizhong, L., Yongxiang, F., Taylor, R. R., Rogers, J., Brass, A. and Cossins, A. R. (2004). Coping with cold: An integrative, multitissue analysis of the transcriptome of a poikilothermic vertebrate. *Proceedings of the National Academy of Science of the United States of America* **101**, 16970–16975.

Gracey, A. Y., Troll, J. V. and Somero, G. N. (2001). Hypoxia-induced gene expression profiling in the euryoxic fish *Gillicthys mirabilis*. *Proceedings of the National Academy of Sciences of the United States of America* **98**, 1993–1998.

Gronostajski, R. M., Pardee, A. B. and Goldberg, A. L. (1985). The ATP dependence of the degradation of short- and long-lived proteins in growing fibroblasts. *The Journal of Biological Chemistry* **260**, 3344–3349.

Han, K. K. and Martinage, A. (1992). Post-translational chemical modification of proteins. *International Journal of Biochemistry* **24**, 19–28.

Hansen, P. E., Lied, E. and Børresen, T. (1989). Estimation of protein synthesis in fish larvae using an *in vitro* polyribosome assay. *Aquaculture* **79**, 85–89.

Haschemeyer, A. E. V. (1968). Compensation of liver protein synthesis in temperature acclimated toadfish, *Opsanus tau*. *Biological Bulletin* **135**, 130–140.

Haschemeyer, A. E. V. (1973). Control of protein synthesis in the acclimation of fish to environmental changes. *In* "Responses of Fish to Environmental Changes" (W. Chavin, ed.), pp. 3–31. Charles C. Thomas, Springfield, Illinois.

Haschemeyer, A. E. V. (1978). Protein metabolism and its role in temperature acclimation. *Biochemical and Biophysical Perspectives in Marine Biology* **4**, 29–84.

Haschemeyer, A. E. V. (1983). A comparative study of protein synthesis in Nototheniids and icefish at Palmer Station, Antarctica. *Comparative Biochemistry and Physiology* **76B**, 541–543.

Haschemeyer, A. E. V. and Mathews, R. W. (1980). Antifreeze glycoprotein synthesis in the Antarctic fish *Trematomus hansoni* by constant infusion *in vivo*. *Physiological Zoology* **53**, 383–393.

Haschemeyer, A. E. V. and Mathews, R. W. (1982). Effects of temperature extremes on protein synthesis in liver of toadfish, *Opsanus tau*, *in vivo*. *Biological Bulletin* **162**, 18–27.

Haschemeyer, A. E. V. and Mathews, R. W. (1983). Temperature dependency of protein synthesis in isolated hepatocytes of Antarctic fish. *Physiological Zoology* **56**, 78–87.

Haschemeyer, A. E. V. and Smith, M. A. K. (1979). Protein synthesis in liver, muscle and gill of mullet (*Mugil cephalus* L.) *in vivo*. *Biological Bulletin* **156**, 93–102.

Haschemeyer, A. E. V. and Williams, R. C. (1982). Temperature dependency of cell-free protein synthetic systems from Antarctic fish. *Marine Biology Letters* **3**, 81–88.

Haschemeyer, A. E. V., Persell, R. and Smith, M. A. K. (1979). Effect of temperature on protein synthesis in fish of the Galapagos and Perlas Islands. *Comparative and Biochemical Physiology* **64B**, 91–95.

Hawkins, A. J. S. (1985). Relationships between the synthesis and breakdown of protein, dietary absorption and turnovers of nitrogen and carbon in the blue mussel, *Mytilus edulis* L. *Oecologia* **66**, 42–49.

Hawkins, A. J. S. (1991). Protein turnover: A functional appraisal. *Functional Ecology* **5**, 222–233.

Hawkins, A. J. S. (1996). Temperature adaptation and genetic polymorphism in aquatic animals. *In* "Animals and Temperature: Phenotypic and Evolutionary Adaptation" (I. A. Johnston and A. F. Bennett, eds). Vol. 59, pp. 103–125. Society for Experimental Biology. Cambridge University Press, Cambridge.

Hawkins, A. J. S. and Bayne, B. L. (1991). Nutrition of marine mussels: Factors influencing the relative utilisations of protein and energy. *Aquaculture* **94**, 177–196.

Hawkins, A. J. S. and Day, A. J. (1996). The metabolic basis of genetic differences in growth efficiency among marine animals. *Journal of Experimental Marine Biology and Ecology* **203**, 93–115.

Hawkins, A. J. S. and Day, A. J. (1999). Metabolic interrelations underlying the physiological and evolutionary advantages of genetic diversity. *American Zoologist* **39**, 401–411.

Hawkins, A. J. S., Bayne, B. L. and Day, A. J. (1986). Protein turnover, physiological energetics and heterozygosity in the blue mussel, *Mytilus edulis*: The basis of variable age-specific growth. *Proceedings of the Royal Society of London B* **229**, 161–176.

Hawkins, A. J. S., Rusin, J., Bayne, B. L. and Day, A. J. (1989a). The metabolic/physiological basis of genotype-dependent mortality during copper exposure in *Mytilus edulis*. *Marine Environmental Research* **28**, 253–257.

Hawkins, A. J. S., Widdows, J. and Bayne, B. L. (1989b). The relevance of whole-body protein metabolism to measured costs of maintenance and growth in *Mytilus edulis*. *Physiological Zoology* **62**, 745–763.

Hazel, J. R. and Prosser, C. L. (1974). Molecular mechanisms of temperature compensation in poikilotherms. *Physiological Reviews* **54**, 620–677.

Hershey, J. W. B. (1989). Protein phosphorylation controls tranlation rates. *Journal of Biological Chemistry* **264**, 20823–20826.

Hershey, J. W. B. (1991). Translational control in mammalian cells. *Annual Review of Biochemistry* **60**, 717–755.

Hershko, A. and Ciechanover, A. (1982). Mechanisms of intracellular protein breakdown. *Annual Review of Biochemistry* **51**, 335–364.

Hershko, A. and Ciechanover, A. (1998). The ubiquitin system. *Annual Review of Biochemistry* **67**, 425–479.

Hewitt, D. R. (1992). Response of protein turnover in the brown tiger prawn *Penaeus esculentus* to variation in dietary protein content. *Comparative Biochemistry and Physiology* **103A**, 183–187.

Hofmann, G. E. (2005). Patterns of gene expression in ectothermic marine organisms on small to large biogeographic scales. *Integrative and Comparative Biology* **45**, 247–255.

Hofmann, G. E., Buckley, B. A., Place, S. P. and Zippay, M. L. (2002). Molecular chaperones in ectothermic marine animals: Biochemical function and gene expression. *Integrative and Comparative Biology* **42**, 808–814.

Hogstrand, C., Balesaria, S. and Glover, C. N. (2002). Application of genomics and proteomics for study of the integrated response to zinc exposure in a non-model fish species, the rainbow trout. *Comparative Biochemistry and Physiology Part B* **133**, 523–535.

Houghton, J. T., Ding, Y., Griggs, D. J., Noguer, M., van der Linden, P. J., Dai, X., Maskell, K. and Johnson, C. A. (2001). Climate change 2001: The scientific basis. Contribution of working group I to the third assessment report of the Intergovernmental Panel on Climate Change. Cambridge University Press, Cambridge.

Houlihan, D. F. (1991). Protein turnover in ectotherms and its relationships to energetics. *Advances in Comparative and Environmental Physiology* **7**, 1–43.

Houlihan, D. F. and El Haj, A. J. (1985). Muscle growth. *In* "Crustacean Growth" (A. Wenner, ed.), pp. 15–30. AA Balkema Press, Netherlands.

Houlihan, D. F. and Laurent, P. (1987). Effects of exercise training on the performance, growth, and protein turnover of the rainbow trout (*Salmo gairdneri*). *Canadian Journal of Fisheries and Aquatic Science* **44**, 1614–1621.

Houlihan, D. F., McMillan, D. N. and Laurent, P. (1986). Growth rates, protein synthesis, and protein degradation rates in rainbow trout: Effects of body size. *Physiological Zoology* **59**, 482–493.

Houlihan, D. F., Hall, S. J., Gray, C. and Noble, B. S. (1988a). Growth rates and protein turnover in Atlantic cod, *Gadus morhua*. *Canadian Journal of Fisheries and Aquatic Science* **45**, 951–964.

Houlihan, D. F., Agnisola, C., Lyndon, A. R., Gray, C. and Hamilton, N. M. (1988b). Protein synthesis in a fish heart: Responses to increased power output. *Journal of Experimental Biology* **137**, 565–587.

Houlihan, D. F., Hall, S. J. and Gray, C. (1989). Effects of ration on protein turnover in cod. *Aquaculture* **79**, 103–110.

Houlihan, D. F., McMillan, D. N., Agnisola, C., Genoino, I. T. and Foti, L. (1990a). Protein synthesis and growth in *Octopus vulgaris*. *Marine Biology* **106**, 251–259.

Houlihan, D. F., Waring, C. P., Mathers, E. and Gray, C. (1990b). Protein synthesis and oxygen consumption of the shore crab *Carcinus maenas* after a meal. *Physiological Zoology* **63**, 735–756.

Houlihan, D. F., Wieser, W., Foster, A. R. and Brechin, J. (1992). *In vivo* protein synthesis rates in larval nase (*Chondrostoma nasus* L.). *Canadian Journal of Zoology* **70**, 2436–2440.

Houlihan, D. F., Pannevis, M. C. and Heba, H. (1993a). Protein synthesis in juvenile tilapia, *Oreochromis mossambicus*. *Journal of the World Aquaculture Society* **24**, 145–151.

Houlihan, D. F., Mathers, E. M. and Foster, A. (1993b). Biochemical correlates of growth rates in fish. *In* "Fish Ecophysiology" (J. C. Rankin and F. B. Jensen, eds), pp. 45–71. Chapman & Hall, London.

Houlihan, D. F., Costello, M. J., Secombes, C. J., Stagg, R. and Brechin, J. (1994). Effects of sewage sludge exposure on growth feeding and protein synthesis of dab (*Limanda limanda*). *Marine Environmental Research* **37**, 331–353.

Houlihan, D. F., Carter, C. G. and McCarthy, I. D. (1995a). Protein turnover in animals. *In* "Nitrogen Metabolism and Excretion" (P. J. Walsh and P. Wright, eds), pp. 1–32. CRC Press, Boca Raton.

Houlihan, D. F., Pedersen, B. H., Steffensen, J. F. and Brechin, J. (1995b). Protein synthesis, growth and energetics in larval herring (*Clupea harengus*) at different feeding regimes. *Fish Physiology and Biochemistry* **14**, 195–208.

Houlihan, D. F., McCarthy, I. D., Carter, C. G. and Marttin, F. (1995c). Protein turnover and amino acid flux in fish larvae. *ICES Marine Science Symposium* **201**, 87–99.

Houlihan, D. F., Carter, C. G. and McCarthy, I. D. (1995d). Protein synthesis in fish. *In* "Biochemistry and Molecular Biology of Fishes" (P. Hochachka and T. Mommsen, eds), Vol. 4, pp. 191–220. Elsevier Science, Amsterdam.

Houlihan, D. F., Kelly, K. and Boyle, P. R. (1998). Correlates of growth and feeding in laboratory-maintained *Eledone cirrhosa* (Cephalopoda: Octopoda). *Journal of the Marine Biological Association of the United Kingdom* **78**, 919–932.

Jackim, E. and La Roche, G. (1973). Protein synthesis in *Fundulus heteroclitus* muscle. *Comparative and Biochemical Physiology* **44A**, 851–866.

Jahoor, F., Zhang, X.-J., Baba, H., Sakurai, Y. and Wolfe, R. R. (1992). Comparison of constant infusion and flooding dose techniques to measure muscle protein synthesis rate in dogs. *Journal of Nutrition* **122**, 878–887.

Jin, W., Riley, R. M., Wolfinger, R. D., White, K. P., Passador-Gurgel, G. and Gibson, G. (2001). The contribution of sex, genotype and age to transcriptional variance in *Drosophila melanogaster*. *Nature Genetics* **29**, 389–395.

Jobling, M. (1985). Growth. *In* "Fish Energetics: New Perspectives" (P. Tytler and P. Calow, eds), pp. 213–230. Croom Helm, London.

Jobling, M. (1995). The influence of environmental temperature on growth and conversion efficiency in fish. *ICES CM* 1–26.

Jobling, M. (2002). Environmental factors and rates of development and growth. *In* "Handbook of Fish Biology and Fisheries" (P. J. Hart and J. Reynolds, eds), Vol. 1, pp. 97–122. Blackwell Science, Oxford.

King, J. C. and Harangozo, S. A. (1998). Climate change in the western Antarctic Peninsula since 1945: Observations and possible causes. *Annals of Glaciology* **27**, 571–575.

Knight, J. C. (2004). Allele-specific gene expression uncovered. *Trends in Genetics* **20** (3), 113–116.

Krasnov, A., Koskinen, H., Pehkonen, P., Rexroad, C. E., III, Afanasyev, S. and Mölsä, H. (2005). Gene expression in the brain and kidney of rainbow trout in response to handling stress. *BMC Genomics* **6**, 3.

Krawielitzki, K. and Schadereit, R. (1992). Estimation of protein synthesis rates using the flooding method and [^{15}N] lysine. *Isotopenpraxis* **28**, 8–12.

Kreeger, D. A., Hawkins, A. J. S., Bayne, B. L. and Lowe, D. M. (1995). Seasonal variation in the relative utilisation of dietary protein for energy and biosynthesis by the mussel *Mytilus edulis*. *Marine Ecology Progress Series* **126**, 177–184.

Land, S. C. and Hochachka, P. W. (1994). Protein turnover during metabolic arrest in turtle hepatocytes: Role and energy-dependence of proteolysis. *American Journal of Physiology* **266**, C1028–C1036.

Langar, H. and Guillaume, J. (1994). Effect of feeding pattern and dietary protein source on protein synthesis in European sea bass (*Dicentrarchus labrax*). *Comparative Biochemistry and Physiology* **108A**, 461–466.

Langar, H., Guillaume, J., Metailler, R. and Fauconneau, B. (1993). Augumentation of protein synthesis and degradation by poor amino acid balance in European sea bass (*Dicentrarchus labrax*). *Journal of Nutrition* **123**, 1754–1761.

Langenbuch, M. and Pörtner, H. O. (2003). Energy budget of hepatocytes from Antarctic fish (*Pachycara brachycephalum* and *Lepidonotothen kempi*) as a function of ambient CO_2: pH-dependent limitations of cellular protein biosynthesis? *Journal of Experimental Biology* **206**, 3895–3903.

Larade, K. and Storey, K. B. (2002). Reversible suppression of protein synthesis in concert with polysome disaggregation during anoxia exposure in *Littorina littorea*. *Molecular and Cellular Biochemistry* **232**, 121–127.

Laurence, G. C. (1975). Laboratory growth and metabolism of the winter flounder *Pseudopleuronectes americanus* from hatching through metamorphosis at three temperatures. *Marine Biology* **32**, 223–229.

Lee, D. H. and Goldberg, A. L. (1998). Proteasome inhibitors: Valuable new tools for cell biologists. *Trends in Cell Biology* **8**, 397–403.

Leipoldt, M., Kellner, H. G. and Stark, S. (1984). Comparative analysis of ribosomal RNA in various fish and other vertebrate species: Hidden breaks and ribosomal function in phylogenetically tetraploid species of cyprinidae. *Comparative Biochemistry and Physiology* **77B**, 769–777.

Levitus, S., Antonov, J. I., Boyer, T. P. and Stephens, C. (2000). Warming of the world ocean. *Science* **287**, 2225–2229.

Lied, E., Lie, O. and Lambertsen, G. (1985). Nutritional evaluation in fish by measurement of *in vitro* protein synthesis in white trunk muscle tissue. *In* "Nutrition and Feeding in fish" (C. B. Cowey, A. M. Mackie and J. G. Bell, eds), pp. 169–176. Academic Press, London.

Lilley, K. S. and Friedman, D. B. (2004). All about DIGE: Quantification technology for differential-display 2D-gel proteomics. *Expert Review of Proteomics* **1**, 401–409.

Liu, Z. and Barrett, E. J. (2002). Human protein metabolism: Its measurement and regulation. *American Journal of Physiology* **263**, E1105–E1112.

Lobley, G. E., Milne, V., Lovie, J. M., Reeds, P. J. and Pennie, K. (1980). Whole body and tissue protein synthesis in cattle. *British Journal of Nutrition* **43**, 491–502.

London, I. M., Levin, D. H., Matts, R. L., Thomas, N. S. B. and Petryshyn, R. (1987). Regulation of protein synthesis. *In* "The Enzymes" (P. D. Boyer and E. G. Krebs, eds), Vol. XVIII, pp. 259–380. Academic Press, New York.

Loughna, P. T. and Goldspink, G. (1985). Muscle protein synthesis rates during temperature acclimation in a eurythermal (*Cyprinus carpio*) and a stenothermal (*Salmo gairdneri*) species of teleost. *Journal of Experimental Biology* **118**, 267–276.

Love, R. M. (1980). "The Chemical Biology of Fishes", Vol. 2. Academic Press, London.

Lowery, M. S. and Somero, G. N. (1990). Starvation effects on protein synthesis in red and white muscle of the barred sand bass, *Paralabrax nebulifer*. *Physiological Zoology* **63**, 630–648.

Lyndon, A. R. and Brechin, J. G. (1999). Evidence of partitioning of physiological functions between holobranchs: Protein synthesis rates in flounder gills. *Journal of Fish Biology* **54**, 1326–1328.

Lyndon, A. R. and Houlihan, D. F. (1998). Gill protein turnover: Costs of adaptation. *Comparative and Biochemical Physiology* **119A**, 27–34.

Lyndon, A. R., Houlihan, D. F. and Hall, S. J. (1992). The effect of short-term fasting and a single meal on protein synthesis and oxygen consumption in cod, *Gadus morhua*. *Journal of Comparative Physiology* **B162**, 209–215.

Marsh, A. G., Maxson, R. E. and Manahan, D. T. (2001). High macromolecular synthesis with low metabolic cost in Antarctic sea urchin embryos. *Science* **291**, 1950–1952.

Martin, N. B., Houlihan, D. F., Talbot, C. and Palmer, R. M. (1993). Protein metabolism during sexual maturation in female Atlantic salmon (*Salmo salar* L.). *Fish Physiology and Biochemistry* **12**, 131–141.

Martin, S. A. M., Cash, P., Blaney, S. and Houlihan, D. F. (2001). Proteome analysis of rainbow trout (*Oncorhynchus mykiss*) liver proteins during short term starvation. *Fish Physiology and Biochemistry* **24**, 259–270.

Martin, S. A. M., Blaney, S., Bowman, A. and Houlihan, D. F. (2002). Ubiquitin-proteasome-dependent proteolysis in rainbow trout (*Oncorhynchus mykiss*): Effect of food deprivation. *European Journal of Physiology* **445**, 257–266.

Martinez, J. A. (1987). Validation of a fast, simple and reliable method to assess protein synthesis in individual tissues by intraperitoneal injection of a flooding dose of [^{3}H] phenylalanine. *Journal of Biochemical and Biophysical Methods* **14**, 349–354.

Mathers, E. M., Houlihan, D. F. and Cunningham, M. J. (1992a). Estimation of saithe *Pollachius virens* growth rates around the Beryl oil platforms in the North Sea: A comparison of methods. *Marine Ecology Progress Series* **86**, 31–40.

Mathers, E. M., Houlihan, D. F. and Cunningham, M. J. (1992b). Nucleic acid concentrations and enzyme activities as correlates of growth rate of the saithe *Pollachius virens*: Growth rate estimates of open sea fish. *Marine Biology* **112**, 363–369.

Mathers, E. M., Houlihan, D. F., McCarthy, I. D. and Burren, L. J. (1993). Rates of growth and protein synthesis correlated with nucleic acid content in fry of rainbow trout, *Oncorhynchus mykiss*: Effects of age and temperature. *Journal of Fish Biology* **43**, 245–263.

Mathews, R. W. and Haschemeyer, A. E. V. (1978). Temperature dependency of protein synthesis in toadfish liver *in vivo*. *Comparative Biochemistry and Physiology* **61B**, 479–484.

Maurizi, M. R. (1993). ATP-dependent proteases. *In* "Proteolysis and Protein Turnover" (J. S. Bond and A. J. Barrett, eds), Proceedings of the Ninth ICOP Meeting, pp. 65–71. Portland Press, London.

Mayer, R. J. and Doherty, F. (1986). Intracellular protein catabolism: State of the art. *FEBS Letters* **198**, 181–193.

McCarthy, I. D. and Houlihan, D. F. (1996). The effect of temperature on protein metabolism in fish: The possible consequences for wild Atlantic salmon (*Salmo salar* L.) stocks in Europe as a result of global warming. *In* "Global Warming: Implications for Freshwater and Marine Fish" (C. M. Wood and D. G. McDonald, eds), pp. 51–77. Cambridge University Press, Cambridge.

McCarthy, I. D., Carter, C. G. and Houlihan, D. F. (1992). The effect of feeding hierarchy on individual variability in daily feding of rainbow trout, *Oncorhynchus mykiss* (Walbaum). *Journal of Fish Biology* **41**, 257–263.

McCarthy, I. D., Houlihan, D. F., Carter, C. G. and Moutou, K. (1993). Variation in individual food consumption rates of fish and its implications for the study of nutrition and physiology. *Proceedings of the Nutrition Society* **52**, 427–436.

McCarthy, I. D., Houlihan, D. F. and Carter, C. G. (1994). Individual variation in protein turnover and growth efficiency in rainbow trout, *Oncorhynchus mykiss* (Walbaum). *Proceedings of the Royal Society B* **257**, 141–147.

McCarthy, I. D., Moksness, E. and Pavlov, D. A. (1998). The effects of temperature on growth efficiency of juvenile common wolfish. *Aquaculture International* **6**, 207–218.

McCarthy, I. D., Moksness, E., Pavlov, D. A. and Houlihan, D. F. (1999). Effects of water temperature on protein synthesis and protein growth in juvenile Atlantic wolfish (*Anarhichas lupus*). *Canadian Journal of Fisheries and Aquatic Science* **56**, 231–241.

McMillan, D. N. and Houlihan, D. F. (1988). The effect of re-feeding on tissue protein synthesis in rainbow trout. *Physiological Zoology* **61**, 429–441.

McMillan, D. N. and Houlihan, D. F. (1989). Short-term responses of protein synthesis to re-feeding in rainbow trout. *Aquaculture* **79**, 37–46.

McMillan, D. N. and Houlihan, D. F. (1992). Protein synthesis in trout liver is stimulated by both feeding and fasting. *Fish Physiology and Biochemistry* **10**, 23–34.

McNurlan, M. A., Essén, P., Heys, S. D., Buchan, V., Garlick, P. J. and Wernerman, J. (1991). Measurement of protein synthesis in human skeletal muscle: Further investigation of the flooding dose technique. *Clinical Science London* **81**, 557–564.

Meek, D. W. (1994). Post-translational modification of p53. *Seminars in Cancer Biology* **5**, 203–210.

Melzner, F., Forsythe, J. W., Lee, P. G., Wood, J. B., Piatkowski, U. and Clemmesen, C. (2005). Estimating recent growth in the cuttlefish *Sepia officinalis*: Are nucleic acid-based indicators for growth and condition the method of choice? *Journal of Experimental Marine Biology and Ecology* **317**, 37–51.

Mente, E., Houlihan, D. F. and Smith, K. (2001). Growth, feeding frequency, protein turnover, and amino acid metabolism in European lobster *Homarus gammarus* L. *Journal of Experimental Zoology* **289**, 419–432.

Mente, E., Coutteau, P., Houlihan, D. F., Davidson, I. and Sorgeloos, P. (2002). Protein turnover, amino acid profile and amino acid flux in juvenile shrimp *Litopenaeus vannamei*: Effects of dietary protein source. *Journal of Experimental Biology* **205**, 3107–3122.

Mente, E., Legeay, A., Houlihan, D. F. and Massabuau, J.-C. (2003). Influence of oxygen partial pressures on protein synthesis in feeding crabs. *American Journal of Physiology* **284**, R500–R510.

Meridith, M. P. and King, J. C. (2005). Rapid climate change in the ocean west of the Antarctic Peninsula during the second half of the 20th century. *Geophysical Research Letters* **32**, L19604, doi:10.1029/2005GLO24042.

Meyer-Burgdorff, K.-H. and Günther, K. D. (1995). N-turnover of carp in relation to protein supply. *Journal of Applied Ichthyology* **11**, 378–381.

Meyer-Burgdorff, K.-H. and Rosenow, H. (1995a). Protein turnover and energy metabolism in growing carp. 1. Method of determining N-turnover using a ^{15}N-labelled casein. *Journal of Animal Physiology and Animal nutrition* **73**, 113–122.

Meyer-Burgdorff, K.-H. and Rosenow, H. (1995b). Protein turnover and energy metabolism in growing carp. 2. Influence of feeding level and protein: Energy ratio. *Journal of Animal Physiology and Animal nutrition* **73**, 123–133.

Meyer-Burgdorff, K.-H. and Rosenow, H. (1995c). Protein turnover and energy metabolism in growing carp. 3. Energy cost of protein deposition. *Journal of Animal Physiology and Animal nutrition* **73**, 134–139.

Millward, D. J. (1989). The nutritional regulation of muscle growth and protein turnover. *Aquaculture* **79**, 1–28.

Millward, D. J., Garlick, P. J., James, W. P. T., Nnanyelugo, D. O. and Ryatt, J. S. (1973). Relationship between protein synthesis and RNA content in skeletal muscle. *Nature* **241**, 204–205.

Millward, D. J., Garlick, P. J., Stewart, R. J. C., Nnanyelugo, D. O. and Waterlow, J. C. (1975). Skeletal-muscle growth and protein turnover. *Biochemical Journal* **150**, 235–243.

Millward, D. J., Bates, P. C., Brown, J. G., Cox, M. and Rennie, M. J. (1981). Protein turnover and the regulation of growth. *In* "Nitrogen Metabolism in Man" (J. C. Waterlow and J. M. L. Stephen, eds), pp. 409–418. Applied Science Publishers, London.

Millward, D. J., Bates, P. C., de Benoist, B., Brown, J. G., Cox, M., Halliday, D., Odedra, B. and Rennie, M. J. (1983). Protein turnover the nature of the phenomenon and its physiological regulation. *In* "Proceedings of Fourth International Symposium on Protein Metabolism and Nutrition" (M. Arnal, R. Pion and D. Bonin, eds), Vol. 1, pp. 69–96. INRA Publications, Paris.

Moldave, K. (1985). Eukaryotic protein synthesis. *Annual Review of Biochemistry* **54**, 1109–1149.

Møller, A. P. (1997). Developmental stability and fitness: A review. *The American Naturalist* **149**(5), 916–932.

Mommsen, T. P. (2004). Salmon migration and muscle protein metabolism: The August Krogh principle at work. *Comparative Biochemistry and Physiology* **B139**, 383–400.

Mommsen, T. P., French, C. J. and Hochachka, P. W. (1980). Sites and patterns of protein and amino acid utilisation during the spawning migration of salmon. *Canadian Journal of Zoology* **58**, 1785–1799.

Morgan, I. J., McCarthy, I. D. and Metcalfe, N. B. (2000). Life-history strategies and protein metabolism in overwintering juvenile Atlantic salmon: Growth is enhanced in early migrants through lower protein turnover. *Journal of Fish Biology* **56**, 637–647.

Müller, M., Siems, W., Buttgereit, F., Dumdey, R. and Rapoport, S. M. (1986). Quantification of ATP-producing and consuming processes of Ehrlich ascites tumour cells. *European Journal of Biochemistry* **161**, 701–705.

Muramatsu, T. and Okumura, J. I. (1983). Effect of ageing on the whole body protein turnover in chicks. *In* "Mètabolisme et nutrition azotès" (M. Arnal, R. Pion and D. Bonin, eds), Number 16, Vol. 2, pp. 57–60. Institut National de la Recherche Agronomique, Paris, France.

Negatu, Z. and Meier, A. H. (1993). Daily variation of protein synthesis in several tissues of the gulf killifish, *Fundulus grandis* Baird and Girard. *Comparative and Biochemical Physiology* **106A**, 251–255.

Nevo, E. (2001). Evolution of genome-phenome diversity under environmental stress. *Proceedings of the National Academy of Sciences of the United States of America* **98** (11), 6233–6240.

Nyachoti, C. M., de Lange, C. F. M. and McBride, B. W. (1998). The effect of a flooding dose of phenylalanine on indicators of metabolic status in pigs. *Canadian Journal of Animal Science* **78**, 715–718.

Oddy, V. H. (1999). Protein metabolism and nutrition in farm animals: An overview. *In* "Proceedings of the VIIIth International Symposium on Protein Metabolism and Nutrition" (G. E. Lobley, A. White and J. C. MacRae, eds), pp. 7–23. EAAP Publication No. 96, Wageningen Pers, Wageningen.

Oleksiak, M. F. and Crawford, D. L. (2006). Functional genomics in fishes: Insights into physiological complexity. *In* "The Physiology of Fishes" (D. H. Evans and J. B. Claiborne, eds), 3rd edn., pp. 523–549. Taylor and Francis Group, Boca Raton, Florida.

Oleksiak, M. F., Churchill, G. A. and Crawford, D. L. (2002). Variation in gene expression within and among natural populations. *Nature Genetics* **32**, 261–266.

Oleksiak, M. F., Roach, J. L. and Crawford, D. L. (2005). Natural variation in cardiac metabolism and gene expression in *Fundulus heteroclitus*. *Nature Genetics* **37**(1), 67–72.

Oñate, S., Amthauer, R. and Krauskopf, M. (1987). Differences in the tRNA population between summer and winter acclimatized carp. *Comparative Biochemistry and Physiology* **86B**, 663–666.

Owen, S. F., McCarthy, I. D., Watt, P. W., Ladero, V., Sanchez, J. A., Houlihan, D. F. and Rennie, M. J. (1999). In vivo rates of protein synthesis in Atlantic salmon (*Salmo salar* L.) smolts determined using a stable isotope flooding dose technique. *Fish Physiology and Biochemistry* **20**, 87–94.

Pace, D. A. and Manahan, D. T. (2001). Differential protein accretion rates and energetic efficiency of protein synthesis during sea urchin development. *American Zoologist* **41**, 1548–1549.

Pace, D. A. and Manahan, D. T. (2006). Fixed metabolic costs for highly variable rates of protein synthesis in sea urchin embryos and larvae. *The Journal of Experimental Biology* **209**, 158–170.

Pace, D. A., Maxson, R. E. and Manahan, D. T. (2004). High rates of protein synthesis and rapid ribosomal transit times at low energy cost in Antarctic echinoderm embryos. *Integrative and Comparative Biology* **43**, 1078.

Place, S. P. and Hofmann, G. E. (2005). Constitutive expression of a stress-inducible heat shock protein gene, HSP70, in phylogenetically distant Antarctic fish. *Polar Biology* **28**, 261–267.

Place, S. P., Zippay, M. L. and Hofmann, G. E. (2004). Constitutive roles for inducible genes: Evidence for the alteration in expression of the inducible HSP70 gene in Antarctic notothenioid fishes. *American Journal of Physiology; Integrative and Comparative Physiology* **287**, R429–R436.

Pannevis, M. C. and Houlihan, D. F. (1992). The energetic cost of protein synthesis in isolated hepatocytes of rainbow trout (*Oncorhynchus mykiss*). *Journal of Comparative Physiology* **B162**, 393–400.

Paul, A. J., Paul, J. M. and Smith, R. L. (1994). Energy and ration requirements of juvenile Pacific halibut (*Hippoglossus stenolepis*) based on energy consumption and growth rates. *Journal of Fish Biology* **44**, 1023–1031.

Peck, L. S. (1998). Feeding, metabolism and metabolic scope in Antarctic marine ectotherms. In "Cold Ocean Physiology" (H. O. Pörtner and R. C. Playle, eds), pp. 365–390. Cambridge University Press, Cambridge.

Peck, L. S. (2002). Ecophysiology of Antarctic marine ectotherms: Limits to life. *Polar Biology* **25**, 31–40.

Peck, L. S. and Veal, R. (2001). Feeding, metabolism and growth in the Antarctic limpet, *Nacella concinna* (Strebel 1908). *Marine Biology* **138**, 553–560.

Peck, L. S., Pörtner, H. O. and Hardewig, I. (2002). Metabolic demand, oxygen supply, and critical temperatures in the Antarctic bivalve *Laternula elliptica*. *Physiological and Biochemical Zoology* **75**, 123–133.

Peck, L. S., Webb, K. E. and Bailey, D. M. (2004). Extreme sensitivity of biological function to temperature in Antarctic marine species. *Functional Ecology* **18**, 625–630.

Pedersen, B. H. (1997). The cost of growth in young fish larvae, a review of new hypotheses. *Aquaculture* **155**, 259–269.

Pedersen, K. S., Kristensen, T. N. and Loeschcke, V. (2005). Effects of inbreeding and rate of inbreeding in *Drosophila melanogaster*—Hsp70 expression and fitness. *Journal of Evolutionary Biology* **18**, 756–762.

Peterson, I. and Wroblewski, J. S. (1984). Mortality rate of fishes in the pelagic ecosystem. *Canadian Journal of Fisheries and Aquatic Science* **41**, 1117–1120.

Pocrnjic, Z., Mathews, R. W., Rappaport, S. and Haschemeyer, A. E. V. (1983). Quantitative protein synthetic rates in various tissues of a temperate fish in vivo by the method of phenylalanine swamping. *Comparative Biochemistry and Physiology* **74B**, 735–738.

Podrabsky, J. E. and Somero, G. N. (2004). Changes in gene expression associated with acclimation to constant temperatures and fluctuating daily temperatures in an annual killifish *Austrofundulus limnaeus*. *The Journal of Experimental Biology* **207**, 2237–2254.

Pörtner, H. O., Hardewig, I. and Peck, L. S. (1999). Mitochondrial function and critical temperature in the Antarctic bivalve, *Laternula elliptica*. *Comparative Biochemistry and Physiology Part A* **124**, 179–189.

Preedy, V. R., Smith, D. M. and Sugden, P. H. (1985). The effects of 6 hours of hypoxia on protein synthesis in rat tissues *in vivo* and *in vitro*. *Biochemical Journal* **228**, 179–185.

Preedy, V. R., Paska, L., Sugden, P. H., Schofield, P. S. and Sugden, M. C. (1988). The effects of surgical stress and short-term fasting on protein synthesis *in vivo* in diverse tissues of the mature rat. *Biochemical Journal* **250**, 179–188.

Quayle, W. C., Peck, L. S., Peat, H., Ellis-Evans, J. C. and Harrigan, P. R. (2002). Extreme responses to climate change in Antarctic lakes. *Science* **295**, 645.

Raae, A. J., Opstad, I., Kvenseth, P. and Walther, B. T. (1988). RNA, DNA and protein during early development of feeding and starved cod (*Gadhus morhua* L.) larvae. *Aquaculture* **73**, 247–259.

Reddy, P. S. and Bhagyalakshmi, A. (1994). Modulation of protein metabolism in selected tissues of the crab, *Oziotelphusa senex senex* (Fabricius), under fenvalerate-induced stress. *Ecotoxicology and Environmental Safety* **27**, 214–219.

Reeds, P. J. and Davis, T. A. (1992). Hormonal regulation of muscle protein synthesis and degradation. *In* "Control of Fat and Lean Composition" (K. N. Boorman, P. J. Buttery and D. B. Lindsay, eds), pp. 1–26. Butterworth-Heinemann, Oxford.

Reeds, P. J., Fuller, M. F. and Nicholson, B. A. (1985). Metabolic basis of energy expenditure with particular reference to protein. *In* "Substrate and Energy Metabolism" (J. S. Garrow and D. Halliday, eds), pp. 46–57. Libbey, London.

Reeds, P. J., Wahle, K. W. J. and Haggarty, P. (1982). Energy costs of protein and fatty acid synthesis. *Proceedings of the Nutrition Society* **41**, 155–159.

Reid, S. D., Linton, T. K., Dockray, J. J., McDonald, D. G. and Wood, C. M. (1998). Effects of chronic sublethal ammonia and a simulated summer global warming scenario: Protein synthesis in juvenile rainbow trout (*Oncorhynchus mykiss*). *Canadian Journal of Fisheries and Aquatic Science* **55**, 1534–1544.

Rennie, M. J. and MacLennan, P. (1985). Protein turnover and amino acid oxidation: The effects of anaesthesia and surgery. *In* "Substrate and Energy Metabolism in Man" (J. S. Garrow and D. Halliday, eds), pp. 213–221. Libby, London.

Rennie, M. J., Smith, K. and Watt, P. W. (1994). Measurement of human protein synthesis: An optimal approach. *American Journal of Physiology* **266**, E298–E307.

Rhoads, R. E. (1993). Regulation of eukaryotic protein synthesis by initiation factors. *Journal of Biological Chemistry* **268**, 3017–3020.

Robertson, R. F., El-Haj, A. J., Clarke, A. and Taylor, E. W. (2001a). Effects of temperature on specific dynamic action and protein synthesis rates in the Baltic isopod crustacean, *Saduria entomon*. *Journal of Experimental Marine Biology and Ecology* **262**, 113–129.

Robertson, R. F., El-Haj, A. J., Clarke, A., Peck, L. S. and Taylor, E. W. (2001b). The effects of temperature on metabolic rate and protein synthesis following a meal in the isopod *Glyptonotus antarcticus* Eights (1852). *Polar Biology* **24**, 677–686.

Rolfe, D. F. S. and Brown, G. C. (1997). Cellular energy utilization and molecular origin of standard metabolic rate in mammals. *Physiological Reviews* **77**, 731–758.

Saez, L., Goicoechea, O., Amthauer, R. and Krauskopf, M. (1982). Behaviour of RNA and protein synthesis during the acclimatization of the carp. Studies with isolated hepatocytes. *Comparative Biochemistry and Physiology* **72B**, 31–38.

Sayegh, J. F. and Lajtha, A. (1989). *In vivo* rates of protein synthesis in brain, muscle and liver of five vertebrate species. *Neurochemical Research* **14**, 1165–1168.

Schaefer, A. L. and Scott, S. L. (1993). Amino acid flooding doses for measuring rates of protein synthesis. *Amino Acids* **4**, 5–19.

Schalburg, K. R., Rise, M. L., Cooper, G. A., Brown, G. D., Gibbs, A. R., Nelson, C. C., Davidson, W. S. and Koop, B. F. (2005). Fish and chips: Various methodologies demonstrate utility of a 16,006-gene salmonid microarray. *BMC Genomics* **6**, 126.

Schoenheimer, R. (1942). "Dynamic State of Body Constituents". Harvard University Press, Cambridge, MA.

Schofield, P. J. (2004). Influence of salinity, competition and food supply on the growth of *Gobiosoma robustum* and *Microgobius gulosus* from Florida Bay, U.S.A. *Journal of Fish Biology* **64**, 820–832.

Scornik, O. A. (1984). Role of protein degradation in the regulation of cellular protein content and amino acid pools. *Federation Proceeding* **43**, 1283–1287.

Shepard, J. L. and Bradley, B. P. (2000). Protein expression signatures and lysosomal stability in *Mytilus edulis* exposed to graded copper concentrations. *Marine Environmental Research* **50**, 457–463.

Shepard, J. L., Olsson, B., Tedengren, M. and Bradley, B. P. (2000). Protein expression signatures identified in *Mytilus edulis* exposed to PCBs, copper and salinity stress. *Marine Environmental Research* **50**, 337–340.

Siems, W., Dubiel, W., Dumdey, R., Müller, M. and Rapoport, S. M. (1984). Accounting for the ATP-consuming processes in rabbit reticulocytes. *European Journal of Biochemistry* **139**, 101–107.

Simon, E. (1987). Effect of acclimation temperature on the elongation step of protein synthesis in different organs of rainbow trout. *Journal of Comparative Physiology B* **157**, 201–207.

Simon, O. (1989). Metabolism of proteins and amino acids. *In* "Protein Metabolism in Farm Animals: Evaluation, Digestion, Absorption and Metabolism" (H.-D. Bock, B. O. Eggum, A. G. Low, O. Simon and T. Zebrowska, eds), pp. 273–366. Oxford University Press, VEB Deutscher Landwirtschnaftsverlag, Berlin.

Skinner, D. M. (1968). Isolation and characterization of ribosomal ribonucleic acid from the crustacean *Gecarcinus lateralis*. *Journal of Experimental Zoology* **169**, 347–356.

Skjermo, J., Salvesen, I., Conceição, L. E. C., Skjåk-Bræk, G. and Vadstein, O. (1999). Microbial control in larval culture: Effects of non-specific immunostimulation of larval turbot *Scophthalmus maximus* L. *EAS Special Publication* **27**, 221–222.

Smith, K., Downie, S., Barua, J. M., Watt, P. W., Scimgeour, C. M. and Rennie, M. J. (1994). Effect of a flooding dose of leucine in stimulating incorporation of constantly infused valine into albumin. *American Journal of Physiology* **266**, E640–E644.

Smith, M. A. K. (1981). Estimation of growth potential by measurement of tissue protein synthetic rates in feeding and fasting rainbow trout, *Salmo gairdnerii* Richardson. *Journal of Fish Biology* **19**, 213–220.

Smith, M. A. K. and Haschemeyer, A. E. V. (1980). Protein metabolism and cold adaptation in Antarctic fishes. *Physiological Zoology* **53**, 373–382.

Smith, M. A. K., Mathews, R. W., Hudson, A. P. and Haschemeyer, A. E. V. (1980). Protein metabolism of tropical reef and pelagic fish. *Comparative Biochemistry and Physiology* **65B**, 415–418.

Smith, R. W. and Houlihan, D. F. (1995). Protein synthesis and oxygen consumption in fish cells. *Journal of Comparative Physiology B* **165**, 93–101.

Stammatoyannopoulos, J. A. (2004). The genomics of gene expression. *Genomics* **84**, 449–457.

Storch, D. and Pörtner, H.-O. (2003). The protein synthesis machinery operates at the same expense in eurythermal and cold stenothermal pectinids. *Physiological and Biochemical Zoology* **76**, 28–40.

Storch, D., Heilmayer, O., Hardewig, I. and Pörtner, H.-O. (2003). *In vitro* protein synthesis capacities in a cold stenothermal and a temperate eurythermal pectinid. *Journal of Comparative Physiology* **B173**, 611–620.

Storch, D., Lannig, G. and Pörtner, H.-O. (2005). Temperature-dependent protein synthesis capacities in Antarctic and temperate (North sea) fish (Zoarcidae). *Journal of Experimental Biology* **208**, 2409–2420.

Sugden, P. H. and Fuller, S. J. (1991). Regulation of protein turnover in skeletal and cardiac muscle. *Biochemical Journal* **273**, 21–37.

Sveier, H., Raae, A. J. and Lied, E. (2000). Growth and protein turnover in Atlantic salmon (*Salmo salar* L.); the effect of dietary protein level and particle size. *Aquaculture* **185**, 101–120.

Talbot, C. and Higgins, P. J. (1983). A radiographic method for feeding studies on fish using metallic iron powder as a marker. *Journal of Fish Biology* **23**, 211–220.

Tesseraud, S., Larbier, M., Chagneau, A. M. and Geraert, P. A. (1992). Effect of dietary lysine on muscle protein turnover in growing chickens. *Reproduction Nutrition Development* **32**, 163–175.

Thorpe, J. E., Talbot, C., Miles, M. S., Rawlings, C. and Keay, D. S. (1990). Food consumption in 24 hours by Atlantic salmon (*Salmo salar* L.) in a sea cage. *Aquaculture* **89**, 41–47.

Tirard, C. T., Grossfeld, R. M., Levine, J. F. and Kennedy-Stoskopf, S. (1997). Effect of osmotic shock on protein synthesis of oyster hemocytes *in vitro*. *Comparative Biochemistry and Physiology* **116A**, 43–49.

Tomas, F. M., Pym, R. A. and Johnson, R. J. (1991). Muscle protein turnover in chickens selected for increased growth rates, food consumption or efficiency of food utilisation: Effects of genotype and relationship to plasma IGF-I and growth hormone. *British Journal of Poultry Science* **32**, 363–376.

Treberg, J. R., Hall, J. R. and Driedzic, W. R. (2005). Enhanced protein synthetic capacity in Atlantic cod (*Gadus morhua*) is associated with temperature-induced compensatory growth. *American Journal of Physiology* **288**, 205–211.

Utting, S. D. and Millican, P. F. (1997). Techniques for the hatchery conditioning of bivalve broodstocks and the subsequent effect on egg quality and larval viability. *Aquaculture* **155**(1–4), 45–54.

Vasemägi, A. and Primmer, C. R. (2005). Challenges for identifying functionally important genetic variation: The promise of combining complementary research strategies. *Molecular Ecology* **14**, 3623–3642.

Vavra, J. and Manahan, D. T. (1999). Protein metabolism in lecithotrophic larvae (Gastropoda: *Haliotis rufescens*). *Biological Bulletin* **196**, 177–186.

Venier, P., Pallavicini, A., De Nardi, B. and Lanfranchi, G. (2003). Towards a catalogue of genes transcribed in multiple tissues of *Mytilus galloprovincialis*. *Gene* **314**, 29–40.

Viarengo, A., Moore, M. N., Pertica, M., Mancinelli, G. and Accomando, R. (1992). A simple procedure for evaluating the protein degradation rate in mussel (*Mytilus galloprovincialis* Lam.) tissues and its application in a study of phenanthrene effects on protein catabolism. *Comparative Biochemistry and Physiology* **103B**, 27–32.

von der Decken, A. and Lied, E. (1992). Dietary protein levels affect growth and protein metabolism in trunk muscle of cod, *Gadus morhua*. *Journal of Comparative Physiology B* **162**, 351–357.

Vrede, T., Dobberfuhl, D. R., Kooijman, S. A. L. M. and Elser, J. J. (2004). Fundamental connections among organism C:N:P stoichometry, macromolecular composition, and growth. *Ecology* **85**, 1217–1229.

Waterlow, J. C. (1995). Whole-body protein turnover in humans—past, present and future. *Annual Review of Nutrition* **15**, 57–92.

Waterlow, J. C. and Millward, D. J. (1989). Energy cost of turnover of protein and other cellular constituents. *In* "Energy Transformations in Cells and Organisms" (W. Wieser and E. Gnaiger, eds), pp. 277–282. Thieme, Stuttgart.

Waterlow, J. C., Garlick, P. J. and Millward, D. J. (1978). "Protein Turnover in Mammalian Tissues and in the Whole Body". North Holland Publishing Company, Amsterdam.

Watson, J. D. (2004). "Molecular Biology of the Gene", 5th edn. Pearson/Benjamin Cummings, San Francisco.

Watt, P. W., Marshall, P. A., Heap, S. P., Loughna, P. T. and Goldspink, G. (1988). Protein synthesis in tissues of fed and starved carp, acclimated to different temperatures. *Fish Physiology and Biochemistry* **4**, 165–173.

Wheatley, D. N. (1984). Intracellular protein degradation: Basis of a self-regulating mechanism for the proteolysis of endogenous proteins. *Journal of Theoretical Biology* **107**, 127–149.

Whitehead, A. and Crawford, D. L. (2005). Variation in tissue-specific gene expression among natural populations. *Genome Biology* **6**, R13.

Whiteley, N. M. and Faulkner, L. S. (2005). Temperature influences whole-animal rates of metabolism but not protein synthesis in a temperate intertidal isopod. *Physiological and Biochemical Zoology* **78**, 227–238.

Whiteley, N. M., Robertson, R. F., Meagor, J., El Haj, A. J. and Taylor, E. W. (2001). Protein synthesis and specific dynamic action in crustaceans: Effects of temperature. *Comparative Biochemistry and Physiology* **128**, 595–606.

Whiteley, N. M., Taylor, E. W. and El Haj, A. J. (1996). A comparison of the metabolic cost of protein synthesis in stenothermal and eurythermal isopod crustaceans. *American Journal of Physiology* **271**, R1295–R1303.

Wieser, W. and Krumschnabel, G. (2001). Hierarchies of ATP-consuming processes: Direct compared with indirect measurements, and comparative aspects. *Biochemical Journal* **355**, 389–395.

Wolfe, R. R. (1992). "Radioactive and Stable Isotope Tracers in Biomedicine". Wiley-Liss, New York, NY.

Yacoe, M. E. (1983). Protein metabolism in the pectoralis muscle and liver of hibernating bats, *Eptesicus fuscus*. *Comparative Physiology* **152**, 137–144.

Yamahira, K. and Conover, D. O. (2002). Intra- vs. interspecific latitudinal variation in growth: Adaptation to temperature or seasonality? *Ecology* **83**, 1252–1262.

Young, V. R. and Munro, H. N. (1978). N^T-Methylhistidine (3-methylhistidine) and muscle protein turnover: An overview. *Federation Proceedings* **37**, 2291–2300.

Young, V. R., Yong-Ming, Y. and Krempf, M. (1991). Protein and amino acid turnover using the stable isotopes ^{15}N, ^{13}C, and ^2H as probes. *In* "New Techniques in Nutritional Research" (R. G. Whitehead and A. Prentice, eds), pp. 17–72. Academic Press Inc., London.

Zuvic, T., Brito, M., Villanueva, J. and Krauskopf, M. (1980). *In vivo* levels of aminoacyl-tRNA species during acclimization of the carp *Cyprinus carpio*. *Comparative Biochemistry and Physiology* **67B**, 167–170.

TAXONOMIC INDEX

SUBJECT INDEX